W0231913

Cell Regulation by Intracellular Signals

NATO ADVANCED STUDY INSTITUTES SERIES

A series of edited volumes comprising multifaceted studies of contemporary scientific issues by some of the best scientific minds in the world, assembled in cooperation with NATO Scientific Affairs Division.

Series A: Life Sciences

Recent Volumes in this Series

This series is published by an international board of publishers in conjunction with NATO Scientific Affairs Division

A Life Sciences	Plenum Publishing Corporation
B Physics	London and New York
C Mathematical and Physical Sciences	D. Reidel Publishing Company Dordrecht, The Netherlands and Hingham, Massachusetts, USA
D Behavioral and Social Sciences	Martinus Nijhoff Publishers The Hague, The Netherlands
E Applied Sciences	

Cell Regulation by Intracellular Signals

Edited by
Stéphane Swillens
and
Jacques E. Dumont

University of Brussels Medical School
Brussels, Belgium

SPRINGER SCIENCE+BUSINESS MEDIA, LLC
Published in cooperation with NATO Scientific Affairs Division

Library of Congress Cataloging in Publication Data

NATO Advanced Study Institute on Control of the Growth and Function of Differentiated
Cells by Intracellular Signals (1980: Nivelles, Belgium)
    Cell regulation by intracellular signals.

    (Nato advanced study institutes series. Series A, Life sciences; v. 44)
    "Proceedings of a NATO Advanced Study Institute on Control of the Growth and
Function of Differentiated Cells by Intracellular Signals, held July 14–23, 1980, in
Nivelles, Belgium" — T.p. verso.
    "Published in cooperation with NATO Scientific Affairs Division."
    Includes bibliographical references and index.
    1. Cellular control mechanisms — Congresses. I. Swillens, Stéphane. II. Dumont,
Jacques E., 1931–   . III. North Atlantic Treaty Organization. Division of Scientific
Affairs. IV. Title. V. Series. [DNLM: 1. Cell communication — Drug effects — Congresses.
2. Nucleotides, Cyclic — Physiology — Congresses. 3. Calcium — Physiology — Congresses.
QH 604.2 N279c 1980]

QH604.N35 1980	574.87′6	82-483
ISBN 978-1-4684-7720-7	ISBN 978-1-4684-7718-4 (eBook)	AACR2
DOI 10.1007/978-1-4684-7718-4		

Proceedings of a NATO Advanced Study Institute on Control of the Growth and
Function of Differentiated Cells by Intracellular Signals, held July 14 — 23, 1980,
in Nivelles, Belgium

© 1982 Springer Science+Business Media New York
Originally published by Plenum Press, New York
Softcover reprint of the hardcover 1st edition 1982

All rights reserved

No part of this book may be reproduced, stored in a retrieval system, or transmitted,
in any form or by any means, electronic, mechanical, photocopying, microfilming,
recording, or otherwise, without written permission from the Publisher

PREFACE

In 1980, the IVth International Cyclic Nucleotide Conference
was held in Brussels. As this meeting attracted many investigators
involved in cyclic nucleotides and calcium role in intracellular
regulation, it was thought that this opportunity could be used to
organize, prior to the Congress, an in-depth introductory course on
the subject. This was carried out as a NATO Advanced Study Institute.
The participants included Ph.D. students and M.D.s engaged in a
research training, but also fully trained and well known researchers
who wanted a refresher course on the whole subject. During the
course, most of the participants and lecturers asked to be provided
with a text summarizing the basic lectures of the course. This
book was therefore conceived as a basic textbook on the regulation
and action of intracellular signal molecules, concentrating mainly
on cyclic nucleotides and calcium. It was deliberately kept at a
basic level. We would therefore be happy if it could be used as an
introduction for interested M.D.s or Ph.D.s working in other fields
or entering this field and as a general refresher for researchers
interested in the subject. For this reason, very general schemes
have been asked of the authors, along with reading lists of available
reviews rather than extensive bibliographies.

The editors should like to thank the NATO Scientific Affairs
Committee for having supported the course and Mrs. Gh. Wilmes who
prepared the manuscripts.

                                        S. SWILLENS
                                        J.E. DUMONT

CONTENTS

CONTENTS

INTRODUCTION : CELL CONTROLS AND SIGNAL MOLECULES

Jacques E. Dumont

Institut de Recherche Interdisciplinaire

Université Libre de Bruxelles

All cells are submitted to a variety of specific and unspecific extracellular signals (Fig. 1.): photons, ions, metabolites, neurotransmitters, and local or general hormones. At all moments, the number, size, and activity of the cells result from their complex present and past network of interactions with these signals. The action of these signals may be defined by several of their characteristics : their nature, their kinetics, their biochemical mechanisms. Signals may control the level of activity of the target cell, either stimulating this activity (positive control) or inhibiting it (negative control). They can influence the nature of the cell activity, e.g., by changing the program of the cell; this is a differentiating action. They may also increase the number of functional units (e.g. organelles) per cell or the number of cells (hypertrophy and hyperplasia) or decrease it; these are positive or negative trophic actions. The kinetics of the action of signals varies from a few microseconds to days. The biochemical mechanisms of action also vary from the simple opening of an ion channel to the complex activation of sets of genes. It is obvious that these three characteristics are related; the opening of an ionic channel is almost immediate and can directly influence very defined enzymatic or transport functions. Differentiation which implies the alteration of the expression of sets of genes, and the consequent modification of the cell protein pattern, will necessarily take at least hours and will involve controls at the level of transcription.

Different extracellular signals acting on the same cell may interact at several levels : the primary site of interaction, the generation or action of intracellular signals, the concentration of receptors or intracellular signal-generating systems. Sometimes they activate the same biochemical system (e.g., adenylate cyclase),

1

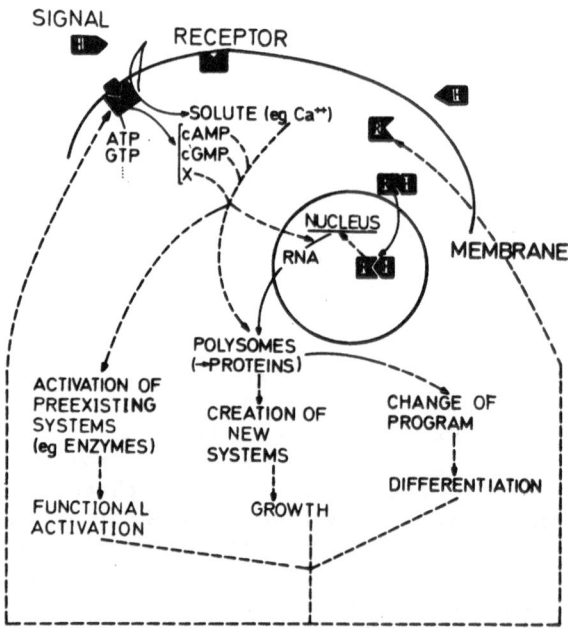

Fig. 1. Mechanisms of action of extracellular signals.

in which case their actions will only differ in amplitude and duration. They can also act in opposite ways on the same system
(e.g., the α-adrenergic inhibition of the β-adrenergic-activated
adenylate cyclase). It was hypothesized a few years ago that cyclic
AMP on the one hand and cyclic guanosine monophosphate (cyclic GMP)
and Ca^{++} on the other hand would act similarly in cells submitted
to one type of external control, positive or negative (unidirectional system), whereas they would have opposite effects in systems
submitted to both positive and negative controls (bidirectional
systems). In fact, when they act on different biochemical systems,
these different intracellular signals may cause similar, antagonistic,
or complementary effects or combinations of such effects. A modification of the response to a signal may be called a modulation. On
the long term, signals may modulate their own action or the action
of other signals by increasing or depressing the level of specific

receptors. This type of action may involve the downward (e.g. insulin) or upward regulation of their receptors in their target cells, the induction or the repression of receptors to other hormones (e.g. steroid induction of insulin receptors), or even the regulation of the level of intracellular target enzyme systems.

In general, it must be emphasized that, whereas the nature of the extracellular signals and for the most part the accessory signals controlling a cell and the sign of the effect (positive or negative) of intracellular or extracellular signals may vary from species to species, the biochemical mechanism of action of these signals (e.g. cyclic AMP through protein kinase) and the overall pattern of control remain constant. All thyroids are activated by thyrotropin and inhibited by iodide.

Among the various regulation systems, this book is concerned only with extracellular signals molecules acting on the plasma membrane of the target cell to elicit intracellular signal molecules which, by affecting their effector proteins, induce their terminal physiological or pharmacological actions. The two best known systems at the moment, involve cyclic AMP and Ca^{++} as intracellular signal molecules. It is interesting that these systems have many properties suggesting that a general framework may apply to these and to other still to be discovered intracellular signals (Fig. 2). It may be useful to summarize these concepts in the introduction before the chapters describing the different elements of those systems.

The concept of the intracellular signal molecule or "secondary messenger" was proposed and developed by Sutherland, Rall, Butcher, Hardman, Krebs and their coworkers on the basis of the work which led to and followed their discovery of cyclic AMP. The model proposed can now be applied to a great number of hormones, local hormones and neurotransmitters. As it stands now this model can be described as follows :

1) Activation of adenylate cyclase. The extracellular signal molecule binds to specific receptor proteins floating in the two dimensional plane of the membrane. This binding leads to a conformational change in the receptor which allows it to activate a transducing protein (N, G/F, GTPase) which will then activate a catalytic unit that will convert ATP to cyclic AMP. Whether the three units of adenylate cyclase are bound together or more probably floating independently in the membrane remains controversial. Catabolism of cyclic AMP involves its hydrolysis to 5'AMP by specific phosphodiesterases.

2) Action of cyclic AMP on effector protein. Cyclic AMP dependent protein kinase is constituted of 2 catalytic units C and 2 regulatory units R_2. The regulatory units block the activity of the catalytic units. By binding to R_2 as $R_2(cAMP)_4$, cyclic AMP causes the release of the 2 catalytic units which thus become operational.

The protein kinases catalyse the phosphorylation of their specific
protein substrates by ATP.

3) <u>Activated proteins and effects</u>. The phosphorylation of the target
enzyme of the kinase may turn "on" or "off" its activity. This positive
or negative effect will more or less directly determine the known
physiological or pharmacological effects of the extracellular and
intracellular signal molecules. Termination of the action is due to
disappearance of the extracellular signal, loss of response to this
signal (desensitization), and the action of cyclic nucleotide phosphodiesterases and protein phosphatases. Direct negative control of the
cyclase by extracellular signal molecules through receptors and transducing proteins has also been demonstrated.

At about the same time that Sutherland developed his model,
neurochemists working on the action of acetylcholine on nicotinic
receptors showed that this agent acts directly on the membrane ionophores. The consequence of this effect in muscle was to elicit
release of calcium from the sarcoplasmic reticulum and thus to induce
contraction (stimulus-contraction coupling). This model was later
extended to other types of muscles and to secretory tissues (stimulus-
secretion coupling). As it now stands the model can be described as
follows:

1) <u>Opening of calcium channel</u>. The extracellular signal molecule
binds to a receptor protein in the plasma membrane. The activated
receptor turns on an ionophore for calcium, thus allowing extracellular calcium to flow down its chemical and electrical gradient in the
cytosol.

2) <u>Effector protein</u>. Calcium combines to a specific protein calmodulin (very similar to skeletal muscle troponin) which is then able to
bind to specific enzymes and to activate them.

3) <u>Activated protein-effect</u>. The binding of calcium containing
calmodulin turns on specific enzymes, the activity of which, through
a more or less extended sequence, causes the physiological or pharmacological response of the target cell. In one such sequence, calmodulin activates specific protein kinases which, by catalyzing the
phosphorylation by ATP of specific protein substrates, thus activates
or inhibits them. Termination of the effect results from desensitization of the receptor, release of the extracellular signal molecule
and from calcium removal mechanisms : plasma membrane extrusion pumps
and sequestration pumps.

The Ca^{++} and cyclic AMP system thus both involve plasma membrane
exterior receptors, membrane catalytic units (cyclase, ionophore),
intracellular signal molecules and their effector proteins (regulatory

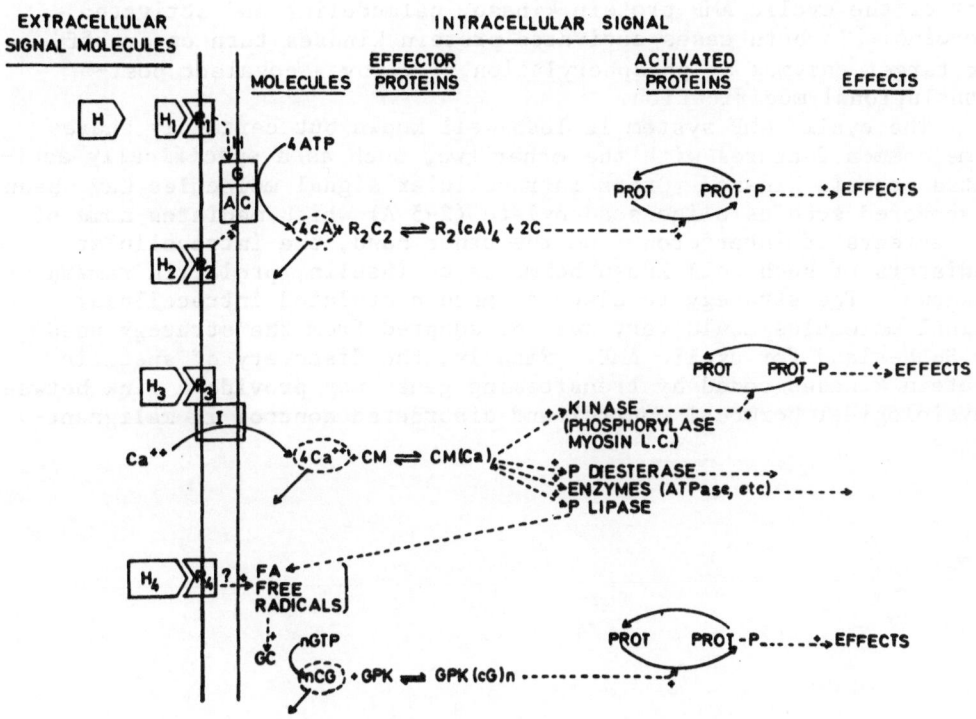

Fig. 2. Cellular control by intracellular signal molecules

H   = extracellular signal molecules
R   = receptor
G   = G/F, N, unit of adenylate cyclase (GTPase)
AC = catalytic unit of adenylate cyclase
I   = ionophore
CA = cyclic AMP
R_2C_2= cyclic AMP dependent protein kinase (R : regulatory
         C : catalytic units)
Prot, Prot P = dephospho and phosphorylated protein
CM = calmoduline
LC = light chains of myosin
P diesterase = phosphodiesterase
P lipase = phospholipase
FA = fatty acids
GC = guanylate cyclase
CG = cyclic GMP
GPK = cyclic GMP activated protein kinase
$\xrightarrow{\pm}$ = positive control
$\dashrightarrow$ = negative control

unit of the cyclic AMP protein kinase, calmodulin) and activated
proteins. In both cases activated protein kinases turn on or off
the target enzymes by phosphorylation, i.e. by a covalent post-
translational modification.

     The cyclic GMP system is less well known but certainly shares
some common features with the other two, such as a specifically activated protein kinase. Other intracellular signal molecules have been
discovered such as oligoisoadenylate (2-5 A) which mediates some of
the effects of interferon. On the other hand, the intracellular
mediators of such well known hormones as insulin, prolactin remain
unknown. The strategy to discover such postulated intracellular
signal molecules could very well be adapted from the strategy used
by Sutherland for cyclic AMP. Finally, the discovery of specific
protein kinases coded by transforming genes may provide a link between
physiological hormonal control and disordered control in malignant
cells.

REGULATION OF THE RESPONSIVENESS OF ADENYLATE CYCLASE TO CATECHOLAMINES

John P. Perkins

Department of Pharmacology
University of North Carolina at Chapel Hill
Chapel Hill, North Carolina 27514

INTRODUCTION:

Cells possess within the plasma membrane a variety of proteins involved in the reception, transduction and amplification of extra-cellular signals. Neurotransmitters and many hormones serve as intercellular signal molecules that alter target cell function upon interaction with cell surface, structure-specific receptors. Beta adrenergic receptors (BAR) can be included in this class of plasma membrane proteins that mediate transmembrane processes.

The idea that target cell response is a function of the extra-cellular concentration of the signal molecule is well accepted and needs no elaboration here. However, the observation that cells, tissues and animals can exhibit refractoriness or tachyphylaxis to the effects of administered neurotransmitters, hormones, or their analogs, indicates that variation in the cellular response to similar doses of these agents is possible. Similarly, states of "super responsiveness" can be induced in tissues of animals after surgical or pharmacological manipulation of neural pathways or hormone-se-creting organs. Thus, it seems reasonable to conclude that the capacity of cells to respond to molecular signals is not a static but a dynamic process and that such processes must be regulated.

One aspect of such regulation is observed when certain target cells are exposed to catecholamines. Not only does such a challenge lead to the activation of adenylate cyclase but it also sets in motion a complicated series of events designed to down-regulate responsiveness of the cell to any subsequent challenge with a catecholamine.

The hormone-sensitive adenylate cyclase system is known to be composed of at least three separate moieties: a hormone receptor, a catalytic protein and a guanine nucleotide binding protein, the latter being involved in coupling the effects of hormone binding to enzyme activation. The response of adenylate cyclase to hormones could thus be regulated at any of the steps that are involved in the sequence of interaction of these proteins. For example, there is now good evidence that the number of hormone receptors can be regulated independently of the other components of the adenylate cyclase system. Such changes appear to involve chronic adaptive or developmental changes in cells and have been observed during ontogenesis and as a result of chronic alterations in the exposure of target cells or tissues to receptor agonists. In addition to changes in receptor number, changes in the functional properties of receptors theoretically could be expected to alter their capacity for interaction with the guanine nucleotide binding protein. Other cell types exhibit a tachyphylaxis to the effects of hormones that involves nonspecific changes in adenylate cyclase; under such a condition basal, NaF-stimulated, and guanine nucleotide-stimulated activities, as well as hormone-stimulated adenylate cyclase activity are all reduced. Losses in enzyme responsiveness that are nonspecific in nature suggest that changes occur in components of the adenylate cyclase system other than in hormone specific receptor sites.

Our discussion will focus on the mechanisms of 1st messenger-induced decreases in the responsiveness of adenylate cyclase, these include agonist-specific and non-specific processes.

THE STRUCTURE AND FUNCTION OF ADENYLATE CYCLASE

Since the discussion will focus on the regulation of the response of adenylate cyclase to hormones it will be useful to have in mind some idea of the structure and function of this complex enzyme system. Furthermore, the regulatory phenomena we will discuss involve relationships derived for the most part from studies of the catecholamine-sensitive adenylate cyclase; thus, regulation of the BAR-linked adenylate cyclase will be emphasized. Excellent and more exhaustive reviews have recently appeared (Maguire et al., 1977; Stadel et al., 1980; Ross and Gilman, 1980; also see Neer and Swillens in this volume).

In Figure 1 a model of the catecholamine-sensitive adenylate cyclase is illustrated. This formulation is based on the "steady-state" model of Cassel and Selinger and incorporates the idea put forth by Lefkowitz and Williams that the receptor can exhist in a form that has relatively low affinity for agonists, (R), or a form that has relatively high affinity for agonists, (RN). The basic premise of the steady-state model is that a GTP-liganded coupling

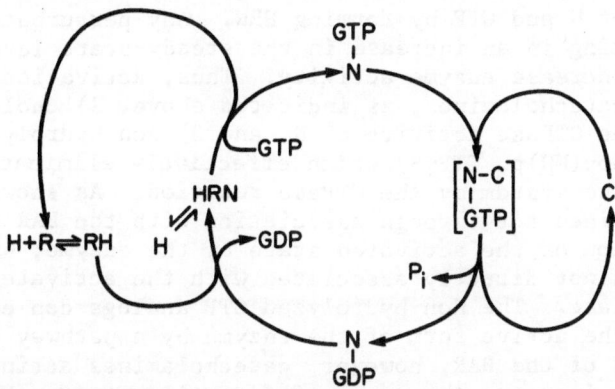

Fig. 1.   This model indicates the hypothetical interactions of a
catecholamine (H), the β-adrenergic receptor (R), a guanine
nucleotide binding protein (N), the catalytic protein (C),
and guanine nucleotides (GDP and GTP). The model indicates
that RH is able to bind to N, resulting in the release of
free GDP and the formation of HRN. The formation of HRN is
rate-limiting in the activation process and, in the presence
of GTP, HRN is rapidly converted to N-GTP and HR; thus, the
role of the hormone-receptor system is to effect the conversion of N-GDP to N-GTP. Once formed, N-GTP interacts
with C to form the enzymically active complex C-N-GTP. The
lifetime of the active complex is determined by the activity
of a GTPase (probably an integral part of N) which hydrolyzes
the bound GTP to release P_i, with the subsequent regeneration
of C and N-GDP. In the absence of GTP, addition of H leads
to the formation of HRN in amounts sufficient to change the
apparent K_a of the system for H. Thus, in the absence of
GTP, agonists (H) exhibit binding characteristic of inter-
reaction at two sites (R and RN). In the intact cells, or
upon addition of GTP to membranes, the amount of HRN would
be small because its rate of formation is postulated to be
the limiting step in the intact system. Under these conditions, agonists (H) would exhibit binding characteristic
of the reaction R + H⇌RH, namely low affinity binding to
a single type of site. Modified from Su et al., 1980.

factor (N-GTP) is the actual activator of adenylate cyclase (C);
inactivation results from hydrolysis of the bound GTP by a GTPase
that is intrinsic to N. Catecholamines stimulate enzyme activity
by accelerating the formation of N-GTP either by increasing the rate
of dissociation of GDP from N-GDP as shown, or by directly increasing
the association of N and GTP by forming HRN. Any perturbation of
the system resulting in an increase in the steady-state level of
(GTP-N-C) would increase enzyme activity. Thus, activation can be
effected by: 1) catecholamines, as indicated above, 2) cholera toxin,
which inhibits the GTPase activity of N, and 3) non-hydrolyzable
analogs of GTP (Gpp(NH)p, GTPγS) which effectively eliminate the
inactivation of the system by the GTPase reaction. As shown in the
model, catecholamines act through association with the BAR to accelerate the formation of the activated state of the enzyme; but, the
BAR apparently is not directly associated with the activated form
of adenylate cyclase. The non-hydrolyzed GTP analogs can effect an
accumulation of the active form of the enzyme by a pathway that circum-
vents involvement of the BAR; however, catecholamines acting through
the receptor will increase the <u>rate</u> of formation of (Gpp(NH)p-N-C)
albeit, not the <u>extent</u>. The effect of cholera toxin, which apparently involves ADP-ribosylation of N, is to markedly reduce GTPase
activity; thus, the kinetics of activation of the enzyme by GTP in
the presence of cholera toxin become similar to those of the analog
Gpp(NH)p, i.e., pseudo-irreversible.

It is clear from the model that a guanine nucleotide is required
for activation of the system by catecholamines. Such absolute dependence on GTP has been shown experimentally under rigorously defined
experimental conditions. Purine nucleotides also decrease the apparent affinity of catecholamines for the BAR as measured by direct
binding techniques. It was initially considered paradoxical that
GTP would on the one hand be required for stimulation of enzyme
activity by catecholamines, but on the other hand markedly decrease
the apparent binding affinity of these agonists. The shift in the
binding affinity caused by GTP is observed only with agonists, not
antagonists, which suggests an effect of the activation process on
the receptor binding reaction. The model accounts for such observations by proposing that the RH complex, which readily dissociates
H, can combine with N to form a complex, HRN. The agonist-selective
reaction, RH + N $\rightleftharpoons$ HRN, could account for the increased agonist
affinity.

However, in the presence of GTP the HRN complex is dissociated
to form N-GTP + RH. If the formation of NRH is slow relative to
the rate of its dissociation in the presence of GTP, then the BAR
would exist primarily in the form (RH) that exhibits lower affinity
for H. In the absence of GTP, the addition of catecholamines would
shift the equilibria such that a significant amount of the high
affinity form (HRN) would be present. The capacity of nucleotides
to decrease agonist binding thus reflects association of the BAR and

a component of adenylate cyclase. Evidence in support of this conclusion includes: 1) the effect of GTP is lost when membranes are
treated with sulfhydryl agents at concentrations that destroy N and
C. 2) the effect of GTP is absent when binding studies are conducted
with membranes from the UNC or cyc- variants of S49 lymphoma cells
but present if HC-1 hepatoma cell membranes are used (the S49 variants
are deficient in N; HC-1 lacks C) and 3) uncoupling of activity from
hormonal effects by agents like filipin or by detergents also results
in loss of the GTP effect on binding.

Even though the evidence suggests that other components of the
enzyme system can affect BAR function, and vice-versa, it is clear
that the receptor is a protein distinct from the other components
of the complex. The evidence in support of this concept can be
summarized briefly as follows: 1) certain cells express hormone-
sensitive adenylate cyclase activity but no BAR function; 2) developmental or growth-related changes in adenylate cyclase activities and
BAR function can exhibit different patterns; 3) BAR function and
adenylate cyclase activity can exhibit different sensitivities to
chemical perturbation; 4) genetic variants of the S49 mouse lymphoma
cell line can differentially express the various components of the
enzyme system; 5) HC-1 hepatoma cells have been shown to exhibit
classical BAR ligand binding properties but are devoid of the catalytic protein; and 6) physical separation of adenylate cyclase and
the BAR has been accomplished after solubilization by detergents.

Recently, procedures for the purification of the BAR and description of its physical and chemical properties have appeared. Caron
and Lefkowitz were first to identify BAR in a soluble form. Using
digitonin extracts of purified frog erythrocyte membranes, it was
demonstrated that ^{3}H-dihydroalprenolol (^{3}H-DHA) could be used as a
radioligand to identify a soluble binding site which expressed the
binding properties of the BAR. The molecular weight of the solubilized receptor was estimated by gel chromatography to be 130,000 -
150,000. Lubrol PX has been used to solubilize radio-labelled receptors from S49 cells after pretreating membranes with ^{125}I-iodohydroxy-
benzylpindolol (^{125}IHYP). The hydrodynamic properties of the receptor
were compared to those of adenylate cyclase, which was clearly shown
to be a separable protein by both gel chromatography and sucrose
density gradient centrifugation. The calculated molecular weights
in the absence of detergent for the receptor and enzyme were 75,000
and 250,000 respectively. Others have separated the receptor from
adenylate cyclase using gel exclusion chromatography or affinity
chromatography.

Limbird and Lefkowitz observed that pre-incubation of frog
erythrocyte membranes with a BAR agonist (hydroxybenzylisoproterenol;
HBI) but not an antagonist (IHYP) changed the apparent stokes radius
of the solubilized receptor as determined by gel exclusion chromatography. One interpretation of such results is that agonist-

induced "coupling" of the receptor to another membrane component occurs, leading to an increase in the molecular weight of the complex
associated with agonist binding. Since the same phenomenon did not
occur with antagonist, Limbird and Lefkowitz suggested that the agonist-induced complex might be the result of an interaction of the components normally involved in the activation process. This phenomenon
could be related to the previously mentioned effect of GTP to decrease
the apparent affinity of the receptor for agonists but not antagonists.

Recently, Vauquelin et al. and Caron et al. have reported affinity chromatography techniques for extensive purification of the BAR
from turkey and frog erythrocytes, respectively. The procedure of
Caron et al., appears to be straightforward and if it can be generally
applied, could make available highly purified preparations of receptors.
For example, these workers have reported a 15,000-fold purification
of receptors using this procedure. These results suggest that the BAR
is now amenable to purification, and thus, it is now feasible to suggest that antibodies against the receptor protein can be generated.
Moreover, the availability of preparations highly enriched in receptors makes attractive the use of the monoclonal antibody - hybridoma
technology for the preparation of specific antibodies. This technology
could potentially allow the preparation of antibodies directed at different structural determinants of the BAR, e.g., antibodies that would
selectively block association with agonists and antagonists, but not
block the association with coupling factors, and vice-versa.

One approach to identifying the component entities of hormone-
responsive adenylate cyclase and understanding the mechanisms involved
in their interaction within the plasma membrane involves in vitro reconstitution studies. In this regard, recent advances in the resolution by physical and genetic means of the catalytic moiety of adenylate cyclase and the BAR have led to successful reconstitution of
catecholamine-stimulated adenylate cyclase. Combination of detergent
extracts of plasma membranes which exhibit adenylate cyclase activity
with membranes of an S49 lymphoma cell variant (cyc-) which is deficient in adenylate cyclase activity (but retains BAR) resulted in reconstitution of catecholamine responsiveness to a functionally uncoupled S49 cell variant (UNC) also has been accomplished. Results from
these studies have led to the resolution of the catalytic component
and guanine nucleotide regulatory component of the adenylate cyclase
system and an identification of the specific lesions of the adenylate
cyclase system that are phenotypically expressed in HC-1 hepatoma cells
(C$^-$) and in two variants of S49 lymphoma (cyc$^-$ = N$^-$; UNC = altered N).
To date reconstitution has only been accomplished using the guanine
nucleotide binding protein as the soluble donor component. In this
regard, it has been proposed that the guanine nucleotide regulatory
protein(s) is an endoprotein probably residing at the inner face of
the plasma membrane. It can be partially solubilized by urea in the
absence of detergent and appears to bind little detergent.

Cholera toxin had long been known to "irreversibly" activate adenylate cyclase activity by a mechanism demonstrated to be due to an inhibition of the GTP hydrolysis function of the adenylate cyclase system. Following exposure to cholera toxin, GTP behaves like the hydrolysis resistant Gpp(NH)p analog in producing a persistent acti-vation of adenylate cyclase. Gill and Meren demonstrated that in-cubation of broken cells with cholera toxin and ^{32}P NAD$^+$ resulted in a transfer of ^{32}P-ADP-ribose to a 42,000 Mr Peptide which corre-lated with the extent of cholera toxin-promoted increase in GTP-sensitive adenylate cyclase activity and inhibition of catecholamine stimulated GTPase activity. Thus, an important consequence of the discovery of the catalytic function of cholera toxin is the ability to specifically identify N with a radiolabelled, covalent marker, ^{32}P-ADP-ribose.

Gilman and co-workers have recently succeeded in purifying N from rabbit liver plasma membranes as well as from turkey erythrocyte plasma membranes. The N-protein isolated from rabbit liver membranes appears to consist of three non-identical subunits of approximately 55,000, 45,000 and 35,000 Mr associated in an unknown stoichiometry. The N-protein from turkey erythrocytes contains only the 45,000 and 35,000 Mr species. The 45,000 Mr subunit is responsible, at least in part, for reconstitution of hormone, Gpp(NH)p and NaF-stimulated activity to S49 cyc- membranes. The 45,000 Mr subunit is a substrate for ADP-ribosylation by cholera toxin in native membranes or following reconstitution of purified N into cyc- acceptor membranes. Interest-ingly, purified N, in the absence of membranes, is not a substrate for cholera toxin-catalyzed ADP-ribosylation, but requires the pres-ence of additional "factors" to be covalently modified.

The 55,000 Mr subunit of N is also ADP-ribosylated by cholera toxin. This subunit is not found in turkey erythrocyte membranes or human erythrocyte membranes. Since, N from both human and turkey erythrocytes reconstitutes NaF and guanine nucleotide sensitivity to cyc- acceptor membranes, the 55,000 Mr subunit apparently is not essential for N activity.

The role of the 35,000 Mr subunit is not understood, but it is apparently an integral part of N since it cannot be resolved from the other two subunits except by conditions which inactivate the reconstituting ability of the 45,000 and 55,000 Mr subunits.

The stoichiometry of the three subunits of the G-protein com-prising a functional complex is not clear; however, hydrodynamic studies of partially purified N suggest that the complex has an apparent protein mass of 130,000 daltons which would be consistent with a 1:1:1 relationship.

A cell fusion approach also has been utilized to investigate the mechanisms of interaction of components of the adenylate cyclase system. For example, Orly and Schramm (1976) have demonstrated that catecholamine-stimulated adenylate cyclase activity could be constituted in heterokaryons formed between BAR deficient Friend erythroleukemia cells and receptor-containing turkey erythrocytes in which adenylate cyclase had been irreversibly inactivated. Similar techniques have been utilized to fuse cyc- cells (which possess BAR but not the guanine nucleotide regulatory protein(s) with BAR-deficient B-82 cells. Similarly, cyc- membranes were fused with intact B-82 cells to generate a hormone responsive system. This approach possesses the obvious advantages of little disruption of membranes and little inactivation of component proteins. Results using this technique also demonstrate that, at least in its membrane environment, the BAR can be fused into a functional catecholamine responsive adenylate cyclase by combination with receptor deplete membranes.

In summary, although the hormone receptor component of adenylate cyclase is involved in the regulation of enzyme activity, its properties clearly distinguish it from the other components of the system. Furthermore, sensitive methods are available for its quantitative analysis in both membrane and solubilized preparations and for its extensive purification. Experiments have been initiated in several laboratories concerning the purification of the components of the BAR-adenylate cyclase and the reconstitution in vitro of these components into a hormone responsive system. These studies have already added significantly to our understanding of the mechnisms of regulation of hormone-sensitive adenylate cyclase. One pertinent and, as yet, unanswered question concerning this multicomponent enzyme system involves the in vivo assembly of the requisite proteins in the functional complex. That is, how are these components synthesized and processed within the cell, and what mechanisms are involved in their assembly and turnover in the plasma membrane environment. The existence of such knowledge would clearly complement the approaches discussed above concerning both the characterization of the individual components of this system and attempts at in vitro reconstitution of functional activity from the isolated individual components.

AGONIST-INDUCED, ADAPTIVE CHANGES IN THE RESPONSE OF ADENYLATE CYCLASE TO CATECHOLAMINES OBSERVED WITH INTACT ANIMALS

It has been proposed that states of variable sensitivity of the neuronal cell membrane could represent an important aspect of the homeostatic control of neuronal activity. More specifically it has been suggested that when nerve activity is chronically increased or decreased, the sensitivity of the innervated tissue is altered in a manner that compensates for the altered neural input. The BAR mediates, at least in part, the effects of neurally-released norepinephrine on postsynaptic cells. Of pertinence to this discussion is the observation that when noradrenergic neural activity is chronically increased or decreased the sensitivity to catecholamines of the BAR-linked adenylate cyclase of postsynaptic structures is decreased or increased, respectively. The speculation has been put forth that adaptive changes in this enzyme system mediate the associated adaptive changes in the function of postsynaptic systems. Evidence in support of this concept has come primarily from experiments designed to determine if modification of noradrenergic neuronal activity in vivo leads to compensatory changes in the BAR-adenylate cyclase system.

Pharmacological perturbation of norepinephrine synthesis, storage and reuptake has been used as one strategy to change the steady-state level of norepinephrine to which post-synaptic receptors are exposed. Kalisker et al., destroyed central noradrenergic pathways in rats by intraventricular injection of 6-hydroxydopamine (6-HDA) and then examined the effect of norepinephrine on cAMP formation in slices of cerebral cortex. Four days after 6-HDA treatment, a two-fold increase occurred in the maximal accumulation of cAMP inducible by norepinephrine. These results and similar results of others were interpreted as being consistent with a postsynaptic adaptation of the adenylate cyclase system in response to the cessation of noradrenergic input that resulted upon destruction of noradrenergic neurons by 6-HDA.

Reserpine blocks the storage of norepinephrine in nerve endings and thereby reduces noradrenergic neuronal activity. Like 6-HDA, reserpine has been shown to cause hyperresponsiveness to norepinephrine of the cAMP generating system of slices from rat limbic forebrain, cortex, hypothalamus and hippocamus. Conversely, when the turnover of norepinephrine was increased by chronic administration of amphetamine to mice, slices of cerebral cortex exhibited a submaximal rise in cAMP upon challenge with norepinephrine. Similarly agents like imipramine, which inhibit reuptake of neuronally-released norepinephrine, would be expected to increase the mean synaptic concentration of norepinephrine. Treatment of rats with imipramine for 6-14 days caused a decrease in responsiveness of cortical slices to norepinephrine as well as a loss of BAR as measured by loss of ^{125}I-iodohydroxybenzylpindolol (IHYP) binding sites.

Physical denervation studies have generated results that are consistent with chemical denervation studies utilizing 6-HDA or reserpine. The median forebrain bundle is the primary noradrenergic innervation of the cerebral cortex in rats. Electrolytic lesions of this nerve tract resulted in enhanced responsiveness to norepinephrine of slices from the innervated brain regions.

A large amount of experimental evidence, some of which is summarized above, supports the idea that chronic, excessive noradrenergic nerve activity results in a compensatory decrease in the responsiveness of postsynaptic adenylate cyclase to norepinephrine;

whereas, prolonged periods of reduced nerve activity result in an
increase in enzyme responsiveness.

Adaptive changes, similar in certain ways to those observed in
the intact nervous system, have been observed in non-neural cells in
culture and in erythrocytes; in fact, agonist-induced changes in the
responsiveness of adenylate cyclase appears to be a general response
of the enzyme system. It follows that if the compensatory adaptations
of the adenylate cyclase system observed in cultured cells occur as a
result of mechanisms similar to those operating in the intact animal,
such cells might serve as easily manipulated model systems for examining the molecular basis of a physiologically important regulatory
process.

The rationale for using non-neural tissue to study putative
neural events is based on two premises. First, since the focus is
on postsynaptic adaptation, only the postsynaptic components may be
required in a model system. Second, compensatory regulation of the
responsiveness of adenylate cyclases is a general phenomenon not
limited to adenylate cyclase of neural tissue or adenylate cyclases
that respond to neurotransmitters.

Agonist-induced, Adaptive Changes in Adenylate Cyclase Observed in Isolated Cell Preparations

There appear to be two basic mechanisms subserving catecholamine-induced desensitization of BAR-linked adenylate cyclase systems.
One process exemplified in the work of Brooker, Terasaki and co-
workers appears to involve changes in the activation process distal
to the receptor that reduce the capacity of the enzyme to respond
to any of its normal stimulators. Another process results in receptor
specific alterations that change only the capacity of the system to
catecholamines.

Catecholamine-induced Desensitization in C6-2B Glioma Cells

Brooker, de Vellis, and co-workers have studied catecholamine-
induced refractoriness using the 2B clone of the rat C6 glioma cell
line. Incubation of this cell with catecholamines results in a
transient elevation of intracellular cAMP levels. Rechallenge of
catecholamine-preincubated cells with the agonist fails to elicit a
full cAMP response. The development of refractoriness appears to
occur through a decrease in the synthetic capacity of the cells since
there is no evidence for an induction of phosphodiesterase activity.
The refractoriness that develops in 2B cells appears to be, at least
in part, dependent upon the synthesis of a protein which affects the
rate of synthesis of cAMP. For example, if cells are incubated with
cycloheximide or actinomycin D, as well as norepinephrine during the
desensitization reaction, the degree of loss of responsiveness is
substantially reduced. The protein involved in refractoriness

appears to exhibit a high rate of turnover. That is, little recovery of catecholamine responsiveness is observed if cells are desensitized and then placed in catecholamine-free medium; however, if C6-2B cells are desensitized and placed in agonist-free medium plus cycloheximide the degree of loss of responsiveness is rapidly reduced.

The "refractoriness protein" is synthesized by transcription of a messenger RNA that appears to be relatively long-lived. For example, if cells are incubated for a long period of time with catecholamine plus cycloheximide and then washed and incubated for a short period of time in their absence, the degree of refractoriness upon subsequent challenge is almost identical to that of cells that were preincubated with catecholamine alone. Similarly, during the recovery process actinomycin D does not effect a reversal of refractoriness as does cycloheximide, and actinomycin D is much more effective in preventing desensitization during the first hour of catecholamine incubation than at later times.

The induction of refractoriness in C6-2B cells appears to involve cAMP as an obligatory mediator. The concentration-effect relationship for induction of refractoriness by catecholamines is similar to the concentration-effect relationship for elevation of intracellular cAMP levels. Moreover, preincubation of cells with a phosphodiesterase inhibitor results in the development of refractoriness to a subsequent catecholamine challenge, and phosphodiesterase inhibitors potentiate the effects of submaximal concentrations of isoproteranol on induction of desensitization. Analogs of cAMP also induce refractoriness. Interestingly, the refractoriness that occurs under these conditions is completely blocked by protein synthesis inhibition whereas catecholamine-induced refractoriness is only partially blocked by cycloheximide.

The BAR-adenylate cyclase system of C6-2B cells also has been examined in membrane fractions prepared from desensitized cells. Although data were not presented, Terasaki et al. reported that there was no change in BAR number in membranes prepared from desensitized cells. Basal, isoproterenol- and GTP-stimulated enzyme activities decreased in membranes from desensitized cells with a time course similar to that of loss of whole cell responsiveness. Interestingly fluoride-stimulated activity did not appear to be significantly changed in desensitized cells.

Since C6-2B cells contain only one known hormone receptor that is linked in a stimulatory manner to adenylate cyclase it is not possible to define the catecholamine-induced changes that occur in C6-2B cells as agonist-specific or agonist-nonspecific. However, the fact that cAMP induces refractoriness would suggest that an agonist-nonspecific change occurs in these cells that may be analogous to that reported for cells that respond to more than one type of hormone agonist (e.g. see Su et al., 1976). The hormone

nonspecific nature of the refractoriness induced by catecholamines
is supported by the fact that the cAMP response to cholera toxin in
C6-2B cells is also potentiated by protein synthesis inhibition.

Several lines of evidence suggest that the refractoriness that
develops in C6-2B cells may be of more than one type. For example
although protein synthesis inhibition is effective in blocking
refractoriness induced by dibutyryl cAMP alone, this blockade is
not complete if refractoriness is induced by norepinephrine. Our
own observations indicate that at least during long-term (2-12 hr)
incubation of C6-2B cells with isoproterenol there is a decrease
(down to 20% of control) in the number of BAR. Thus, as with several
other cell lines (see below), it appears that multiple mechanisms
may be responsible for the regulation of catecholamine responsiveness
C6-2B cells. The data of Brooker and co-workers would suggest that
in C6-2B cells a hormone-nonspecific type of regulation is of major
importance.

Agonist-induced, Receptor-specific Changes in Adenylate Cyclase Activity

Based on our perception of the adenylate cyclase system, it was
reasonable to expect that agonist-specific desensitization would
involve a change in the number, or the functional state of the receptor protein specific for the agonist. With the use of direct
radioligand binding procedures it is possible to directly assess
the role of receptor modification as a basis for desensitization of
cells to catecholamine.

Lefkowitz and co-workers applied the ligand binding approach
to a study of catecholamine-induced desensitization of the BAR-
linked adenylate cyclase system of frog erythrocytes. They demonstrated that administration of catecholamines to frogs caused not
only a loss in responsiveness of adenylate cyclase to catecholamines,
but also a significant loss in binding sites for ^{3}H-dihydroalprenolol,
a BAR antagonist. Similar changes could be induced in vitro simply
by incubating erythrocytes with catecholamines. Although these
original studies emphasized the similarity in time course and extent
of loss of BAR and catecholamine-responsive adenylate cyclase, it subsequently became apparent that a quantitative discrepancy usually existed between these two parameters. For example, it was shown that
incubation of human astrocytoma cells with isoproterenol resulted in
an 80-90% loss of catecholamine responsiveness in intact cells but
only a 20% decrease in the number of BAR. Isoproterenol-stimulated
adenylate cyclase activity of homogenates of catecholamine-treated
cells was reduced by 40-50%. The effects on receptor number and catecholamine-stimulated adenylate cyclase activity were shown to be
agonist specific.

Shear and co-workers carried out similar experiments using S49 mouse lymphoma cells. Incubation of cells for two hr with isoproterenol resulted in a 60 to 80% reduction in isoproterenol-stimulated adenylate cyclase activity, but the number of BAR as measured by ^{125}IHYP binding decreased only by 25-45%. In experiments with frog erythrocytes Lefkowitz and co-workers also documented that the catecholamine-induced loss of isoproterenol-stimulated adenylate cyclase activity is substantially greater than the loss of BAR as measured with radioligands that are BAR antagonists. The observed differences between agonist-induced changes in receptor number and loss of response of adenylate cyclase to agonists called into question the role of receptor loss as the sole basis of desensitization.

Recent studies of the adenylate cyclase system during catecholamine-induced desensitization have focused on an analysis of the kinetics of alterations in BAR and catecholamine-stimulated adenylate cyclase activity and on attempts to define the molecular basis of these alterations. Su et al., demonstrated that incubation of 132INI astrocytoma cells with isoproterenol results in a rapid ($t_{\frac{1}{2}}$= 3min) decrease of isoproterenol-stimulated adenylate cyclase activity to a new, essentially steady state level of 40-50% of control. During this time there was no alteration in the number of BAR or in basal or PGE_1-, NaF-, or $Gpp(NH)_p$-stimulated adenylate cyclase activity. Since the amounts of adenylate cyclase or BAR per se do not appear to be altered during this condition it was suggested that an uncoupling occurs between BAR and adenylate cyclase during this phase of desensitization. Recovery from this uncoupled state was rapid ($t_{\frac{1}{2}}$= 7min) upon removal of isoproterenol from the medium (see Figure 2).

The conclusion that short-term desensitization of 132INI astrocytoma cells involves an uncoupling of BAR and adenylate cyclase also was indicated by receptor binding studies. Work from a number of laboratories suggests that in well coupled BAR-adenylate cyclase systems, agonists bind with high affinity to the receptors in the absence of GTP. In the presence of GTP this high affinity complex does not accumulate and only low affinity binding of agonists is observed. In systems that have been uncoupled by chemical means or by genetic selection agonists bind to BAR with only low affinity and there is no effect of guanine nucleotides on agonist binding. The agonist binding properties of BAR of astrocytoma cells that have been incubated with isoproterenol for 15 min are similar to those of uncoupled receptors. That is, although there is no change in the number of receptors, the apparent affinity of isoproterenol for inhibition of ^{125}IHYP binding is reduced by 10 fold; similarly, there is little effect of GTP on the apparent affinity of isoproterenol (Figure 3). The time course of the uncoupling reaction as detected by changes in the affinity of agonist binding is similar to the time course of loss of isoproterenol-stimulated adenylate cyclase activity. Furthermore, the alteration in agonist binding affinity and the reduction of isoproterenol-stimulated adenylate cyclase activity appear to return to control levels with similar time courses upon removal of isoproterenol.

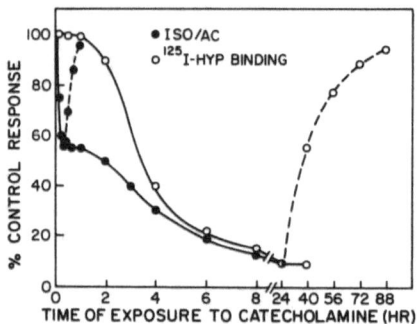

Fig. 2. Time courses of desensitization and Reversal of Desensitis-
         ation. Cells were incubated with 1 μM isoproterenol. At the
         times indicated, isoproterenol-stimulated adenylate cyclase
         activity (●) and β-adrenergic receptor density (o) were de-
         termined in membrane preparations. The symbols connected by
         dashed lines indicate the recovery of enzyme activity or β-
         receptor number after washing the cells free of isoproter-
         enol at the indicated times.

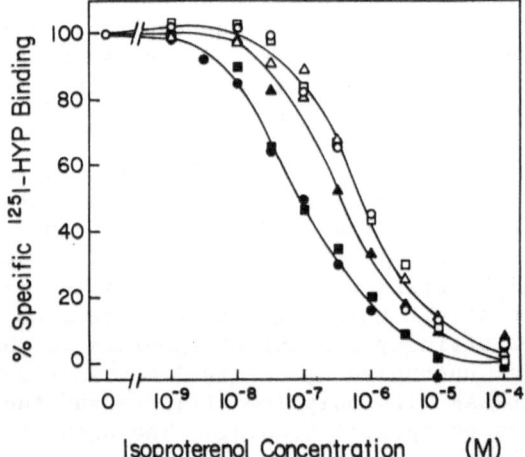

Fig. 3. Competition by isoproterenol for specific [125]IHYP binding
         sites in membranes from control or desensitized cells. Cells
         were incubated for 15 min in the absence or presence of iso-
         proterenol. Other cells were incubated for 15 min with iso-
         proterenol, the cells washed, and the incubation continued
         in the presence of fresh agonist-free medium for an add-
         itional 15 min. Binding assays were carried out on washed
         membranes prepared from each group of cells. Membranes from
         control cells (-●-,-O-) desensitized cells (-▲-Δ-) or re-
         covered cells (-■-,-□-) were incubated in the absence (closed
         symbols) or presence (open symbols) of 100 μM GTP.

The uncoupling reaction exhibits the same concentration-effect relationship for isoproterenol as does stimulation of cAMP accumulation in intact cells ($K_{0.5}$ values approximately 0.03μM). Also, partial agonists have roughly equivalent partial effects on the degree of uncoupling and the degree of activation of adenylate cyclase.

Although no change in the density of BAR occurs during a 30-60 min incubation of astrocytoma cells with isoproterenol, continued exposure to agonist results in a precipitous loss of receptors after 2 to 4 hr. By 24 hr the loss of BAR and isoproterenol-stimulated enzyme activity are usually similar (0-15% of control). After long-term incubation with catecholamine, recovery of hormone responsiveness or receptor number occurs slowly ($t_{\frac{1}{2}}$ = 7-12 hr) upon removal of isoproterenol from the medium (see Figure 2).

Since an essentially irreversible reaction appears to be involved in the second stage of catecholamine-induced changes of the BAR-adenylate cyclase system of astrocytoma cells, the concentration-effect relationship for induction of desensitization changes as a function of time. As stated above, at short times (i.e., during the uncoupling reaction) the concentration-effect curve for desensitization is essentially the same as that for activation of adenylate cyclase. With time, however, the concentration effect curve is shifted to the left. For example, the $K_{0.5}$ for loss of responsiveness is approximately 300 nM at 30 min compared to 2.0 nM at 24 hr. Similarly, although partial agonists cause only a partial uncoupling (compared to isoproterenol) at short times, continued incubation with partial agonists can lead to a virtually complete loss of receptors from the cell.

The results of experiments presented thus far suggest the following kinetic model for agonist-specific desensitization.

$$BAR_N \rightleftharpoons BAR_U \rightleftharpoons BAR_D$$

The initial process can be envisioned to involve the formation of an altered form of the receptor, BAR_U. Although [125]IHYP still binds to this form of the receptor, the interaction of agonists with BAR_U does not elicit activation of adenylate cyclase. During a short term incubation with catecholamine, the receptor exists either as BAR_N or BAR_U, and the steady state levels of these two forms are determined by the balance of two reactions. The extent of formation of BAR_U appears to depend on the concentration of isoproterenol. With time, the receptor is converted to a form (BAR_D) that is not detected by [125]IHYP.

Using genetic mutants of the S49 lymphoma cell line it was shown that occupancy alone of BAR by agonists was insufficient to cause a decline in BAR number. Incubation of wild type S49 cells (which

contain a fully responsive BAR-linked adenylate cyclase) with isoproterenol resulted in a decrease in receptor number and catecholamine responsiveness. Also, experiments with an S49 cell variant
that is deficient in cAMP dependent protein kinase yielded similar
results. In contrast, incubation with isoproterenol of an S49 cell
variant (cyc-) that lacks hormone responsiveness resulted in no change
in receptor number. Combinations of isoproterenol and Bt2cAMP also
had no effect on β-receptor number in cyc- cells. It was concluded
that receptor occupancy alone or in combination with elevated cAMP
levels was insufficient to cause refractoriness; it was thus proposed
that the development of refractoriness required an intact adenylate
cyclase. Since it is now known that cyc- cells lack the guanine
nucleotide binding protein while retaining the catalytic unit, these
studies suggest that coupling per se may be involved in agonist-induced changes of the BAR-linked adenylate cyclase system. Our recent
studies support this contention. For example, incubation of HC-1
hepatoma cells with isoproterenol results in an initial uncoupling
reaction as measured by agonist competition binding curves and in an
eventual large decrease in the number of BAR. Since this cell line
lacks the catalytic protein but possesses both BAR and the guanine
nucleotide binding protein (i.e., it will couple) the catalytic unit
per se would not appear to be involved in the desensitization process.

Our experiments with cyc- S49 lymphoma cells confirm the results
of Shear et al. That is, even long term incubation (24 hr) of cyc-
cells with isoproterenol had no effect on BAR number. Surprisingly,
a 24 hr incubation of the UNC clone of S49 lymphoma cells with isoproterenol resulted in a 20-30% loss of BAR. Since these cells appear
to be uncoupled in terms of the interaction of hormone receptors and
the guanine nucleotide binding protein, an agonist-induced change in
receptor number was somewhat surprising. However, as has been recently pointed out by Ross and Gilman, UNC cells appear to undergo a small
degree of hormone induced coupling of components of the adenylate
cyclase; thus, a small loss of BAR during long term incubations with
catecholamine is not unexpected.

Iyengar and co-workers have obtained results indicating that the
nucleotide binding protein (N) involved in coupling is not altered
during the uncoupling phase of catecholamine-induced desensitization.
Wildtype S49 cells were incubated with isoproterenol and then the
N-protein was extracted from the desensitized cells and compared to
N extracted from control cells for its ability to reconstitute
hormone-sensitive adenylate cyclase activity in membranes from S49
cyc- cells. No difference was observed in the capacity of N from
either control or experimental cells to reconstitute enzyme activity.
These results are consistent with the idea that it is the BAR that
is altered during the uncoupling phase of desensitization.

Recent work by Lefkowitz and co-workers suggests that the
mechanisms involved in agonist-induced desensitization of the

adenylate cyclase system of frog erythrocytes is in general similar to that of astrocytoma cells. Incubation of erythrocytes with iso-proterenol results in a rapid and parallel loss (70%) of both iso-proterenol-stimulated adenylate cyclase activity and receptor binding of the agonist hydroxybenzyl isoproterenol (^{3}H-HBI). In contrast BAR number as measured by the binding of the antagonist ^{3}H-DHA was de-creased by only 35% over the same time period. The affinity of iso-proterenol for inhibition of ^{3}H-DHA binding also was reduced in de-sensitized membranes, as was the affinity of ^{3}H-HBI binding. The authors concluded that an alteration occurred during desensitization which preferentially interfered with agonist versus antagonist bin-ding. Recently, this group has carried out a computer analysis of ligand binding to BAR of frog erythrocytes. The capacity of agonists to form a high affinity complex in membrane preparations was found to correlate closely with the efficacy of these agonists for stim-ulation of adenylate cyclase activity. Analysis of ligand binding with membrane from desensitized frog erythrocytes has led to the suggestion that during agonist-induced refractoriness there is an impairment of the capacity of agonists to form a high affinity binding state. This conclusion is similar to that based on studies of agonist-induced desensitization of astrocytoma cells and supports the general conclusion that the process of BAR-specific desensitization involves a change in the receptor that prevents its association with the nucleotide binding protein.

     In light of the properties of agonist-induced desensitization described above it is apparent that catecholamine specific desensit-ization of a number of cell lines occurs through a multistep process. Although the relative rates of the involved reactions may be consider-ably different, kinetic data would suggest that at least in two astro-cytoma cell lines, in HC-1 hepatoma cells, in frog erythrocytes, and in human fibroblasts, a common mechanism of catecholamine-induced desensitization is involved. That is, catecholamine agonists induce a rapid uncoupling reaction which is followed by another reaction(s) which leads to a loss of BAR from the cell. The molecular basis of these changes is not known.

Separation of Native and Desensitized β-adrenergic Receptors

     Using sucrose density gradients Harden and co-workers recently demonstrated that short-term incubation of astrocytoma cells with isoproterenol results in an alteration in the membrane form of BAR. Isoproterenol or NaF-stimulated adenylate cyclase activities are recovered from gradients as a single peak in preparations from both control and desensitized cells. BAR from control cells are also pre-dominantly recovered in this fraction. In contrast, receptors from desensitized cells migrate as two peaks (30-35% and 45-50% sucrose). (Figure 4). The "light" peak of BAR that is generated during short-term incubation with catecholamines exhibits binding properties

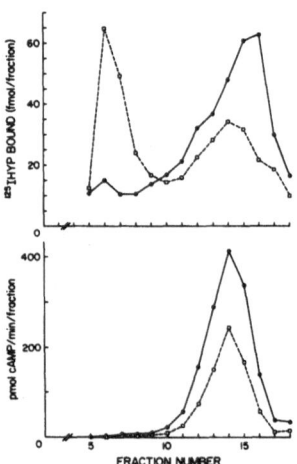

Fig. 4.   Sucrose Density Gradient Distribution of β-Adrenergic Recep-
          tors and Adenylate Cyclase after Short-Term Incubation of
          Cells with Isoproterenol.   Cells were incubated with 1 mM
          sodium ascorbate (●) or 1 mM sodium ascorbate plu- 1 M iso-
          proterenol (o) for 15 minutes.   The cells were treated with
          Concanavalin A then lysed and centrifuged in a sucrose den-
          sity gradient. ^{125}IHYP binding (upper panel) and isopro-
          terenol-stimulated adenylate cyclase activity (lower panel)
          were determined in each gradient fraction.   Taken from
          Harden et al., 1980.

similar to those of uncoupled receptors. That is, only low affinity
agonist binding is observed and there is no effect of GTP on agonist
affinity. The time course of appearance and disappearance of the new
receptor form upon addition or removal of isoproterenol from the
medium correlates with the time courses of decrease or recovery of
isoproterenql-stimulated adenylate cyclase activity. At present,
the significance of this physical form of receptors that appears
during desensitization is not known. Clearly, its appearance co-
incides on a temporal basis with the agonist-induced uncoupling
reaction. Whether an alteration in physical form of the receptor
is casual in the uncoupling process or occurs as a result of such an
event is not known. Also, the cellular localization of the altered
receptor form has not been determined. These receptors do not appear
to be associated with the plasma membrane per se since markers of this
membrane migrate to heavier equilibrium densities. Possibly, the
altered receptors could be associated with the surface membrane initially but upon cell lysis become altered in such a way as to not
associate with the plasma membrane. Such an idea would require the
further qualification that a domain of receptors exists in the plasma
membrane which, due to an agonist-induced aggregation or some other

alteration, are affected differently by cell lysis than is the rest
of the plasma membrane. A more plausible explanation is that these
uncoupled receptors exist in cytoplasmic vesicles that have resulted
from an agonist-induced selective endocytosis of receptor protein.
Such a phenomenon has been demonstrated with receptors for a number
of polypeptide hormones (see Pastan and Willingham, 1981). Our recent
studies indicate that the "light" receptor peak sediments at an equilibrium density similar to that for Golgi apparatus, lyosomes, and
endoplasmic reticulum. A number of compounds, methylamine, quinacrine
and dansylcadavarine have been used to block the processes of endocytosis or degradation of polypeptide hormone/receptor complexes.
Although BAR loss in astrocytoma cells can be at least partially
antagonized by agents such as methylamine or quinacrine, the lack of
specificity of these drugs limits the conclusions that can be drawn
from such results. In contrast to our experience, Robison and co-
workers have shown that dansylcadavarine can block the catecholamine-
induced loss of BAR in BHK cells.

Recently, Chuang and Costa reported that incubation of bullfrog
erythrocytes with isoproterenol results in the appearance of a "soluble" fraction of BAR in cell lysates. Although only about 25% of
the lost receptors could be accounted for in this fraction, these
results do indicate that catecholamines can induce a change in the
physical form of BAR.

Mallorga and co-workers have utilized a different approach to
assess the mechanisms of agonist-induced desensitization in C6 astrocytoma cells. Incubation of cells with isoproterenol for 2 hr resulted in an 80% decrease in responsiveness of intact cells to catecholamines and a 20-30% decrease in the number of BAR. These changes
were also paralleled by increased activity of phospholipase A_2 and
turnover of methylated phospholipids as measured by an increase in
the release of arachidonic acid. Quinacrine, which inhibits phospholipase A_2, blocked the decrease in responsiveness and the reduction
in receptor number. In light of these results, Mallorga proposed
that activation of phospholipase A_2 may be intimately involved in the
process of desensitization. However, the type of desensitization
that was studied in these experiments is not clear. It has been
shown that catecholamine agonists induce a nonspecific type of desensitization in C6 cells that, at least during the first several
hours, does not involve loss of BAR. Thus, it is unlikely that only
agonist-specific desensitization was studied under the conditions
reported by Mallorga et al.

Agonist-induced Desensitization in Cell-free Preparations

Bockaert et al., first established that desensitization to the
effects of hormones on adenylate cyclase could be induced in vitro
using broken cell preparations. Desensitization to the effects of

luteinizing hormone (LH) in this in vitro preparation from pig graaf-
ian follicles required ATP and Mg^{++}; the process was stimulated by,
but did not require, LH. Hormone-stimulated activity was selectively
affected in that basal activity did not change. Neither GTP nor App
(NH)p would substitute for ATP in the desensitization reaction suggesting an involvement of phosphorylation of one or more of the components of the adenylate cyclase system. This idea was indirectly
supported by a subsequent study demonstrating reversal of in vitro
desensitization to LH upon activation of endogenous phosphatase activity with Mn^{++} or dithiothreitol or by the addition of a partially
purified phosphatase preparation. Both studies provided evidence
for a lack of involvement of cAMP in the desensitization process observed in vitro.

Recently, Ezra and Salmon have studied what may be a similar
reaction in rat ovarian plasma membrane preparations. In this study,
LH caused desensitization of the response of adenylate cyclase to LH
but not to NaF. The enzyme system would reconvert to its fully responsive state if the desensitized preparations were incubated in
GTP-free medium. Alterations in LH receptor number or binding properties were not detected during in vitro desensitization. The authors
proposed that GTP serves as a substrate in a phosphotransferase
reaction that leads to phosphorylation of adenylate cyclase in its
inactive (GDP-bound) state (see Fig. 1).

To date only one study of catecholamine-induced desensitization
in cell-free systems has appeared. Anderson and Jaworski incubated
crude membranes from normal rat kidney cells (NRK-S) with isoproterenol and observed a selective loss in the response of adenylate cyclase to isoproterenol; basal and PGE_1-stimulated activities were not
altered. An absolute requirement for ATP, GTP and Mg^{++} was established. Also, increasing concentrations of Mg^{++}, in the presence
of ATP and GTP led to desensitization even in the absence of isoproterenol. It may be relevant to successful demonstration of
desensitization in cell-free preparations that NRK-S cells exhibit
a very rapid rise and fall in intracellular cAMP when exposed to
isoproterenol. It is possible that in these cells the capacity of
the desensitization reaction is exceptionally great, and thus the
probability of observing this reaction in diluted cell free preparations is favourable. Iyengar (personal communication) has shown
that incubation of membranes from S49 lymphoma cells with isoproterenol also leads to agonist-specific desensitization. In this study,
ATP was required and its nonphosphorylating analog App(NH)p would
not substitute. Neither basal activity not NaF-, Gpp(NH)p- or PGE_1-
stimulated activities were altered by a prior 45 min incubation of
S49 membranes with isoproterenol. Changes in the number of BAR or
in their binding properties were not determined in either study.

Although only these few studies of agonist-induced desensitization in cell-free systems have appeared they exhibit the common

feature of requiring phosphorylating conditions for the observance of desensitization. In addition, the reaction does not appear to involve cAMP, eliminating the obvious possibility that cAMP-dependent protein kinases could be functioning in a feed-back regulatory mechanism. Although the data are too limited to determine if these cell-free systems are valid models of the agonist-specific desensitization observed with whole cells there are at least two positive correlations. First, at least in some of the studies the desensitization appears agonist-specific. Second, the "phosphorylation" is not cAMP dependent consistent with the lack of involvement of cAMP in agonist-specific desensitization in whole cells. Careful comparative studies are needed to firmly establish any relation between whole cell and cell-free agonist-induced desensitization reactions.

In summary, although some progress has been made in understanding the mechanisms of catecholamine-induced desensitization in vitro, the physiological significance of these processes is still not clear. Nonetheless, a growing body of evidence indicates the existence of cellular mechanisms for modification of the number of functional BAR as an adaptive response to fluctuating levels of catecholamines in vivo. The study of these processes in model systems may help to provide the details of what appears to be an elaborate set of regulatory processes that modulate catecholamine-mediated intercellular communication.

REFERENCES

Anderson, W.B. and Kaworski, C.J., 1979, Isoproterenol-induced Desensitization of Adenylate Cyclase Responsiveness in a Cell-free System, J. Bio. Chem., 254:4596.

Barber, R., Clark, R.B., Kelly, L.A., and Butcher, R.W., 1978, A Model of Desensitization in Intact Cells, Adv. Cyclic Nucleotide, 9:507.

Bockaert, J., Hunzicker-Dunn, M., and Birnbaumer, L., 1976, Hormone-stimulated Desensitization of Hormone-dependent Adenylyl Cyclase, J. Bio. Chem., 251:2653.

Caron, M.G., Srinivasan, Y., Pitha, J., Kociolek, K., and Lefkowitz, R.J., 1979, Affinity Chromatography of the β-Adrenergic Receptor, J. Bio. Chem., 254:2923.

Cassel, D. and Selinger, Z., 1977, Mechanism of Adenylate Cyclase Activation by Cholera Toxin: Inhibition of GTP Hydrolysis at the Regulatory Site, Proc. Natl. Acad. Sci. U.S.A., 74:3307.

Chuang, D.M. and Costa, E., 1979, Evidence for Internalization of the Recognition Site of β-Adrenergic Receptors During Receptor Subsensitivity Induced by (-)-Isoproterenol, Proc. Natl. Acad. Sci. U.S.A., 76:3024.

Citri, Y. and Schramm, M., 1980, Resolution, Reconstitution and Kinetics of the Primary Action of a Hormone Receptor, Nature, 287:297.

Ezra, E. and Salomon, Y., 1980, Mechanism of Desensitization of Adenylate Cyclase by Lutropin, J. Bio. Chem., 255:653.

Gill, D.M. and Meren, R., 1978, ADP-ribosylation of Membrane Proteins Catalyzed by Cholera Toxin: Basis of the Activation of Adenylate Cyclase, Proc. Natl. Acad. Sci. U.S.A., 75:3050.

Haga, T., Haga, K., and Gilman, A.G., 1977, Hydrodynamic Properties of the β-Adrenergic Receptor and Adenylate Cyclase From Wild Type and Variant S49 Lymphoma Cells, J. Bio. Chem., 252:5776.

Harden, T.K., Cotton, C.U., Waldo, G.L., Lutton, J.K., and Perkins, J.P., 1980, Catecholamine-induced Alteration in the Sedimentation Behaviour of Membrane-bound β-adrenergic Receptors, Science, 210:441.

Harden, Y.K., Su, Y.F., and Perkins, J.P., 1979, Catecholamine-induced Desensitization Involves an Uncoupling of β-adrenergic Receptors and Adenylate Cyclase, J. Cyclic Nucleotide Res., 5:99.

Howlett, A.C., Sternweis, P.C., Macik, B.A., Van Arsdale, P.M., and Gilman, A.G., 1979, Reconstitution of Catecholamine-sensitive Adenylate Cyclase, J. Bio. Chem., 254:2287.

Iyengar, R., Bhat, M., Riser, M., and Birnbaumer, L., 1981, Receptor Specific Desensitization of the S49 Lymphoma Cell Adenylyl Cyclase: Unaltered Behaviour of the Regulatory Component. J. Bio. Chem.,256.

Kalisker, A., Rutledge, C.O., and Perkins, J.P., 1973, Effect of Nerve Degeneration by 6-hydroxydopamine on Catecholamine-stimulated Adenosine 3', 5'-monophosphate Formation in Rat Cerebral Cortex, Mol. Pharmacol., 9:619.

Kent, R.S., DeLean, A., and Lefkowitz, R.J., 1980, A Quantitative Analysis of β-adrenergic Receptor Interactions: Resolution of High and Low Affinity States of the Receptor by Computer Modeling of Ligand Binding Data, Mol. Pharmacol., 17:14.

Lefkowitz, R.J., Wessels, M.R., and Stadel, J.M., 1980, Hormones, Receptors, and Cyclic AMP: Their Role in Target Cell Refractoriness, Current Topics in Cell Reg., 17:205.

Lefkowitz, R.J. and Williams, L.T., 1978, Molecular Mechanisms of Activation and Desensitization of Adenylate Cyclase Coupled β-adrenergic Receptors, Adv. Cyclic Nucleotide Res., 9:1.

Limbird, L.E., Gill, D.M., and Lefkowitz, R.J., 1980, Agonist-promoted Coupling of the β-adrenergic Receptor with the Guanine Nucleotide Regulatory Protein of the Adenylate Cyclase System, Proc. Nat. Acad. Sci. U.S.A., 77:775.

Limbird, L.E., and Lefkowitz, R.J., 1977, Resolution of β-adrenergic Receptor Binding and Adenylate Cyclase Activity by Gel-exclusion Chromatography, J. Bio. Chem., 252:799.

Limbird, L.E. and Lefkowitz, R.J., 1978, Agonist-induced Increase in Apparent β-adrenergic Receptor Size, Proc. Nat. Acad. Sci. U.S.A. 75:228.

Maguire, M.E., Ross, E.M., and Gilman, A.G., 1977, β-adrenergic Receptor: Ligand Binding Properties and the Interaction with Adenylyl Cyclase, Adv. Cyc. Nucleotide Res., 8:1.

Maguire, M.E., Van Arsdale, P.M., and Gilman, A.G., 1976, An Agonist-specific Effect of Guanine Nucleotides on Binding to the β-adrenergic Receptor, Mol. Pharmacol. 12:335.

Mallorga, P., Tallman, J.F., Henneberry, R.C., Hirata, F., Stritt-
    matter, W.T., and Axelrod, J., 1980, Mepacrine Blocks β-adren-
    ergic Agonist-induced Desensitization in Astrocytoma Cells, Proc.
    Nat. Acad. Sci. U.S.A., 77:1341.
Mickey, J.V., Tate, R., Mullikin, D., and Lefkowitz, R.J., 1976,
    Regulation of Adenylate Cyclase-coupled β-adrenergic Receptor
    Binding Sites by β-adrenergic Catecholamines in Vitro, Mol.
    Pharmacol., 12:409.
Mukherjee, C., Caron, M.G., and Lefkowitz, R.J., 1975, Catecholamine-
    induced Subsensitivity of Adenylate Cyclase Associated with Loss
    of β-adrenergic Receptor Binding Sites, Proc. Nat. Acad. Sci.
    U.S.A., 72:1945.
Nickols, G.A. and Brooker, G., 1979, Induction of Refractoriness to
    Isoproterenol by Prior Treatment of C6-2B Rat Astrocytoma Cells
    with Cholera Toxin, J. Cyc. Nucleotide Res., 5:435.
Northup, J.K., Sternweis, P.C., Smigel, M.D., Schleifer, L.S., and
    Gilman, A.G., 1980, Purification of the Regulatory Component of
    Adenylate Cyclase, Proc. Nat. Acad. Sci., 77:6516.
Orly, J. and Schramm, M., 1976, Coupling of Catecholamine Receptor
    from One Cell with Adenylate Cyclase from Another Cell by Cell
    Fusion, Proc. Nat. Acad. Sci., U.S.A., 73:4410.
Pastan, I.H. and Willingham, M.C., 1981, Receptor-mediated Endocytosis
    of Hormones in Cultured Cells, Ann. Rev. Physiol., 43:239.
Perkins, J.P., Su, Y-F., and Harden, T.K., 1979, Adaptive Changes in
    the Responsiveness of Adenylate Cyclase to Catecholamines, Drug
    and Alcohol Dependence, 4:279.
Reggiani, A, Vernaleone, F., and Robison, G.A., 1980, Loss and Re-
    storation of Sensitivity to Catecholamines in Cultured BHK Cells:
    Influence of Dansylcadavarine, Soc. Neuroscience Abst., 6:183.
Rodbell, M., 1980, The Role of Hormone Receptors and GTP-regulatory
    Proteins in Membrane Transduction, Nature, 284:17.
Ross, E.M. and Gilman, A.G., 1980, Biochemical Properties of Hormone-
    sensitive Adenylate Cyclase, Ann. Rev. Biochem., 49:533.
Ross, E.M., Howlett, A.C., Ferguson, K.M., and Gilman, A.G., 1978,
    Reconstitution of Hormone-sensitive Adenylate Cyclase Activity
    with Resolved Components of the Enzyme, J. Bio. Chem., 253:6401.
Schramm, M. and Rodbell, M., 1975, Persistent Active State of the
    Adenylate Cyclase System Produced by the Combined Actions of
    Isoproterenol and Guanylyl Imidodiphosphate in Frog Erythrocyte
    Membranes, J. Bio. Chem., 250:2232.
Shear, M., Insel, P.A., Melmon, K.L., and Coffino, P., 1976, Agonist-
    specific Refractoriness Induced by Isoproterenol, J. Biol. Chem.
    251:7572.
Sternweis, P.C. and Gilman, A.G., 1979, Reconstitution of Catechol-
    amine-sensitive Adenylate Cyclase, J. Bio. Chem., 254:3333.
Su, Y-F., Cubeddu-Ximenez, L., and Perkins, J.P., 1976a, Regulation
    of Adenosine 3':5'-monophosphate Content of Human Astrocytoma
    Cells: Desensitization to Catecholamines and Prostaglandins,
    J. Cyc. Nucleotide Res., 2:257.

Su, Y-F., Harden, T.K., and Perkins, J.P., 1980, Catecholamine-specific
    Desensitization of Adenylate Cyclase: Evidence for a Multistep
    Process, J. Bio. Chem., 255:7410.
Terasaki, W.L., Brooker, G., de Vellis, J., Inglish, D., Hsu, C-Y.,
    and Moylan, R.D., 1978, Involvement of Cyclic AMP and Protein
    Synthesis in Catecholamine Refractoriness, Adv. Cyc. Nucleotide
    Res., 9:33.
Vauquelin, G., Geynet, P., Hanoune, J., and Strosberg, A.S., 1977,
    Isolation of Adenylate Cyclase-free, β-adrenergic Receptor from
    Turkey Erythrocyte Membranes by Affinity Chromatography, Proc.
    Nat. Acad. Sci. U.S.A., 74:3710.
Wessels, M.R., Mullikin, D., and Lefkowitz, R.J., 1979, Selective
    Alteration in High Affinity Agonist Binding: A Mechanism of
    β-adrenergic Receptor Desensitization, Mol. Pharmacol., 16:10.

STUDIES OF SOLUBILIZED ADENYLATE CYCLASE

Eva J. Neer

Department of Medicine
Harvard Medical School and Brigham and Women's Hospital
Boston, Massachusetts 02115

INTRODUCTION: PROPERTIES OF DETERGENTS

To understand, in detail how adenylate cyclase works, we must understand its structure. One approach to this problem is to solubilize the enzyme in an active state, isolate and identify its components and reassemble the whole in a functional form. The first step of this process requires the use of detergents to solubilize adenylate cyclase since the enzyme is firmly membrane bound in all eukaryotic cells, except those from the mature rat testis[1-4]. Adenylate cyclase occurs in membranes in very small amounts, probably making up about 0.01-0.005% of the membrane protein[4], therefore it is impossible to identify the enzyme by protein determination. This limits the kinds of detergents which one may use for solubilization because it is essential to retain enzymatic activity.

Non-ionic detergents are the most useful in solubilizing active adenylate cyclase from a variety of cells. To understand why, we must examine certain properties of ionic and non-ionic detergents. We will use sodium dodecyl sulfate (SDS) and Lubrol 12A9 as examples. SDS is a powerful ionic detergent which solubilizes membranes, disrupts non-covalent protein-protein interactions and denatures virtually all proteins. Lubrol 12A9, a non-ionic detergent, is a less effective solubilizer of membranes, does not break most protein-

---

1. Abbreviations: Gpp(NH)p, guanosine 5' (β,γ-imino)-triphosphate; C, the catalytic unit of adenylate cyclase; G/F, the component of the adenylate cyclase system which mediates activation of the catalytic unit by guanine nucleotides and fluoride.

protein bonds and preserves the activity of most enzymes. The
structure of the two detergents is given below[5]:

$$CH_3-(CH_2)_{11}-O-\overset{\overset{O}{\|}}{\underset{\underset{O}{\|}}{S}}-O^-Na^+ \quad \text{Sodium Dodecyl Sulfate}$$

$$CH_3-(CH_2)_{11}-O-\left[CH_2-CH_2-O\right]_nH \quad \text{Lubrol 12A9}$$

Like all detergents, these two compounds have a polar region
and a hydrophobic one. The hydrophobic portion of both is dodecanol.
They differ in the polar portions: SDS has a strongly acidic sulfate
group, while Lubrol 12A9 is a condensate of dodecanol and several
ethylene oxide units. Both detergents exist in three states in an
aqueous medium: as a monolayer at the air-water interface, as
roughly spherical micelles (in which the hydrophobic tails are
sequestered from water by being in the interior of the micelle) and
as detergent monomers. These forms are illustrated in Figure 1.

Because of the differences in their polar groups, these two
detergents differ greatly in their distribution between the micellar
and monomeric forms. This is reflected in their different critical
micelle concentration (CMC) and different micelle size. Both these
properties affect the usefulness of the detergents for studies of
active membrane enzymes.

Ionic detergents form smaller micelles than do non-ionic
detergents. The average number of SDS molecules in a micelle is 62
and the micelle has an average molecular weight of 18,000: the
average number of Lubrol 12A9 molecules in a micelle is 106 and the
molecular weight is about 64,000[6,7]. The micelle size is not a
fixed property, but increases if the ionic strength of the medium
increases. Non-ionic detergents are less sensitive than ionic
detergents to the effects of ionic strength. This is an advantage
for studies of membrane proteins because one would like the micelle

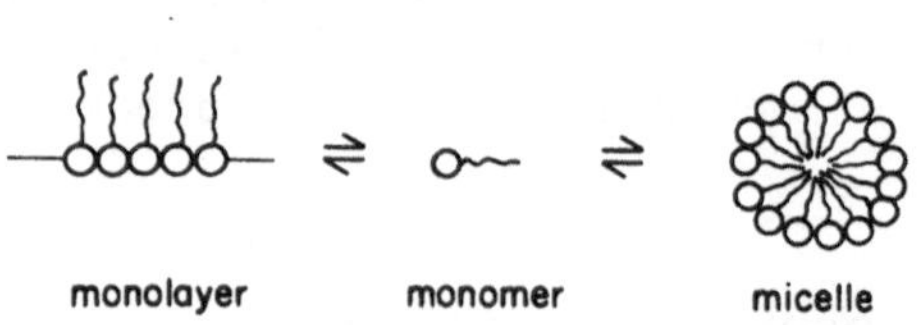

monolayer   monomer   micelle

Fig. 1. States of detergents in aqueous medium.

size to stay relatively constant during further studies with a
solubilized enzyme (for example, ion exchange chromatography with a
salt gradient). The micelle size is very sensitive to temperature
and can decrease 10-fold if the temperature rises from 15° to 45°.
This may be important in electrophoresis (during which gels often
warm) and may account for smearing of bands[6]. The micelle size is
also affected by the pH of the medium.

The critical micelle concentration (CMC) is the highest concentration of detergent monomers in solution. At this concentration
addition of more detergent to the solution causes micelles to form
but there is no further change in the monomer concentration. The
CMC does not have an exact value but is actually a narrow range of
concentrations. Ionic detergents ordinarily have a higher value
for the CMC than do non-ionic detergents. Thus, the CMC of SDS in
water is 8mM while that of the non-ionic detergent Lubrol 12A9 is
less than 0.1mM or about 0.06%[5,8]. The CMC of a detergent varies
with the ionic strength and pH of the medium. This is particularly
true of ionic detergents: the CMC of SDS increases about 10-fold
in 0.2M NaCl compared to water[6].

The CMC is important because of the ways in which detergents
interact with proteins. Makino et al.[9] showed that detergents can
bind to proteins in two ways. One is co-operative and denaturing;
the other is non-co-operative, non-denaturing, and probably confined
to specific hydrophobic sites on the surface of the enzyme. At low
concentration SDS binds to hydrophobic sites on the surface of
proteins. However, the introduction of the negative charge denatures
the protein and exposes interior hydrophobic regions to the detergent.
The process continues until the protein structure is abolished, and
the protein converted into a rigid rod binding a constant amount of
SDS/mass of protein. Non-ionic detergents bind only to the surface
regions of proteins. Their CMC is so low that the monomer concentration is not high enough to produce the massive binding seen with
SDS.

The non-ionic detergents can be classified according to the
hydrophile-lipophile balance number (HLB), an empirical classification
which reflects a detergent's solubility in standardized solvents.
The HLB number has values of 0 through 20 with higher numbers
corresponding to greater water solubility. In practice, detergents
with an HLB number of 12-14 are usually best for solubilizing active
proteins from biological membranes[10,11].

For a more extensive discussion of the properties of detergents
and their interactions with proteins, the reader is referred to
references[5-8]. A useful compendium of the properties of commercial
detergents is McCutcheon's Detergents and Emulsifiers[12].

PROPERTIES OF ADENYLATE CYCLASE SOLUBILIZED WITH NON-IONIC DETER-
GENTS

In most tissues, other than brain, detergents inhibit adenylate
cyclase at concentrations which effectively solubilize it.   In
contrast the enzyme from brain is activated 2-3-fold by Lubrol 12A9
or by Triton X100, another non-ionic detergent[13]. Soluble adenylate
cyclase from most tissues can be stimulated by guanine nucleotides
and by fluoride, but no longer responds to hormone. A number of
reports describe the restoration of hormone responsiveness to
detergent-solubilized enzyme[1], but the techniques have not yet proven
to be generally and reproducibly useful.

Once the enzyme is solubilized, one would like to determine
its size, its shape and how much detergent it binds. Since non-
ionic detergents bind to specific hydrophobic regions on proteins,
the amount of detergent bound may reveal how much of the protein's
surface is hydrophobic and available for interaction with membrane
lipids or hydrophobic proteins. Helenius and Simons[8] and Clarke[14]
showed that a number of proteins which were known on independent
grounds to span the lipid bilayer of the membrane, bind large amounts
of non-ionic detergent (0.2-1.1mg detergent/mg protein). Soluble
proteins bind virtually no detergent since their hydrophobic amino
acids are located in the interior of the molecule. Membrane proteins
which associate which a large amount of detergent can do so in two
ways: they can insert into a detergent micelle as does glycophorin[14]
or they can bind individual detergent molecules, like the Band 3
protein of the erythrocyte membrane[14].

How does adenylate cyclase compare with other membrane proteins
in this respect? Because the enzyme has not yet been purified to
homogeneity, one cannot measure detergent binding directly using
radioactively labeled detergent. However, one can exploit the
partial specific volume $\bar{v}$, of detergents to estimate the amount of
detergent bound. The partial specific volume of most proteins is
0.73-0.74ml/g, while the partial specific volume of Lubrol 12A9 or
Triton X-100 is 0.94-0.98ml/g[7,15]. A complex of protein and detergent has a partial specific volume intermediate between these and
proportional to the fraction of detergent and protein in the complex.
If one measures the partial specific volume of the adenylate cyclase
in the presence of detergent, one can calculate the amount of
detergent bound, assuming the protein itself has a value for $\bar{v}$
similar to a "typical" protein. The partial specific volume of an
impure enzyme can be determined by measuring its rate of sedimentation
(relative to marker proteins of known $\bar{v}$) in sucrose density gradients
made up in H_2O or D_2O. The methods for performing such analysis
and calculating the results have been described previously[14,16].

The results of this kind of analysis for adenylate cyclase from
rat renal medulla and bovine cerebral cortex are given in Table 1.

It is evident that adenylate cyclases solubilized from different tissues by non-ionic detergents are not alike. The solubilized enzyme from the rat kidney has the partial specific volume of a "typical" protein, indicating that it binds no detectable detergents[16]. Since the rat renal enzyme requires detergent for solubilization, it undoubtedly has some hydrophobic regions on its surface, but they must account for a very small fraction of the total surface. Adenylate cyclase solubilized from bovine thyroid has similar properties[17]. In contrast, the enzyme from bovine brain has a partial specific volume of 0.79ml/g in detergent solution indicating that about 28% of its mass represents bound detergent. We do not know whether brain adenylate cyclase inserts into a detergent micelle like glycophorin, or binds individual detergent molecules like Band III from erythrocytes. It is difficult with present methods to distinguish these two modes of binding. Like the brain enzyme, adenylate cyclase from S49 lymphoma cells[18], rat liver[19] and Fasciola hepatica[20] all bind a substantial amount of detergent.

From sucrose density gradient experiments we can measure not only the partial specific volume but also the sedimentation coefficient of adenylate cyclase. By eluting the enzyme from Sepharose columns together with calibrating enzymes of known Stokes radius, we can determine the Stokes radius of the enzyme in detergent. From these three empirically determined values, one can calculate the molecular weight of the detergent-protein complex and then the molecular weight of its protein component. These values are given in Table 1.

The methods which were used to determine the physical properties of adenylate cyclase from rat renal medulla and bovine brain are generally applicable to detergent-solubilized enzymes. Certain precautions, however, must be observed. The density gradient centrifugations and the gel filtration of the enzyme must be carried out under identical conditions, in so far as possible (i.e. constant salt concentration, pH, reducing agents, temperature). The detergent used must have a partial specific volume very different from that of a "typical" protein. Otherwise the change in $\bar{v}$ due to the contribution of the detergent is very hard to detect. One can only determine the molecular weight of the detergent-protein complex in such a detergent, and not the fraction of that complex which is protein.

Lubrol 12A9 and Triton X-100 have both a high $\bar{v}$ and a low CMC so they are useful for binding studies. However, they have the disadvantage of having a large micelle size which elutes close to adenylate cyclase from gel filtration columns. Thus, it is difficult to free the enzyme of detergent by this means or by dialysis. These detergents also precipitate in high salt so that one cannot free the enzyme of detergent by precipitating the protein with ammonium sulfate.

Table 1. Physical Properties of Adenylate Cyclase from Rat Renal
         Medulla and Bovine Cerebral Cortex[a]

Physical Property	Bovine Brain	Rat Renal Medulla
Sedimenation coefficient, $s_{20,w}(s)$	8.1 ± 0.1 (14)	5.9 ± 0.2 (4)
Stokes radius, a (A)	70 ± 2 (3)	62 ± 3 (3)
Partial specific volume, $\bar{v}$ (ml/g)	0.79 ± 0.01 (7)	0.74 ± 0.01 (3)
Molecular weight		
protein plus detergent	305,000	
protein alone	220,000	159,000

[a]The values given are the mean ± SEM for the number of determinations
shown in parentheses. The methods for solubilizing the enzyme and
determining the physical properties have been described in
references 13 and 16.

POSSIBLE ARRANGEMENT OF THE ADENYLATE CYCLASE COMPLEX IN THE PLASMA MEMBRANE

The adenylate cyclase system is known to contain at least two
other separable components in addition to the catalytic unit: the
hormone receptor and the guanine nucleotide regulatory unit. The
latter is a protein thought to couple the hormone receptor to the
catalytic unit. Both of these components have been studied by
techniques similar to those described for adenylate cyclase.

The hormone receptor has, in all cases, proven to be a hydrophobic protein which binds a large amount of detergent. In some
cases, it has been possible to show that hormone receptors are
glycoproteins[21]. The guanine nucleotide regulatory unit from S49
lymphoma cells and from human erythrocytes has a mass of 126–130,000
daltons and binds little[22] or no[23] detergent.

Putting these studies together allows us to propose the following, very speculative, arrangement of the adenylate cyclase system
within the plasma membrane (Fig. 2.). The adenylate cyclase catalytic
unit (C) must face the inside of the cell since it uses intracellular
ATP as the substrate. The holoenzyme is probably a hydrophobic

protein like that solubilized from bovine brain and other tissues.
The catalytic site of the holoenzyme may be on a hydrophilic portion
of the molecule since it is possible to solubilize the enzyme in a
hydrophilic, active form from rat kidney[16] or bovine thyroid[17]. In
addition to these hydrophilic membrane-bound forms of adenylate
cyclase, which can only be solubilized in detergents, there is one
water-soluble form of the enzyme which is found in the mature rat
testis. This form of the enzyme is not activated by guanine nucleotides or by fluoride. It differs also from other adenylate cyclases
in it's requirement for Mn++ in the assay for expression of catalytic
activity. This enzyme is smaller than the forms solubilized from
membranes with detergent, and has a molecular weight of 56,000[3].
Its relationship to the other forms of adenylate cyclase is not
known. It may be an entirely independent enzyme, the product of a
different gene. Or, it may be the hydrophilic catalytic portion of
the larger enzyme which is detached from the holoenzyme by proteolysis
or by other means. Thus, Fig. 2. shows adenylate cyclase to be
composed of a hydrophilic catalytic portion and a hydrophobic
membrane anchor. Whether these are on one or more polypeptide chains
is not known nor is it known whether the enzyme spans the membrane.

Hormone receptors (R) must face the outer surface of the cell
to bind the hormone. Their being glycoproteins is consistent with
this location. There is no evidence that they are transmembrane
proteins although most of the glycoproteins which have been studied
do, indeed, span the membrane. Furthermore, hormone receptors
probably have large hydrophobic surfaces[21].

The guanine nucleotide regulatory protein (G/F) has been shown
to be on the inner surface of human erythrocytes[24]. It seems to be
more loosely attached to the plasma membrane than either the hormone
receptor or adenylate cyclase. Recent reports suggest that some of

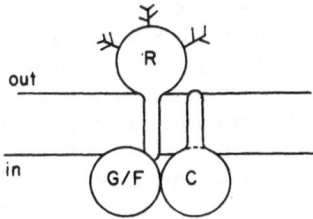

Fig. 2. Possible arrangement of the adenylate cyclase complex in
         the plasma membrane.

this component may be found in the cytoplasm[25]. In the diagram G/F
is shown interacting with both hormone receptors and adenylate
cyclase but not deeply embedded in the lipid layer of the plasma
membrane. This arrangement is consistent with the fact that the
solubilized G/F unit can restore hormone sensitive adenylate cyclase
activity to membranes having hormone receptors and adenylate cyclase
catalytic unit but lacking G/F[26]. This can take place without
permanent insertion of G/F into the membrane. In contrast, it has
been very difficult to directly incorporate catalytic units into
membranes. Hormone receptors and adenylate cyclase catalytic unit
have been incorporated into liposomes[27,28]. In the case of hormone
receptors, these liposomes have been fused with membranes in which
the receptor has been shown to function[27].

EFFECT OF ENZYME ACTIVATION ON PHYSICAL PROPERTIES

        A question which can be approached using solubilized enzyme
preparations is whether the state of activation affects the physical
properties of adenylate cyclase. The studies with adenylate cyclase
from bovine cerebral cortex described above used enzyme which had
been fully activated with Gpp(NH)p. We now asked whether the
properties of the enzyme were the same if the analyses were done
without prior activation. The results of gel filtration of Lubrol
12A9 solubilized adenylate cyclase over Sepharose 6B are shown in
Fig. 3.[29]. After gel filtration, the fractions were assayed either
with no activation (basal activity) or after incubation with 1.6
x10-5M Gpp(NH)p. Also shown in the figure is the ratio of Gpp(NH)p-
stimulated to basal activity. The peak of activity in the void
volume, which was imperfectly solubilized enzyme, was variable and
was not studied further.

        As can be seen from the figure, the ratio of stimulated to
basal activity increases across the included peak. Thus, the basal
activity is partially separated from the activity which can be
stimulated by Gpp(NH)p. Enzyme from the regions marked 1 and 2 in
the figure was rechromatographed on Sepharose 6B. A second gel
filtration separated the basal activity from the Gpp(NH)p-responsive
enzyme and showed that Region 1 contained enzyme which was entirely
unresponsive to activation by guanine nucleotides. Enzyme from
Region 2 had very little basal activity but was very dependent on
stimulation by Gpp(NH)p for activity. Enzyme from Region 1 and
Region 2 was analyzed by sucrose density gradient centrifugation in
H_2O and D_2O to determine the sedimentation coefficient and partial
specific volume. These values are given in Table 2. Both forms of
the enzyme are hydrophobic. The size of the enzyme giving basal
adenylate cyclase activity is about the same as that previously
reported for the Gpp(NH)p-activated enzyme (see Table 1) while the
Gpp(NH)p-responsive enzyme is apparently smaller by about 50,000
daltons.

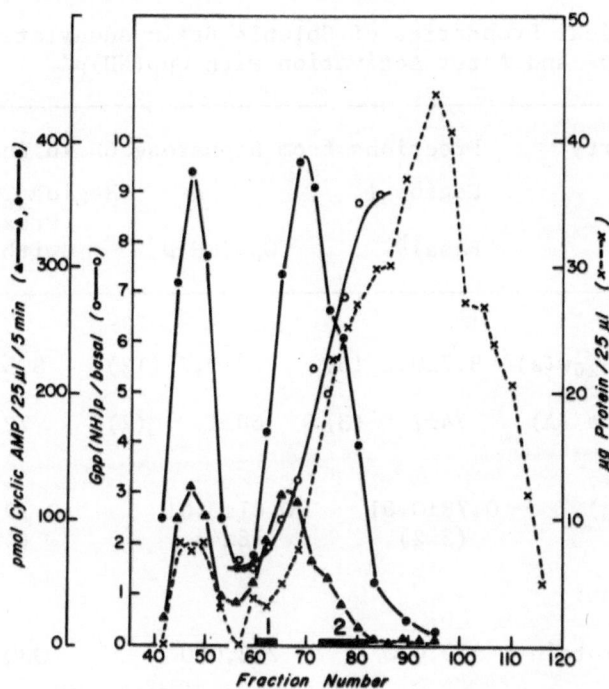

Fig. 3.   Sepharose 6B filtration of Lubrol 12A9 solubilized adenylate
           cyclase. Adenylate cyclase was solubilized with 2% Lubrol
           12A9 as described in reference 30. A 180ml sample containing
           700mg of protein was applied to a 5 x 100cm column of
           Sepharose 6B in 0.1M Tris.Cl, pH 7.6, 0.075 sucrose, 10mM
           MgCl2, 10mM DTT 0.1% Lubrol 12A9 at 4°. The flow rate was
           65ml/hr; 5.5ml fractions were collected. The calibrating
           enzymes were included in the sample applied. Aliquots from
           the fractions were incubated with 1 x 10^{-5}M Gpp(NH)p at 23°
           for 3 hrs then assayed for adenylate cyclase activity (o—o).
           Other aliquots were assayed for enzyme activity without
           Gpp(NH)p (Δ—Δ). The ratio of adenylate cyclase activity
           with/without Gpp(NH)p is indicated by the open circles
           (o—o). β-gal, β-galactosidase, LDH, lactate dehydrogenase,
           MDH, malate dehydrogenase.

    If enzyme from Region 2 is activated by Gpp(NH)p before further
analysis to determine the Stokes radius or sedimentation coefficient,
its size increases to 251,000 daltons. It is not likely that this
increase is due to association of the enzyme with a 50,000 dalton
protein, since Region 2 is well separated from proteins of that size
which would elute farther back on the column.

    One explanation for these observations is that a rapid
equilibrium exists between the catalytic unit of adenylate cyclase

Table 2. Physical Properties of Soluble Brain Adenylate Cyclase
Before and After Activation with Gpp(NH)p[a]

Physical Property	Fractions from Sepharose 6B in Lubrol 12A9[e]		
	Region 1		Region 2
	Basal[b]	Gpp(NH)p[c]	Preactivated with Gpp(NH)p[d]
Sedimentation coefficient, $S_{20}w(s)$	8.7±0.2 (5)	7.3±0.2 (12)	8.5±0.2 (7)
Stokes radius, a (A)	74±2 (3)	68±1 (7)	72±1 (2)
Partial specific volume $\bar{v}$ (ml/g)	0.78±0.01 (3x2)	0.81±0.01 (8x4)	0.79±0.01 (4x3)
Molecular weight:			
detergent + protein	330,000	293,000	330,000
protein	265,000	199,000	251,000

[a]Data are presented as the mean ± SE for the number of determinations
shown in parentheses except where n=2 when the range is given.
These experiments are described in detail in reference 29.
[b]Basal activity refers to activity without Gpp(NH)p in the assay.
[c]Enzyme was activated by $1-1.6 \times 10^{-5}M$ Gpp(NH)p for 3-4 hours at 23°
after gel filtration or sucrose density gradient centrifugation.
[d]Enzyme from region 2 (Fig. 3.) was activated with $1-1.6 \times 10^{-5}M$
Gpp(NH)p for 3-4 hours at 23° before further gel filtration or
sucrose density gradient centrifugation.

and the G/F unit:

$$C + G/F \rightleftharpoons C.G/F$$

If this equilibrium is rapid with respect to the rate of separation
on a column or on a sucrose gradient, one would observe a peak of
adenylate cyclase activity with an apparent size less than that of
the complex but greater than that of either individual component.
We propose that activation might shift such an equilibrium toward
the associated form, increasing the apparent molecular weight.

On the basis of its size, we propose that the basal activity

is given by a fixed C.G/F complex which does not seem to participate
in the equilibrium described above. These experiments show that
most of what is measured as basal activity is not, in fact, activated
by Gpp(NH)p. The increase in measurable enzyme activity after
treatment with the guanine nucleotide analogue seems to be the
result of activation of a separate pool of enzyme which contributed
very little to the basal activity.

A final point should be made from these studies with regard to
the general problem of determining the size of impure multi-component
enzymes. The values obtained are only interpretable if the system
is either fully associated or fully dissociated. In intermediate
states, or where equilibria among components may exist, one cannot
draw any conclusions about the size of the individual components in
the system since the apparent size will be greater than the true
size.

USES OF IONIC DETERGENTS IN STUDIES OF ADENYLATE CYCLASE

At the start of this review, we discussed the difference between
ionic detergents like SDS and non-ionic detergents like Lubrol 12A9.
There exists another category of ionic detergents, the bile salts,
which have a different structure from SDS. One of these, cholate,
is proving useful in studies of adenylate cyclase.

Cholate is a steroid. Instead of having an ionic head group
and a hydrophobic "tail", cholate is hydrophobic on one side of the
ring system and polar on the other. The CMC is high, 13-15 15mM,
and the micelle is very small, only 2-4 cholate molecules[8]. This
means that the free detergent is easily separated from large proteins
by gel filtration or dialysis. Cholate is not precipitated by
ammonium sulfate, so it can also be removed by precipitating the
protein. Cholate is not useful for determining the amount of detergent bound to a solubilized protein, since its $\bar{v}$ is 0.75-0.77ml/g[30].

Cholate is a low ionic strength buffer like that used for
Lubrol 12A9 or Triton X-100 solubilization, will not solubilize
adenylate cyclase from bovine cerebral cortex. However, in the
presence of 1.2M $(NH_4)_2SO_4$, approximately 50% of the enzyme activity
is solubilized[31]. The need for a high ionic strength buffer
emphasizes the fact that solubilization may depend on factors other
than the hydrophobic interactions of the detergent with the protein
in question. The high ionic strength may affect the structure of
adenylate cyclase or it may modify the conformation of the other
proteins in the membrane and so perhaps allow access of the detergent
to otherwise inaccessible sites.

The properties of adenylate cyclase solubilized with cholate
and $(NH_4)_2SO_4$ are very different from those of the enzyme solubilized

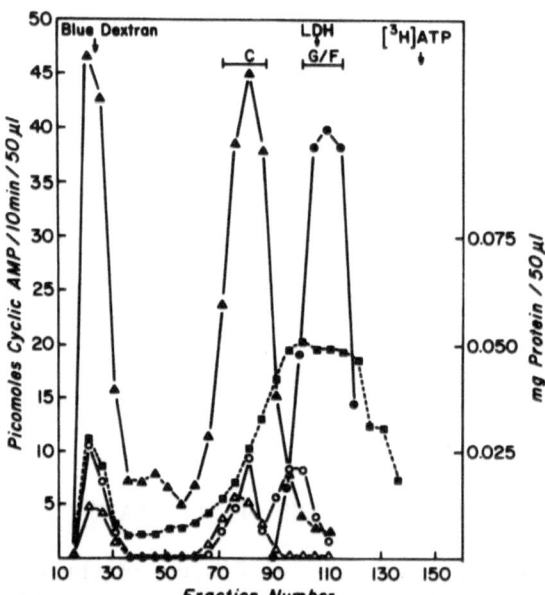

Fig. 4.   Sepharose 6B gel filtration of brain adenylate cyclase
          solubilized with cholate and ammonium sulfate. Eleven ml
          of enzyme prepared as described in reference 31 were applied
          to a 400ml Sepharose 6B column equilibrated at 4° with
          0.05M Tris.Cl, 7mg/ml soybean phospholipids, 14mM cholic
          acid, 25% saturated (4°) ammonium sulfate, 15mM MgCl2,
          0.2M sucrose, 1mM dithiotreitol adjusted to pH 7.7 with
          Tris.OH.   The flow rate was 23ml/hr; fraction size was 2.3ml;
          65ml were collected before the first fraction. Blue Dextran,
          lactate dehydrogenase (LDH) and a trace amount of   3H ATP
          were included in the sample. Fractions were precipitated
          and resuspended in 7/8 their volume 0.05M Tris.Cl, pH 7.7,
          0.2M sucrose, 15mM MgCl2, 1mM dithiothreitol, 0.1% Lubrol
          12A9, 7mg/ml soybean phospholipids. Fifty μl samples were
          assayed for adenylate cyclase activity at 30° for 10 minutes.
          (Δ——Δ) Basal activity; (▲——▲) activity with 5mM Mn^{++};
          (O——O) activity after incubation with 6×10^{-5}M Gpp(NH)p for
          6.5hr at 23°.   The ability of fractions to restore Gpp(NH)p
          responsiveness to adenylate cyclase is indicated by the
          close circles (●——●). Twenty five μl of sample to be
          tested was mixed with 25μl of pooled, precipitated catalytic
          unit. The activity shown is normalized to 50μl of precipi-
          tated catalytic unit for purposes of comparison. Protein
          (■- - - -■) was determined without precipitation.

with non-ionic detergents. Fig. 4. shows the elution pattern of
adenylate cyclase solubilized in this way and filtered over
Sepharose 6B equilibrated with cholate and $(NH_4)_2SO_4$. After gel
filtration the activity was assayed in three ways: with no additions
(basal activity), with Gpp(NH)p, and with Mn^{++}. There was very
little basal activity and virtually no response to Gpp(NH)p in the
column effluent. However, the enzyme was active in the presence of
Mn^{++}. The ratio of activity with Mn^{++} to activity with Mg^{++} as the
only divalent cation was 8-15 in different experiments. In contrast,
the adenylate cyclase activity in the experiment shown in Fig.3. was
only increased 2-3-fold by Mn^{++}.

The requirement for Mn^{++} to express catalytic activity and the
absence of response to Gpp(NH)p, suggested that gel filtration in
cholate and $(NH_4)_2SO_4$ had separated the catalytic unit from the
guanine nucleotide regulatory unit. The combination of a Mn^{++}
requirement and a lack of response to guanine nucleotides occurs in
three other instances: the soluble testicular enzyme described
above, adenylate cyclase from mutant S49 lymphoma cells which lack
a functional G/F unit, and adenylate cyclase from pigeon erythrocytes
separated from G/F by affinity chromatography[1].

This suggestion was confirmed by locating the G/F unit in the
column effluents, using its ability to reconstitute guanine nucleotide responsiveness to the separated catalytic unit. The same factor
restored fluoride responsiveness to the catalytic unit. For the
reconstitution to take place, cholate had first to be removed from
the components by precipitating them with ammonium sulfate[34].

The use of cholate as the solubilizing detergent allowed us to
separate the catalytic and the guanine nucleotide regulatory units
of adenylate cyclase. The resolved components are extremely useful
for further studies of the interaction of the sub-units of adenylate
cyclase. However, as discussed above, cholate is not suitable for
the kinds of studies which we had done with the Lubrol 12A9 and the
Triton X-100 solubilized enzyme.

The hormone-responsive adenylate cyclase system is complex of
at least three proteins: the hormone receptor, the catalytic unit
and the guanine nucleotide regulatory unit. The latter two proteins
each have enzymatic activity, the catalytic unit forming cyclic AMP
from ATP and the G/F unit hydrolyzing GTP to GDP[12]. Other proteins
may also modulate the system and be more or less intimately associated
with it. For example, calmodulin, the adenosine receptor, and a
GTP-dependent inhibition protein[1]. In such a complex system there
is no single "best" detergent for solubilization. Which detergent
is best depends on the reasons for solubilizing the fundamental
properties of detergents and their interactions with membrane
proteins, one can more rationally choose the most efficient strategy
for resolving the problem at hand.

ACKNOWLEDGEMENT

This work was supported by grant AM 19277 from the National Institutes of Health.

REFERENCES

1. E. M. Ross and A. G. Gilman, Ann. Rev. Biochem. 49:533-63 (1980).
2. T. Braun and R. F. Dods, Proc. Natl. Acad. Sci. U.S.A. 72:1097-1101 (1975).
3. E. J. Neer, J. Biol. Chem. 253:5808-5812 (1978).
4. E. J. Neer, in: "Receptors and Hormones Action," B. W. O'Malley and L. Birnbaumer, eds., 463-484, Academic Press, New York (1977).
5. A. Helenius, D. R. McCaslin, E. Fries and C. Tanford, Methods in Enzymology 56:734-749 (1978).
6. L. M. Hjelmeland, D. W. Nebert and A. Chrambach, in: "Electro-phoresis," N. Catsimpplas, ed., Elsevier, North Holland, Amsterdam (1978).
7. C. Tánford and J. A. Reynolds, Biochim. Biophys. Acta. 457:133-170 (1976).
8. A. Helenius and K. Simons, Biochim. Biophys. Acta. 415:29-79 (1975).
9. S. Makino, J. A. Reynolds and C. Tanford, J. Biol. Chem. 248:4926:4932 (1973).
10. J. N. Umbriet and J. L. Strominger, Proc. Natl, Acad. Sci. U.S.A. 70:2997-3001 (1973).
11. G. Guillon, C. Roy and S. Jard, Eur. J. Biochem. 92:341-348 (1978).
12. McCutcheon's Detergents and Emulsifiers, Nowth American Edition, MC Publishing Co., New Jersey (1979).
13. E. J. Neer, J. Biol. Chem. 253:1498-1502 (1978).
14. S. Clarke, J. Biol, Chem. 250:5459-5469 (1975).
15. H. L. Greenwald and G. L. Brown, J. Phys. Chem. 58:825-828 (1954).
16. E. J. Neer, J. Biol. Chem. 249:6527-6531 (1974).
17. R. F. Asbury, G. H. Cook and J. Wolff, J. Biol. Chem. 253:5286-5292 (1978).
18. T. Haga, K. Haga and A. G. Gilman, J. Biol. Chem. 252:5776-5782 (1977).
19. D. Stengel and J. Hanoune, Eur. J. Biochem. 102:21-34 (1979).
20. J. K. Northup, M. F. Renart, J. R. Grove and T. E. Mansour, J. Biol. Chem. 254:11861-11867 (1979).
21. K. J. Catt and M. J. Dufau, Ann. Rev. Physiol. 39:529-557 (1977).
22. A. C. Howlett and A. G. Gilman, J. Biol. Chem. 255:2861-2866 (1980).
23. H. R. Kaslow, G. L. Johnson, V. M. Brothers and H. R. Bourne, J. Biol. Chem. 255:3736-3741 (1980).
24. Z. Farfel, H. R. Kaslow and H. R. Bourne, Biochem. Biophys. Res. Commun. 90:1237-1241 (1979).

25. K. M. Bhat, R. Iyengar, J. Abramowitx, M. E. Bordelon-Riser and L. Birnbaumer, Proc. Natl. Acad. Sci. U.S.A. 77:3836-3840 (1980).
26. E. M. Ross, A. C. Howlett, K. M. Ferguson and A. G. Gilman, J. Biol. Chem. 253:6401-6412 (1978).
27. S. Eimerl, G. Neufeld, M. Korner and M. Schramm, Proc. Natl. Acad. Sci. U.S.A. 77_760-764 (1980).
28. E. Ross, Fed. Proc. Fed. Am. Soc. Exp. Biol. 39:2105 (1980).
29. E. J. Neer, D. Echeverria and S. Knox, J. Biol. Chem. 255:9782-9789.
30. S. Strittmatter and E. J. Neer, Proc. Natl. Acad, Sci. U.S.A. in press (1980).
31. J. C. H. Steele,Jr., C. Tanford and J. A. Reynolds, Methods in Enzymology 48:11-23 (1978).

CONTRIBUTION OF THE THEORETICAL SIMULATION TO THE MECHANISTIC
DESCRIPTION OF A BIOCHEMICAL SYSTEM: APPLICATION TO THE ADENYLATE
CYCLASE SYSTEM

Stéphane Swillens

Institut de Recherche Interdisciplinaire
Free University of Brussels, School of Medicine
B-1000 Brussels, Belgium

INTRODUCTION

     The advance in the mechanistic description of a biochemical
system is the natural consequence of the accumulation of experimental
results. The interpretation of these results usually leads to a
general scheme describing the relations between the different components of the system. Such a scheme is, more often than not, a
set of descriptive sentences, which are sometimes illustrated by a
picture where arrows and conventional signs define biochemical
interactions between the constituents. Less frequently is presented
a complete description of the system in terms of component concentrations and of kinetic parameters of the reactions. Only in this
latter case, the description might be referred to as a model of the
system, since it contains the whole information required for
generating the behaviour of the experimental system.

     The interpretation of the experimental results usually consists
in explaining, on the basis of the model, the behaviour of the
system observed in well-defined conditions. The demonstration of
the validity of the simplest models, which contain only a few
components and reactions, is often evident as the model is self
explanatory, at least in what concerns the qualitative description
of the behaviour. On the other hand, the complexity of the model
which rapidly increases with the number of components and reactions
render the intuitive interpretation quite difficult or just impos-
ible. In any case, the quantitative analysis of the experimental
results always requires demonstrations as rigorous as the experimental protocols used for the study of the system. Thus the unsat-
isfactory intuition must be replaced or at least supported by a
strict theoretical approach which consists of the numerical
simulation of the proposed model. Such a simulation can be

facilitated by using computers, especially when the mathematical equations describing the model cannot be solved due to their number or complexity.

The first and evident aim of the simulation is thus to prove that a model can or cannot account for the experimental results. A second point of interest is the possibility to predict and to define some new experimental conditions which should give rise to a new unknown behaviour. Last but not least, simulation requires models in which all the hypotheses must be explicitly stated. Indeed, contrary to the verbal description of the system in which a lot of hypotheses are masked, the mathematical transformation of the biochemical reactions into mathematical equations compels to select the appropriate assumption between the alternatives or at least to precise that the theoretical result is obtained for a well-defined set of conditions and hypotheses.

A frequent difficulty met in such a work rests on one hand in finding the mathematical solution of the equations, This is in part smoothed away by using computers. The major remaining problem lies in fact in the beginning of the theoretical analysis, that is the elaboration of the model. If the experimental results are incomplete as they usually are, the number of possible models becomes so tremendous that the theoretical conclusions, whatever they might be, loose their potential interest. Indeed too many combinations between the different possibilities at each level of the model reduce the power of the analysis in discriminating between acceptable solutions. The simplicity criterion is even not applicable in this case since the complexity most probably does not differ sufficently. Although in such circumstances theoretical simulation becomes tedious and hopeless, the explicitation of the numerous alternatives during the elaboration of the model leads to propose new protocols which could elucidate the unanswered questions.

In this paper we intend to illustrate how such a theoretical approach can be conducted and applied to a biochemical system which has been widely investigated, namely the adenylate cyclase system. It is not our purpose to demonstrate that the model defined here- after can account for the observed behaviour of this system. These theoretical results are in fact published elsewhere[1,2]. We would prefer to show how such results can be obtained by using the theoretical methodology which consists of four steps:

1. compilation of the observations describing the constituents of the system and the molecular interactions between them.
2. elaboration of the model and explicitation of the hypotheses.
3. simulation of the model and comparison of theoretical results with experimental behaviour.
4. prediction of new behaviours and elaboration of new protocols.

We restrict this presentation to the regulation of adenylate cyclase by β-adrenergic agonists and guanine nucleotides. As suggested in other chapters of this book, the system should essentially take into account the interactions between β-adrenergic agonists, β-adrenergic receptors, the guanine nucleotide binding protein, guanine nucleotides and the adenylate cyclase enzyme.

1. THE EXPERIMENTAL OBSERVATIONS

The role of cyclic AMP as the second messenger of hormone and neurotransmitter stimulations has been widely demonstrated. The transmission of the signal from the extracellular milieu to the inside of the cell has been intensively studied and a detailed minimal molecular mechanism can be proposed on the basis of recent discoveries. The general description of the system can be summarized as follows for a hormone. A membrane bound receptor R can recognise with a high specificity the hormone H. The binding of hormone activates the receptor which in turns acts on a transducer protein G. The G-unit is capable of binding guanine nucleotides and this binding is facilitated when the HR complex and the G-unit interact. The so-called guanine nucleotide binding protein G is active when GTP (or other guanosine triphosphate analogs) is bound to the specific binding site and it is inactive when the site is occupied by GDP, GMP or unoccupied. Finally the active G-unit can activate the enzyme adenylate cyclase (E) which transforms intracellular ATP in cyclic AMP and consequently induces an increase of intracellular cyclic AMP. The three components of the system, i.e. the receptor R, the transducer G and the enzyme E, interact in the plane of the membrane and are considered as independent entities floating in the phospholipidic milieu of the cell membrane.

The finding which has given rise to the greatest progress in the understanding of the activation of adenylate cyclase is the demonstration of the role of GTP. This new development was initiated by the demonstration of an effect of GTP[3] and was pursued by the isolation of the guanine nucleotide binding protein G[4] and by the demonstration of the associated GTPase activity[5]. The G-unit and its interactions with other components of the system thus constitute the transducer mechanism between the agonist receptor and the adenylate cyclase enzyme. The key findings are:

1. the turn on process of adenylate cyclase activation is the binding of GTP on the G-unit;
2. the turn off process is due to the hydrolysis of bound GTP by the G-unit.

This mechanism was described by the cyclic model of Cassel and Selinger[6,7].

On the other hand, recent experimental evidence suggests that the binding of a triphosphate to the G-unit facilitates the association with adenylate cyclase and thus the formation of the active holoenzyme[8].

The agonist action through its specific receptor has been also described in terms of chemical interactions with the G-unit. It has been proposed[9] that the agonist receptor complex interacts with the G-unit and facilitates the binding of GTP as well as the release of GDP as a product of the hydrolysis of GTP by the G-unit. Moreover the direct interaction of the receptor with the G-unit in the presence of agonist is supported by the fact that the receptor size apparently increases due to agonist binding[10]. This suggests the coupling of the receptor R and the G-unit. However this increase of the receptor size is reversed by the addition of guanine nucleotides[10].

The current view of the mechanism of adenylate cyclase activation is thus that the actual effector is GTP which acts through the guanine nucleotide binding protein G, that the system is inactivated due to GTP hydrolysis by the G-unit and that the action of agonist mediated by the receptor is to facilitate the interaction of guanine nucleotide with the G-unit by opening the guanine nucleotide binding site. This view has been nicely expressed by Pfeuffer (see Fig. 3 in[8].

2. ELABORATION OF THE MODEL

The experimental observations presented before should be integrated in a rigorous formulation where the interactions between the β-agonist, the receptor, the transducer unit G, adenylate cyclase E and guanine nucleotides must be explicitly defined. At each step of the elaboration of the model, all the hypotheses made must be expressed.

The main axiom of the proposed model is that the three membrane constituents of the system, namely R, G and E, exist in two conformations, an active form (marked by as asterisk) and an inactive form.

Interactions between β-agonist and Receptor

It has been widely described that most of the β-agonists and antagonists bind to their specific receptor in a reversible manner[11]. Thermodynamic analysis[12] supports the model where the β-agonist binds to the receptor and transforms the receptor in an active conformation R*. Thus the interactions of the β-agonist with the receptor can be represented by a scheme where H binds to both forms of receptors R and R* (Fig. 1.a).

The accumulation of active receptor due to β-agonist action can be a consequence of pathway A, or of pathway B or of both pathways (Fig.1.a).

a)

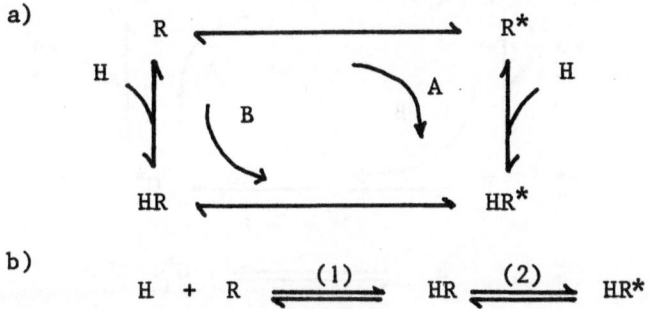

b)

$$H + R \xrightleftharpoons{(1)} HR \xrightleftharpoons{(2)} HR^*$$

Fig. 1. Model of β-adrenergic agonist interactions with receptor.
H : agonist; R : inactive receptor; R* : active receptor.

In order to account for the simplicity criterion, and since it was not demonstrated that both pathways must be considered, the model is restricted to only one pathway. Since pre-equilibrium between the two free forms of the receptor has not been demonstrated so far and since pathway B seems to be necessary for explaining the respective activities of partial agonists and full agonists (see hereafter), the model only consists of pathway B (Fig. 1.b). This first hypothesis is in agreement with experimental results, but is not yet experimentally proved. Likewise, it is supposed that antagonists which can saturate the receptors without eliciting any stimulation, participates only in the first reaction and thus cannot favour the formation of the active receptor[12].

Interactions between G-unit and Adenylate Cyclase

The experimental observation[8] suggested that the binding of adenylate cyclase E with the G-unit is facilitated when GTP is associated to the G-unit whereas GDP-associated G-unit and free G-unit are weakly bound to E. We assume that only the interaction between GTP-bound G-unit (G_T^*) and E must be taken into account. All the other possible interactions are neglected (Fig. 2.a). As shown for the β-agonist - receptor interaction, the activation of E is described by a cyclic equilibrium model. Because of the lack of experimental data on the respective kinetics for pathways A and B (Fig. 2.a), we assume that the concentrations of the intermediate components are negligible, and thus the model shown in Fig. 2.b is obtained. Thus the adenylate cyclase is in the active form only when bound to the active G-unit. The inactivation of E is mainly

due to the hydrolysis of GTP by the G-unit. This enzymatic reaction
must be thus introduced as well as the dissociation of GDP bound
G-unit from E. The reverse reactions are neglected (Fig. 2.c).

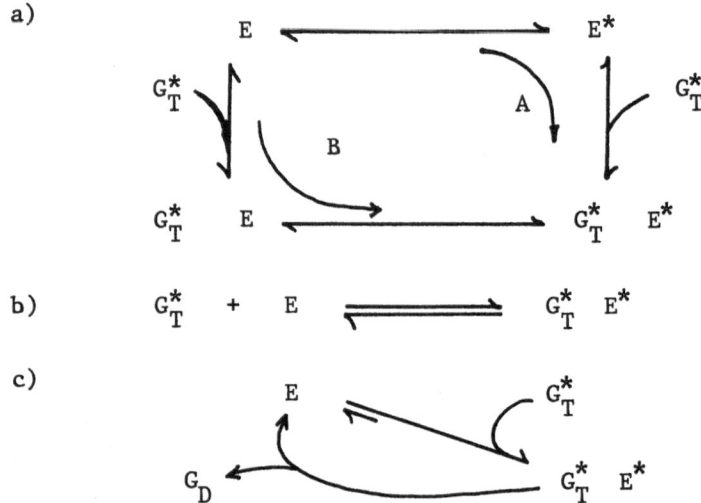

Fig. 2.   Model describing the interactions between G-unit and
          adenylate cyclase; E and E* are inactive and active adenylate
          cyclase; G_T^* is the active GTP bound G-unit; G_D is the
          inactive GDP bound G-unit.

Interactions between Receptor and G-unit

It has been already mentioned that the G-unit can exist in
three physiological states, i.e. the guanine nucleotide binding
site is unoccupied (G), or occupied by GDP (G_D) or by GTP and is
active in this latter form (G_T^*). That only GTP bound G-unit is
active is a hypothesis which needs further experimental investigation. However since it has been suggested that only this form
significantly binds to the adenylate cyclase, the activity of
adenylate cyclase elicited by free G and GDP bound G-unit is
probably of little significance in physiological conditions. Nevertheless, the alternative should be considered[13].

According to the cyclic model of Cassel and Selinger[6,7] hormone
receptor complex binds to the G-unit and facilitates the exchange
of GDP for GTP. In the absence of hormone a pre-equilibrium between
G_D, G and G_T^* could exist. The activation of the system would
result from the following chain of events: binding of HR* with
inactive G-units, facilitated dissociation of GDP, association of
GTP and consequent G-unit activation, dissociation of HR* from the
active G-unit. This equilibrium model must also contain the
enzymatic GTPase reaction which transforms GTP bound G-unit in GDP

bound G-unit (Fig.3.a). Up to now, we have only considered binary
interactions, i.e., agonist with receptor, active receptor with
G-unit, active G-unit with adenylate cyclase.

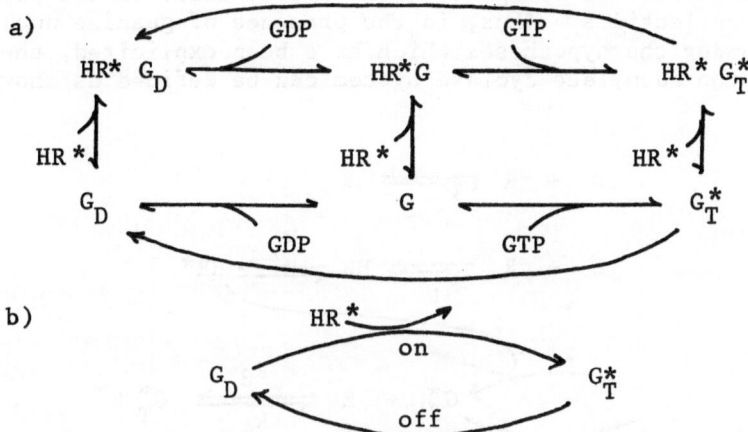

Fig. 3. Model describing the interactions between active agonist
receptor complex HR* and G-unit. G, G_D and G_T^* are respectively inactive free and GDP bound G-units, and active GTP
bound G-unit.

It is evident that the complete model should consider more
interactions as for instance, the interaction of adenylate cyclase
with HR* G_T* complex since this latter complex is active. Fortu-
nately, some experimental evidences can support several simplifi-
cations of the model at least if the model is used in some well-
defined conditions which essentially are mimicking physiological conditions (i.e., presence of guanine nucleotides). Indeed it has been
suggested that HR*G_D and HR*G_T are short lived complexes as guanine
nucleotides destabilize the coupling between receptor and G-unit by
reducing the affinity[10]. Moreover the facilitated interaction of
guanine nucleotide with HR*G complex decreases the concentration of
guanine free HR*G in the presence of guanine nucleotides. Thus, it
can be assumed that, at the steady state, the proportion of G-unit
bound to HR* can be neglected, although this interaction is necessary
for G-unit activation (Fig. 3.b.)

In fact the G-unit activation results from a continuous
recycling process where the on-rate is controlled by the interaction
with HR* and the off-rate is the rate of the GTPase activity. This
hypothesis is in agreement with the model of Cassel and Selinger[6,7]
and with the collision coupling concept[14] which states that
activation of adenylate cyclase does not require a persistent binding of the receptor with the enzymatic components.

Global Model of Adenylate Cyclase Activation

    As proposed above, the functional independence of H-R inter-
actions and R-G interactions allows to consider separately the
equilibrium properties of agonist binding, at least in the presence
of guanine nucleotides. Thus, in the presence of guanine nucleotides and under the hypotheses which have been explicited, the model
describing the adenylate cyclase system can be defined as shown in
Fig. 4.

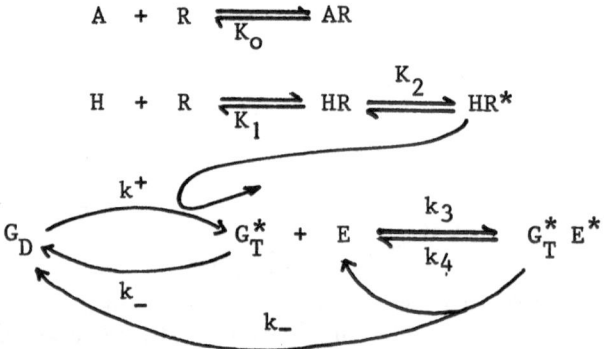

Fig. 4. Model describing the action of β-adrenergic agonist (H) and
        antagonist (A) on the adenylate cyclase system.

    It describes the binding of antagonists to the receptor (K_0),
the binding of the agonist to the receptor (K_1) and the induced
receptor activation (K_2), the activation of the G-unit due to HR*
action (k_+), the interaction between active G-unit and adenylate
cyclase (k_3, k_4), the inactivation of the G-unit due to the GTPase
activity (k_-).

    K_0 and K_1 are equilibrium dissociation constants; K_2 is an
isomerization equilibrium constant; k_+, k_-, k_3 and k_4 are kinetic
constants. All these parameters must be numerically defined as well
as the concentrations (density of molecules per membrane surface
unit) of receptor, G-unit and adenylate cyclase molecules.

3. SIMULATION OF THE MODEL

    Theoretical simulation is essentially a tool for generating
the behaviour of the model which has to be compared to the behaviour
of the experimental system. We shall treat as an example the well-
known property of the adenylate cyclase concerning the non-linear
coupling between receptor occupancy by β-adrenergic agonist and the
adenylate cyclase activation. The original analysis was presented

elsewhere[2]. In this chapter, we would like to emphasize more the
technical aspect than the interpretation of the theoretical results,
in order to show the relative simplicity of such a theoretical
methodology. The application of these results to the experimental
systems is thus neglected in this presentation.

The level of adenylate cyclase activity mainly depends on the
relative amount of active and inactive G-units (Fig. 4). In the
absence of hormone the rate of G-unit activation is slow compared
to the inactivation rate. Thus only a few G-unit molecules are
active. The role of the active hormone-receptor complex is to
accelerate the activation rate in such a way that the on-process
can overwhelm the off-process. This acceleration depends on the
amount of active HR*.

It has been widely reported that β-adrenergic receptors, can
recognize different classes of molecules characterized by their
respective abilities in activating adenylate cyclase despite the
full occupancy of the receptors: full agonists exhibit high efficacy
whereas antagonists cannot activate the system (zero efficacy);
partial agonists exhibit intermediate efficacies. Such properties
can be explained by the respective capabilities of saturating concentrations of these drugs to accelerate the on-process[15,16]. Within
the framework of the proposed model, such an effect results from
different characteristics of the isomerization equilibrium (K_2).
An agonist exhibits a high efficacy if it can induce an accumulation
of HR* high enough for inducing G_T* accumulation. Since it is
hypothesized that the association of HR* to G is only transient
according to the collision coupling concept[14], the HR* concentration
is only governed at saturation by the equilibrium of the isomerization reaction.

In many experimental systems, an apparent discrepancy between
the level of activation and the level of receptor occupancy has
been observed and cannot be explained without further assumptions
(e.g. spare receptors). Such a discrepancy applies to the
β-adrenergic system[17,18]. The more efficient the β-adrenergic
agonist in activating the adenylate cyclase enzyme, the wider the
difference in agonist concentrations necessary for binding and
effect[18]. The aim of the simulation is to demonstrate if the
proposed model can or cannot account for both observations, namely,

1. the non-linear coupling between binding and activation
2. the higher the discrepancy between receptor occupancy level and
   adenylate cyclase activation level, the higher the agonist
   efficacy.

In order to translate these two observations in mathematical
terms, we must introduce some parameters: a is the agonist
efficacy, that is the ratio between the maximal activity induced by

a saturating agonist concentration and the activity obtained when
all the G-units are activated (for instance by using a non-
hydrolyzable analog of GTP which blocks the turn-off process); K_A
is the activation constant, that is the agonist concentration which
produces half of the maximal activation obtained with this type of
agonist. K_D is the dissociation constant, that is the agonist
concentration which produces half maximal occupancy of the receptors.
σ is the similarity index, that is the ratio K_A/K_D. The lower the
σ value, the higher the discrepancy between K_A and K_D, the more
pronounced the non-linear coupling between receptor occupancy and
adenylate cyclase activation.

On the basis of these definitions, the observations become:
1 - $\sigma<1$ (that is $K_A<K_D$)
2 - there exists an inverse correlation between a and σ (that is
    the higher the efficacy, the higher the discrepancy between
    K_A and K_D).

The demonstration of these two mathematical expressions needs
to define the parameters a, K_A, K_D and σ in terms of the parameters
of the model. In the conditions we shall choose, such a demonstration does not require numerical simulation since as we shall see,
the equations can be easily handled. However, in most cases numerical analysis is an obliged tool to solve the problem.

K_D Determination

Since it is assumed that the amount of HR* bound to the G-unit
is always negligible, the agonist binding is governed by the following equations:

$$K_1 = \frac{(H)\,(R)}{(HR)} \tag{1}$$

$$K_2 = \frac{(HR*)}{(HR)} \tag{2}$$

$$(R_{tot}) = (R) + (HR) + (HR*) \tag{3}$$

Dividing eq(1) by eq(2), we get:

$$\frac{K_1}{K_2} = \frac{(H)\,(R)}{(HR*)} \tag{4}$$

or:

$$(R) = \frac{K_1}{K_2} \frac{(HR*)}{(H)} \tag{5}$$

Let us introduce eqs (5) and (2) in eq (3) :

$$(R_{tot}) = (HR^*) \left(\frac{K_1}{K_2(H)} + \frac{1}{K_2} + 1 \right) \tag{6}$$

and thus:

$$\frac{(HR*)}{(R_{tot})} = \frac{K_2 \ (H)}{K_1 + (H) \ (1 + K_2)} \tag{7}$$

Fractional binding is given by:

$$f_B = \frac{(HR) + (HR*)}{(R_{tot})} \tag{8}$$

and thus, using eqs (2) and (7),

$$f_B = \frac{(H)}{\dfrac{K_1}{1 + K_2} + (H)} \tag{9}$$

Eq(9) corresponds to michaelian kinetics : half maximal binding is obtained when $(H) = K_1/(1 + K_2)$ and thus:

$$K_D = \frac{K_1}{1 + K_2} \tag{10}$$

a determination.

_____

The rates of G-unit activation and inactivation are respectively described by:

$$v_{on} = k_+ \ (HR*) \ (G_D) \tag{11}$$

$$v_{off} = k_- \ (G_T* + G_T* \ E*) \tag{12}$$

The conservation law of G-unit is in this case:

$$(G_{tot}) = (G_D) + (G_T*) + (G_T* \ E*) \tag{13}$$

At the steady state, $v_{on} = v_{off}$ and thus

$$k_+ (HR^*) \left[(G_{tot}) - (G_T^*) - (G_T^* E^*) \right] = k_- \left[(G_T^*) + (G_T^* E^*) \right]$$

(14)

Let us define two new kinetic parameters as functions of the model parameters :

$$k_{on} = k_+ (R_{tot}) \frac{K_2}{K_2 + 1}$$

(15)

$$k_{off} = k_-$$

(16)

Introducing k_{on} and k_{off} in eq(14) and using the value of (HR*) (eq(7)), we obtain:

$$\frac{(G_T^*) + (G_T^* E^*)}{(G_{tot})} = \frac{\dfrac{k_{on}(H)}{K_D + (H)}}{k_{off} + \dfrac{k_{on}(H)}{K_D + (H)}}$$

(17)

As this presentation is only used as an illustration of the simulation approach, we restrict the analysis to the particular case where it is assumed that the adenylate cyclase activity parallels the amount of active G-unit. On the basis of this hypothesis and because of the definition of the efficacy a, we have:

$$a = \lim_{(H) \to \infty} \frac{(G_T^* + G_T^* E^*)}{(G_{tot})}$$

(18)

Using eq(17), eq(18) becomes:

$$a = \frac{k_{on}/k_{off}}{1 + k_{on}/k_{off}}$$

(19)

K_A determination.

The calculation of K_A needs to determine the fractional activation of adenylate cyclase f_A that is the adenylate cyclase activity proportioned to the maximal activity obtained with a given agonist:

$$f_A = \frac{(G_T^*) + (G_T^* E^*)}{(G_{tot})} / a$$

(20)

Dividing eq(17) by eq(19), we get:

$$f_A = \frac{(H)}{\dfrac{K_D}{1 + k_{on}/k_{off}} + (H)} \qquad (21)$$

The fractional activity f_A is, just like f_B, a michaelian function of H concentration which is by definition the K_A value:

$$K_A = \frac{K_D}{1 + k_{on}/k_{off}} \qquad (22)$$

σ determination.
―――――――――――――

By definition of σ and using eq(22) we obtain:

$$\sigma = \frac{1}{1 + k_{on}/k_{off}} \qquad (23)$$

On the basis of all these results based on the proposed model, we can demonstrate if the model can or cannot fulfil the two mathematical conditions describing the experimental behaviour of the system.

Because k_{on} and k_{off} are two positive numbers, eq(23) tells us that σ is always lower than 1. Thus, activation is elicited at agonist concentrations lower than those required for the binding. Using eqs(19) and (23), we obtain a relationship between a and σ:

$$a + \sigma = 1 \qquad (24)$$

Thus within the framework of the proposed model and keeping in mind the different hypotheses, we can see that there exists an inverse correlation between a and σ which is linear in this case. In conclusion, the simulation has demonstrated that the proposed model contains enough information for accounting for the experimentally observed behaviour of the system. If some hypotheses are not acceptable because of further experimental evidence, the simulation work must be done again on the basis of the new model.

4. PREDICTIONS

On the basis of theoretical results, obtained thanks to the simulation of the model, the behaviour of the system can be predicted and new experiments can be designed for testing these predictions.

For instance, because of the instability and of the consequent non-accumulation of the ternary complex H-R-G in the presence of guanine nucleotide, it can be predicted that the rate of G-unit activation should be first order with respect to the concentration of G-unit (see eq(11)). A new methodology has recently suggested that the G-unit activation rate would be a zero-order process with respect to G-unit concentration (19). This apparent contradiction which is rigorously pointed out by the theoretical results cannot however be solved by simulation. What simulation can do is to propose new key experiments which could help to find a clue.

CONCLUSIONS

It was our purpose to show how theoretical simulation can contribute to the development of knowledge concerning biochemical systems. The major advantage of such an approach is to oblige us to commit ourselves since some implicit assumptions contained in a verbal description of the system must be explicitly expressed. Modelling and simulation are thus tools which are easily usable and available to anyone. The only possible technical difficulty lies in some cases in the level of the computer numerical analysis. All the other steps of such an approach are based on pure logic.

ACKNOWLEDGMENTS

This work has been realized under Contract of the Ministère de la Politique Scientifique (Actions Concertées). We would like to thank Mrs D. Leemans for typing the manuscript.

REFERENCES

1. S. Swillens and J. E. Dumont, Life Sci. 27:1013-1028 (1980).
2. S. Swillens and J. E. Dumont, Mol. Cell Endocrin. 20:233-242 (1980).
3. M. Rodbell, L. Birnbaumer, S. L. Pohl and H. M. J. Krans, J. Biol Chem. 246:1877-1882 (1971).
4. T. Pfeuffer and E. M. J. Helreich, J. Biol. Chem. 250:867-876 (1975).
5. D. Cassel and Z. Selinger, Biochim. Biophys. Acta 452:538-551 (1976).
6. D. Cassel and Z. Selinger, Proc. Natl. Acad. Sci. 75:4155-4159 (1978).
7. D. Cassel, H. Levkovitz and A. Selinger, J. Cycl. Nucl. Res. 3:393-406 (1977).
8. T. Pfeuffer, FEBS Letters 101:85-89 (1979).
9. D. Cassel and Z. Selinger, J. Cycl. Nucl. Res. 3:11-22 (1977).

10. L. E. Limbird and R. J. Lefkowitz, Proc. Natl. Acad. Sci. 75: 228-232 (1978).
11. L. T. Williams and R. J. Lefkowitz, in: "Receptor Binding Studies in Adrenergic Pharmacology," Raven Press, New York (1978).
12. G. A. Weiland, K. P. Minneman and P. B. Molinoff, Nature 281: 114-117 (1979).
13. L. Birnbaumer, C. F. Bearer and R. Iyengar, J. Biol. Chem. 255: 3552-3557 (1980).
14. A. M. Tolkovsky and A. Levitzki, Biochemistry 17:3795-3810 (1978).
15. J. P. Perkins, T. K. Harden and Y. F. Su in: "Catecholamines : Basic and Clinical Frontiers," E. Usdin, I.J. Kopin and J. Barchas, eds., Pergamon Press, Elmsford, N.Y., Vol.1, pp.542-546 (1979).
16. H. Arad and A. Levitzki, Mol. Pharmacol. 16:749-756 (1979).
17. P. A. Insel and L. M. Stoolman, Mol. Pharmacol. 14:549-561 (1978).
18. W. L. Terasaki and G. Brooker, J. Biol. Chem. 253:5418-5425 (1978).
19. Y. Citri and M. Schramm, Nature 287:297-300 (1980).

GUANYLATE CYCLASE: PROPERTIES AND REGULATION

Joel G. Hardman

Department of Pharmacology
Vanderbilt University School of Medicine
Nashville, Tennessee 37232

INTRODUCTION

Cyclic GMP was discovered in biological material in 1963[1], and guanylate cyclase, the enzyme that catalyzes the formation of the cyclic nucleotide from GTP, was first described in 1969[2-5]. Because of its chemical similarity to cyclic AMP, cyclic GMP has long been assumed to be involved in cellular regulation. This assumption has been reinforced by observations that cellular levels of cyclic GMP rapidly increase in response to agonists that affect such processes as myocardial and smooth muscle contraction, cell division, exocrine and endocrine secretion, platelet aggregation and neuronal activity (see reference 6 for a thorough review of this area). The role of cyclic GMP in these processes, however, is still obscure.

Cyclic GMP and cyclic AMP have several biochemical similarities. Both are formed from the corresponding nucleoside triphosphates by relatively specific cyclases; both are hydrolyzed by one or more phosphodiesterases; both can be extruded from cells; both can activate specific protein kinases; both can be increased in cells and tissues by hormones and neurotransmitters. As Greengard[7] has pointed out, these analogies are superficial and perhaps misleading. Hormones and neurotransmitters that raise cellular levels of cyclic AMP can activate adenylate cyclase in cell-free systems. Hormones and neurotransmitters that increase cellular cyclic GMP levels do not consistently and reproducibly activate guanylate cyclase in cell-free systems. "Physiological" effects of hormones and neurotransmitters that raise cyclic AMP levels often can be reproduced by application to the cells or tissues of exogenous cyclic AMP (or a derivative of it) or of phosphodiesterase inhibitors. Effects of hormones and neurotransmitters that raise cyclic GMP levels seldom can be reproduced by the application of exogenous cyclic GMP (or its derivatives) or of phosphodiesterase inhibitors or both agents that increase cyclic GMP levels (e.g. nitroprusside). It has been suggested that, rather than serving as a mediator of effects of naturally occurring agonists that increase its formation, cyclic GMP may instead serve in a kind of negative feedback capacity, restoring the cell to its prestimulus cytoplasmic Ca^{2+} concentration or redox state[8,9]. Suffice it to say that there is still much to be learned about the role of cyclic GMP in cellular regulation. Our ignorance in this regard is in part the result of inadequate knowledge about mechanisms involved in the physiological regulation of guanylate cyclase.

The purpose of this chapter is to provide an introduction to and brief review of some aspects of guanylate cyclase and potential mechanisms for its regulation. More detailed information on the subject can be found in reviews by Goldberg and Haddox[6], Murad et al.[9] and Garbers et al.[10].

DISTRIBUTION AND PROPERTIES OF GUANYLATE CYCLASE

As with the adenylate cyclase reaction, the other product of the guanylate cyclase reaction (in addition to cyclic GMP) is pyrophosphate[11]. Guanylate cyclase can use dGTP and ITP as substrates, catalyzing the formation of cyclic dGMP and cyclic IMP, respectively[12], and the enzyme can generate cyclic AMP from ATP when it is activated by nitric oxide and other oxidants[9].

Guanylate cyclase is located in both soluble and particulate fractions of homogenates of most mammalian tissues. The proportion of activity associated with the two fractions varies widely from one source to another. In human platelets, for example, the soluble form predominates, whereas in rat small intestinal mucosa the particulate form accounts for virtually all of the activity. The richest source of guanylate known is the sperm of sea urchins, where all of the activity is particulate and associated with plasma membranes of the sperm tail. Among mammalian cells and tissues, the highest activity found is in the rod outer segments of the retina, followed by platelets and then lung and intestinal mucosa; brain, liver and kidney contain somewhat lower activities and heart and skeletal muscle still less. The particulate form of the enzyme is associated with the plasma membrane fraction of several tissues, and some activity has been reported to be associated with several other organelles of some tissues, including endoplasmic reticulum, Golgi and the nucleus. Detergents increase guanylate cyclase activity in particulate fractions by several fold in part at least because of solubilization. Activity in soluble fractions also is increased by some detergents, but generally to a much smaller extent than is particulate activity.

A distinguishing property of both soluble and particulate guanylate cyclases has been their apparent strong dependence upon Mn^{2+} for optimal activity. Activity with Mg^{2+} or Ca^{2+} as the sole divalent cation is ordinarily only 5-10 percent or less of that obtained with Mn^{2+}. When activated by various oxidants, detergents and fatty acids, however, guanylate cyclase can be as active with Mg^{2+} as with Mn^{2+}. Mg^{2+} rather than Mn^{2+} is probably the physiologically important cation for guanylate cyclase activity; Mn^{2+} is present in cells in concentrations much lower than appear to be required to support significant activity. For more complete discussions and original references related to the material covered in the foregoing paragraphs, see references 6, 9, and 10.

Soluble and particulate forms of guanylate cyclase are probably truly different enzymes and not the same protein localized in different parts of the cell. In addition to their solubilities, the two forms have been known for some time to differ in their kinetic behavior, sensitivity to activation and inhibition by divalent cations, apparent molecular weights and some other properties[6,9,10]. The strongest published evidence that they are different proteins comes from studies by Garbers[13], who obtained antibodies to the particulate guanylate cyclase from sea urchin sperm. These antibodies cross reacted with particulate but not with soluble guanylate cyclase from several mammalian tissues. It has been speculated that the soluble enzyme is loosely associated with membranes in intact cells. There is some circumstantial but not conclusive published evidence to support this possibility (see reference 6).

A major unanswered question is: which form of guanylate cyclase is activated by hormones and neurotransmitters that cause rapid increases in cyclic GMP in intact cells? There are a few unconfirmed reports of hormones and neurotransmitters activating guanylate cyclases in cell-free systems. Most investigators, however, have been unable to show activation of either soluble or particulate forms of the enzyme in broken cells by acetylcholine and other agents that increase cellular cyclic GMP levels[6,9,10]. It is nevertheless probably premature to rule out the possibility that such agonists can activate guanylate cyclase more or less directly, perhaps in a manner analogous to that involved in hormonal activation of adenylate cyclase. It is now well-recognized that hormonal activation of adenylate cyclase is a complex process and is highly dependent upon optimal assay conditions. The common inability of investigators to demonstrate activation by hormones and neurotransmitters of guanylate cyclase in cell-free systems could reflect a requirement for cofactors or conditions that have not been identified. There is, however, substantial evidence to indicate that regulation of guanylate cyclase by hormones and neurotransmitters occurs by more indirect processes than are involved in regulation of adenylate cyclase (see following sections).

GUANYLATE CYCLASE REGULATION

Acetylcholine, histamine, alpha-adrenergic agents and several other hormones and neurotransmitters can increase cyclic GMP levels in numerous cells and tissues (see reference 6); these agents generally require Ca^{2+} to increase cyclic GMP levels. Depolarizing agents and the divalent cation ionophore, A-23187 also raise cyclic GMP levels in some cells and tissues in a Ca^{2+} dependant manner, and simply changing the Ca^{2+} concentrations of the incubation medium can cause changes in cyclic GMP levels in some tissues[6,9,14]. These observations led to the suggestion that cyclic GMP elevations in response to hormones and neurotransmitters occur secondarily to increased intracellular Ca^{2+}[14]. Ca^{2+} at rather high concentrations can activate soluble guanylate cyclase in cell-free systems under certain conditions, but this is probably not a physiologically important phenomenon[15]. Low concentrations of Ca^{2+} have been reported to activate particulate guanylate cyclases from some mammalian sources [16-18], and the Ca^{2+}-binding protein, calmodulin, can activate a membrane-associated guanylate cyclase from tetrahymena[19]. Activation by calmodulin of either soluble or particulate guanylate cyclase from a mammalian tissue has not been reported.

Another group of agents that increase cyclic GMP levels in intact cells does not require Ca^{2+}. Included in this group are azide, nitroprusside, nitrosoureas, and nitroso-guanidines, all of which seem to act by liberating nitric oxide, and several other agents that alter the redox state of the cell[6,9]. These agents in general can activate guanylate cyclase in cell-free systems, apparently by causing oxidation of the enzyme.

The first evidence that guanylate cyclase could be activated by an oxidative process came from observations by Böhme et al.[20], who showed that incubation of the soluble enzyme from platelets caused an increase in activity that was prevented by dithiothreitol. Similar phenomena were subsequently seen with other soluble guanylate cyclases, and White et al.[21] showed that this activation process required O_2 and that H_2O_2 could activate the enzyme under some conditions. Oxidative activation of guanylate cyclase subsequently was shown with the nitric oxide-generating agents mentioned above[9], [22,23], with lipid peroxides or hydroperoxides[24,25], with dehydro-ascorbate[26] and apparently with hydroxyl radical[27]. Effects of these agents on soluble guanylate cyclases are pronounced and are in general much more dramatic with Mg^{2+} than with Mn^{2+}[6,9]. Their effects on particulate guanylate cyclase are less clear. Small effects of nitric oxide-related agents on activity on particulate fractions have been reported, but the possibility that contaminating soluble enzyme could have accounted for the responses does not appear to have been ruled out.

Both soluble and particulate forms of guanylate cyclase can be activated by lipids. Guanylate cyclase activity in whole homogenates or in particulate and occasionally in soluble fractions can be increased by phospholipase A_2 or by lysolecithin[9,24,28-32]. Shier et al.[32] have suggested that lysolecithin, formed through phospholipase A_2 activation, could play a role in agonist-induced activation of membrane-associated guanylate cyclase. Particulate guanylate cyclases can be activated by fatty acids, but high concentrations are required and both saturated and unsaturated fatty acids produce effects[33-35]. Soluble guanylate cyclases, on the other hand, can be activated by low concentrations of unsaturated but not saturated fatty acids [6,9,24,25,36,37]. The unsaturated acids probably act in part at least through the formation of lipid peroxides[24,25].

POSSIBLE LINK BETWEEN OXIDATION AND Ca^{2+}-DEPENDENT REGULATION

The requirement for Ca^{2+} in the effects of neurotransmitters and some other agents on cyclic GMP levels in intact cells, the lack of a clear demonstration of physiologically relevant activation of guanylate cyclase by Ca^{2+} or neurotransmitters in cell-free systems, and the remarkable activation of the enzyme by various oxidants in cell-free systems have led to the suggestion that the effects of Ca^{2+} on cyclic GMP levels in intact cells are mediated by an oxidative process, perhaps one involving the formation of oxidized metabolites of arachidonic acid[6,9]. Several lines of evidence support this suggestion. Ca^{2+}-sensitive phospholipases are known to exist[38,39]; arachidonic acid and its metabolites are released from various cells in response to agents that increase cytoplasmic Ca^{2+} concentrations[40]; and lipid peroxides derived from arachidonic acid can activate guanylate cyclase in cell-free systems[24,25].

Clyman et al.[41] and DeRubertis and Craven[42] have shown in human umbilical arteries and in rat kidney slices, respectively, that O_2 is required for Ca^{2+}-dependent agonists and for Ca^{2+} itself to raise cyclic GMP levels. Glass et al.[43] and Anderson et al.[44] have shown that in platelets and in smooth muscle, respectively, inhibitors of prostaglandin cyclooxygenase impair cyclic GMP responses to agonists. Spies et al.[45] demonstrated in the rat ductus deferens that increases in cyclic GMP caused by Ca^{2+} or by Ca^{2+}-dependent agonists were inhibited by mepacrine, a putative inhibitor of phospholipase A_2, and by eicosatetraynoic acid and nordihydroguaiaretic acid, putative inhibitors of lipoxygenases, but not by the cyclooxygenase inhibitor, indomethacin. These findings led these authors to conclude that hormone-induced elevations in cyclic GMP in that tissue are caused by the release of arachidonic acid (or perhaps other unsaturated fatty acids) by Ca^{2+}-activated phospholipases and the ensuing formation of peroxidized fatty acid products which in turn activate guanylate cyclase. Craven and DeRubertis[46] have shown that Ca^{2+}-dependent increases in cyclic GMP in rat colonic mucosa are inhibited by the exclusion of O_2 and by mepacrine or indomethacin, suggesting the participation of an oxidized product of arachidonic acid arising through the cyclooxygenase pathway.

All of these findings are consistent with the concept that agents that increase Ca^{2+} influx into cells or that cause Ca^{2+} release within cells can bring about increased guanylate cyclase activity through the mediation of an oxidized fatty acid, which, depending on the cell type involved, may arise through either the lipoxygenase or the cyclooxygenase pathways for metabolism of arachidonic acid or other unsaturated fatty acids. Much of the evidence supporting this concept, however, relies on the assumption that the drugs used as putative inhibitors of phospholipase, lipoxygenase and cyclooxygenase are indeed specific inhibitors of these enzymes and do not have other actions that could affect cyclic GMP metabolism. The validity of this assumption does not seem to have been proven. Furthermore, the existence of Ca^{2+}-lipid peroxide mechanism would not necessarily exclude other possible mechanisms for the regulation of guanylate cyclase by hormones and neurotransmitters. Other potential mechanisms could include more direct effects of hormones or neurotransmitters on membrane associated gunaylate cyclase, activation of guanylate cyclase by calmodulin, activation of the enzyme by lysolecithin or other lysophosphatides, and activation by oxidative processes unrelated to lipid peroxide formation. It is conceivable that more than one mechanism could be operable in a single cell and that there could be variation in the importance of a single mechanism from cell type to cell type.

PURIFIED GUANYLATE CYCLASES

Soluble guanylate cyclase from mammalian tissues and the particulate enzyme from a non-mammalian source have been purified to homogeneity or near homogeneity. Garbers[47] purified the particulate enzyme from sea urchin sperm to a specific activity of about $12\mu mol/min/mg$ protein. This activity is over 10-fold greater than has been achieved with any apparently homogeneous soluble guanylate cyclase (see below). The degree of homogeneity of the sea urchin enzyme has not been established, but it is likely that it is nearly if not completely homogeneous. The enzyme has an apparent molecular weight of 182,000 as determined by gel filtration, and the preparation yields two protein-staining bands on SDS gel electrophoresis with molecular weights of 118,000 and 75,000. Whether the two bands are subunits of the enzyme or one (or both) of them is a contaminant has not been determined. The enzyme is over 150 times as active with Mn^{2+} as with Mg^{2+}, and unlike the crude enzyme which shows positively cooperative kinetics, the purified one exhibits classical, Michaelis–Menten kinetics with respect to MnGTP, which has an apparent Km of 170μM. After purification the enzyme is soluble in the absence of detergent.

Tsai et al.[48] purified the soluble guanylate cyclase from rat liver to a specific activity or 0.12 μmol/min/mg protein. This preparation was not homogeneous, showing several protein bands on SDS gels. After early purification steps, the activity of the enzyme was no longer proportional to the protein concentration. Proportionality between activity and protein concentration and an increase in total activity could be obtained by assaying the enzyme in the presence of an "activator fraction", which was separated from the enzyme by one of the early purification steps. The activating material was non-dialyzable, heat-labile and alkali-stable.

Soluble guanylate cyclase from rat liver has been purified to apparent homogeneity by Braughler et al.[49]. The preparation had an apparent specific activity of 0.28 μmol/min/mg protein, and a molecular weight as determined on non-denaturing polyacrylamide gels of 150,000. Braughler et al., like Tsai et al., found that their enzyme did not yield proportionality between activity and protein concentration, with higher protein concentrations yielding higher apparent specific activities. This problem makes it difficult to assign a precise value for specific activity to the enzyme. The enzyme purified by Braughler et al. was activated by several agents that can activate cruder preparations of guanylate cyclase, including nitroprusside, nitric oxide and unsaturated fatty acids.

Garbers[50] purified the soluble enzyme from rat lung to apparent homogeneity and a specific activity of 0.7 μmol/min/mg protein when assayed in the absence and over 1 μmol when assayed in the presence of bovine serum albumin. This preparation yielded constant specific activity with enzyme protein concentration when assayed in the presence of bovine serum albumin but not in its absence. The molecular weight of the enzyme was found to be about 151,000 by gel filtration. Two bands were visible after electrophoresis on SDS gels, with molecular weights of 74,000 and 79,400. A recent modification of the procedure (Garbers, personal communication) has yielded a preparation with a single band on SDS gels at 79,400. The purified lung enzyme is highly sensitive to activation by nitroprusside and nitric oxide (S.G. Laychock, D.L. Garbers and J.G. Hardman, unpublished).

Another apparently homogeneous soluble guanylate cyclase recently has been purified from bovine lung by Gerzer et al.[51]. The specific activity is 0.5 μmol/min/mg protein and the apparent molecular weight from gel filtration is 155,000. This preparation also is highly sensitive to activation by nitroprusside.

Zwiller and Mandel[52] have reported the purification of soluble
guanylate cyclase from rat brain to an extent such that only one
protein band was detectable on denaturing and non-denaturing gels.
This preparation, however, had a specific activity of only
0.0007 µmol/min/mg protein and was purified only some 10-fold from
the supernatant fraction of the tissue homogenate (the other soluble
preparations discussed above were purified 5,000 to 17,000-fold from
supernatant fractions). The authors suggest that the low degree of
purification must have been due to the loss of an activator of the
enzyme. Another possibility would seem to be that the single band
of protein detected was a protein other than guanylate cyclase.
This enzyme preparation was not sensitive to activation by nitroprusside.

CONCLUDING REMARKS

In conclusion, two points deserve re-emphasis. First, we do
not know whether Ca^{2+}-dependent agonists that raise cyclic GMP in
cells do so by activating the soluble or the particulate forms of
guanylate cyclase. Second, it is unclear that particulate guanylate
cyclase is as generally sensitive to activation by oxidants as is
the soluble form. If we knew which form of guanylate cyclase,
soluble or particulate (or both), was regulated by Ca^{2+}-dependent
hormones and neurotransmitters and if we knew whether or not the
particulate as well as the soluble form of the enzyme was, in
general, capable of being activated by oxidants, some seemingly
disparate observations might be reconciled.

It has been a fairly commonplace finding that agents such as
nitroprusside increase cyclic GMP in tissues but do not elicit
the "physiological" responses characteristically elicited in the
same tissues by calcium-dependent agonists (such as acetylcholine)
that also increase cyclic GMP. It is possible of course that the
"physiological" responses – for example, acetylcholine-induced
negative inotropism and chronotropism in the heart – have nothing
to do with cyclic GMP. On the other hand it is possible that in a
tissue composed of more than one cell type, Ca^{2+}-dependent agonists
increase cyclic GMP in the cell type that is responsible for the
"physiological" responses, for example myocytes in the heart, while
agents such as nitroprusside raise cyclic GMP in another cell type
that is not involved in the "physiological" responses, for example
capillary endothelium. Such a situation could come about if:
(a) Ca^{2+}-dependent agonists that elicit "physiological" responses
activate only or predominantly membrane associated guanylate cyclase;
(b) oxidants such as nitroprusside activate only or predominantly
soluble guanylate cyclase; (c) the cell type responsible for the
"physiological" response contains only or predominantly membrane
associated guanylate cyclase; and (d) other cell types in the
tissues contain substantial amounts of soluble enzyme. The only

bits of evidence that are consistent with this perhaps unlikely set of circumstances are: (a) activation of guanylate cyclase by various types of oxidants has been in general much more conspicuous with soluble than with particulate enzyme; and (b) there is a wide variation in relative amounts of soluble and particulate enzyme from one cell type to another. The reader should recognize quickly that the suggestion that Ca^+-dependent agonists raise cyclic GMP in cells through the mediation of an oxidative process is not consistent with the above speculation. The possibility that confusing results can arise because of selective effects of various agents on cyclic GMP levels in different cell types has been raised here to draw attention to the following point: too little still is known about mechanisms involved in the physiological regulation of guanylate cyclase to permit straightforward interpretation of results that seem to show either associations or dissociations between changes in tissue levels of cyclic GMP and "physiological" responses to agonists.

REFERENCES

1. D.F. Ashman, R. Lipton, M.M. Melicow, and T.D. Price, Biochem. Biophys. Res. Commun. 11:330-334 (1963).
2. J.G. Hardman, and E.W. Sutherland, J. Biol. Chem. 244:6363-6370 (1969).
3. E. Ishikawa, S. Ishikawa, J.W. Davis, and E.W. Sutherland, J. Biol. Chem. 244:6371-6376 (1969).
4. A.A. White, and G.D. Aurbach, Biochim. Biophys. Acta. 191:686-697 (1969).
5. G. Schultz, E. Böhme, and K. Munske, Life Sci. 8:1323-1332 (1969).
6. N.D. Goldberg, and M.K. Haddox, Ann. Rev. Biochem. 46:823-896 (1977).
7. P. Greengard, Trends in Pharm. Sci. 1:27-29 (1979).
8. G. Schultz, J.G. Hardman, K. Schultz, C.E. Baird and E.W. Sutherland, Proc. Nat. Acad. Sci. U.S.A. 70:3889-3893 (1973).
9. F. Murad, W.P. Arnold, C.K. Mittal, and J.M. Braughler, Advan. Cyc. Nuc. Res. 11:175-204 (1979).
10. D.L. Garbers, T.D. Chrisman, and J.G. Hardman. in: "Eukaryotic Cell Function and Growth: Regulation by Intracellular Cyclic Nucleotides" J.E. Dumont, B.L. Brown and N.J. Marshall, eds., Plenum Press, New York and London, pp. 155-193 (1976).
11. D.L. Garbers, T.D. Chrisman, J.L. Suddath, and J.G. Hardman, Arch. Biochem. Biophys 166:135-138 (1975).
12. D.L. Garbers, J.L. Suddath, and J.G. Hardman, Biochim. Biophys. Acta 377:174-185 (1975).
13. D.L. Garbers, J. Biol. Chem. 253:1898-1901 (1978).
14. G. Schultz, and J.G. Hardman, Advan. Cyc. Nuc. Res. 5:339-351 (1975).

15. D.L. Garbers, T.D. Chrisman, and J.G. Hardman, in: "Molecular Biology and Pharmacology of Cyclic Nucleotides", G. Folco, and R. Paoletti, eds., Elsevier/North Holland Biomedical Press, Amsterdam, pp. 43-55 (1978).

16. D. Wallach, and I. Pastan, Biochem. Biophys. Res. Comm. 72:859-865 (1976).

17. J. Levilliers, F. Lecot, and J. Pairault, Biochem. Biophys. Res. Commun. 84:727-735 (1978).

18. S.N. Levine, A.L. Steiner, H.S. Earp, and G. Meissner, Biochim. Biophys. Acta 566:171-182 (1979).

19. S. Nagao, Y. Suzuki, Y. Watanabe, and Y. Nozawa, Biochem. Biophys. Res. Comm. 90:261-268 (1979).

20. E. Böhme, R. Jung, and I. Mechler, in: "Methods in Enzymology", J.G. Hardman and B.W. O'Malley, eds., Academic Press, New York, Vol. 38, pp. 199-202 (1974).

21. A.A. White, K.M. Crawford, C.S. Patt, and P.J. Lad, J. Biol. Chem. 251:7304-7312 (1976).

22. E. Böhme, H. Graf, and G. Schultz, Advan. Cyc. Nuc. Res. 9:131-143 (1978).

23. F.R. DeRubertis, and P.A. Craven, J. Biol. Chem. 251:4651-4658 (1976).

24. H. Hidaka, and T. Asano, Proc. Nat. Acad. Sci. U.S.A. 74:3657-3661 (1977).

25. G. Graff, J.H. Stephenson, D.B. Glass, M.K. Haddox, and N.D. Goldberg, J. Biol. Chem. 253:7662-7676 (1978).

26. M.K. Haddox, J.H. Stephenson, M.E. Moser, and N.D. Goldberg, J. Biol. Chem. 253:3143-3152 (1978).

27. C.K. Mittal, and F. Murad, Proc. Nat. Acad. Sci. U.S.A. 74:4360-4364 (1977).

28. M. Fujimoto, and T. Okabayashi, Biochem. Biophys. Res. Comm. 67:1332-1336 (1975).

29. J. Zwiller, J. Ciesielski-Treska, and P. Mandel, FEBS Lett., 69:286-290 (1976).

30. S.J. Sulakhe, N.L.-K. Leung, and P.V. Sulakhe, Biochem. J. 157:713-719 (1976).

31. D. Aunis, M. Pescheloche, and J. Zwiller, Neuroscience 3:83-93 (1978).

32. W.T. Shier, J.H. Baldwin, M. Nilsen-Hamilton, R.T. Hamilton, and N.M. Thanassi, Proc. Nat. Acad. Sci. U.S.A. 73:1586-1590 (1976).

33. D. Wallach, and I. Pastan, J. Biol. Chem. 251:5802-5809 (1976).

34. T. Asakawa, I. Scheinbaum, and R-j. Ho, Biochem. Biophys. Res. Comm. 73:141-148 (1976).

35. T. Asakawa, M. Takenoshita, S. Uchida, and S. Tanaka, J. Neurochem. 30:161-166 (1978).

36. A.J. Barber, Biochim. Biophys. Acta. 444:579-595 (1976).

37. D.B. Glass, W. Frey, D.W. Carr, and N.D. Goldberg, J. Biol. Chem. 252:1279-1285 (1977).

38. A. Derksen, and P. Cohen, J. Biol. Chem. 250:9342-9347 (1975).

39. H.Brockerhoff, and R.G.Jensen, "Lipolytic Enzymes", Academic Press, New York (1974)
40. H.R. Knapp, O. Oelz, J. Roberts, B.J. Sweetman, J.A. Oates, and P.W. Reed, Proc. Nat. Acad. Sci. U.S.A. 74:4251-4255 (1977).
41. R.I. Clyman, A.S. Blacksin, V.C. Manganiello, and M. Vaughan, Proc. Nat. Acad. Sci. U.S.A. 72:3883-3887 (1975).
42. F.R. DeRubertis, and P.A. Craven, Metabolism 28:855-868 (1978).
43. D.B. Glass, J.M. Gerrard, D. Townsend, D.W. Carr, J.G. White, and N.D. Goldberg, J. Cyclic Nuc. Res. 3:37-44 (1977).
44. K.-E. Andersson, R.G.G. Andersson, P. Hedner, and C.G.A. Persson, Life Sci. 20:73-78 (1977).
45. C. Spies, K.-D. Schultz, and G. Schultz, Naunyn Schmiedeberg's Arch. Pharmacol. 311:71-77 (1980).
46. P.A. Craven, and F.R. DeRubertis, Fed. Proc. 39:1898 (1980).
47. D.L. Garbers, J. Biol. Chem. 251:4071-4077 (1976).
48. S.-C. Tsai, V.C. Manganiello, and M. Vaughan, J. Biol. Chem. 253:8452-8457 (1979).
49. J.M. Braughler, C.K. Mittal, and F. Murad, Proc. Nat. Acad. Sci. U.S.A. 76:219-222 (1979).
50. D.L. Garbers, J. Biol. Chem. 254:240-243 (1979).
51. R. Gerzer, F. Hofmann, and G. Schultz, Hoppe-Seyler's Z. Physiol. Chem. 361:249 (1980).
52. J. Zwiller, and P. Mandel, C.R. Acad. Sc. Paris. 286:423-426, (Series D) (1978).

α-ADRENERGIC AND CHOLINERGIC CONTROL OF CYCLIC NUCLEOTIDE METABOLISM

Karl H. Jakobs and Günter Schultz

Pharmakologisches Institut
Universität Heidelberg
Im Neuenheimer Feld 366
D-6900 Heidelberg
Germany

INTRODUCTION

A large variety of hormones and neurotransmitters elicits specific cellular responses that reach their maximum within a few minutes. These hormonal effects are mediated by membrane-bound receptors and generally involve intracellular signals or "second messengers". One part of these fast cellular hormone responses is triggered by an increased formation of cyclic AMP. The effects of β-adrenergic agonists, of histamine via H_2-receptors, some effects of prostaglandins (e.g., PGE_1, PGD_2 and PGI_2 in platelets), of adenosine (in platelets) and of some polypeptide hormones, e.g., glucagon, belong to this group[1].

Many of these hormones and neurotransmitters can induce other cellular responses that are not mediated by increased cyclic AMP formation. For some of these hormonal effects, the involvement of receptors other than those triggering cyclic AMP-mediated responses has been established. The group of hormonal effects not mediated by an increase in the intracellular cyclic AMP level includes the effects of α-adrenergic and (muscarinic) cholinergic agonists, of other smooth muscle-contracting factors such as angiotensin II, vasopressin and bradykinin, of some prostaglandins (including PGE_1 and PGE_2 in adipocytes), of adenosine (in adipocytes) and of opiates (Table 1). The nature and the generation of the signals formed under such hormonal stimulations at the plasma membrane and involved in the cellular responses to the above stimuli will be described in the following.

75

Table 1. Intracellular Mediation of Cellular Responses to
          Hormonal Factors

Hormonal factor	Hormonal effects mediated by	
	increased cAMP	other mechanisms
noradrenaline, adrenaline	β (various)	α_1, α_2 (various)
dopamine	D_1 (CNS)	D_2 (CNS)
histamine	H_2 (stomach)	H_1 (smooth muscle)
serotonin	+ (CNS)	+ (smooth muscle)
acetylcholine (muscarinic)		+ (heart, smooth muscle)
adenosine	+ (platelets)	+ (adipocytes)
ADP		+ (platelets)
prostaglandins		
$\quad I_2$	+ (platelets)	
$\quad F_{2\alpha}$		+ (smooth muscle)
$\quad E_2$	+ (fibroblasts)	+ (adipocytes)
polypeptides		
$\quad$ glucagon	+ (liver)	
$\quad$ enkephalines		+ (CNS)
$\quad$ vasopressin	+ (kidney)	+ (smooth muscle)

Receptor subtypes involved in different mechanisms of intracellular mediation are given so far identified. Typical
examples of cell or tissue, for which a type of mediation
has been established, are given in parentheses.

INTRACELLULAR SIGNALS IN HORMONAL EFFECTS NOT MEDIATED BY INCREASED
CYCLIC AMP FORMATION

Cholinergic and α-adrenergic stimuli cause several cellular
changes that may serve as intracellular signals, transferring the
information from the membrane-bound receptors to the sensitive
intracellular systems (Table 2). In some cell types, α-adrenergic
agonists (via α_2-adrenoceptors) and cholinergic agonists (via
muscarinic receptors) elicit cellular responses that are accompanied
by decreased intracellular cyclic AMP levels. Binding of these
hormonal agonists to their respective receptors is generally
affected by guanine nucleotides and sodium ions[2].

In other cells and tissues, responses to α-adrenergic stimuli
(acting via α_1-adrenoceptors) and (muscarinic) cholinergic agonists
involve increases in the cytoplasmic calcium ion and cyclic GMP
concentrations. It has been proposed by Michell and co-workers[3,4]

Table 2.   Cellular Changes Observed in Hormone Effects Not
Mediated by Increased Cyclic AMP Formation

---

1. Decreased cyclic AMP concentration
2. Increased cyclic GMP concentration
3. Increased cytoplasmic calcium concentration
4. Increased phosphatidylinositol metabolism
5. Increased arachidonic acid release and peroxidation

---

that the increased influx of calcium from the extracellular space
(and from intracellular storage sites) into the cytoplasm is preceded
by and the consequence of hormone-induced changes in membrane
phosphatidylinositol metabolism. Binding of agonists to their
receptors appears to induce a rapid, clacium-independent degradation
of phosphatidylinositol to inositolphosphates and diacylglycerol,
which is converted to phosphatidate (Fig. 1). It is not clear which
of these products is responsible for the calcium gating. The
increased release and peroxidation of arachidonate observed with
α-adrenergic, cholinergic and other hormonal stimuli appears to be
secondary to increased cytoplasmic calcium. The receptor-binding
of hormonal agonists whose actions are mediated by increased

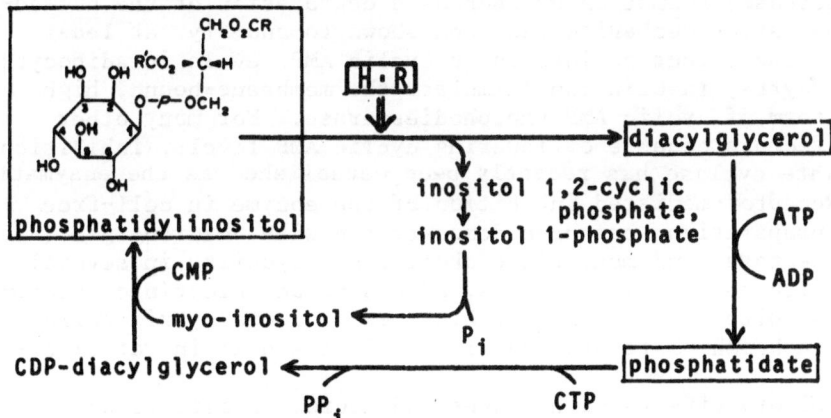

Fig. 1.   Phosphatidylinositol metabolism. Interaction of hormones
with their receptors (H·R) is assumed to stimulate a phospholipase catalyzing the conversion of phosphatidylinositol to
inositol-1,2-cyclic phosphate and diacylglycerol. The latter
product is rapidly converted to phosphatidate. It is not
known which one of the products formed during this "PI
response" is responsible for calcium gating.

calcium concentrations is generally not affected by guanine nucleotides.

In some cell types, hormonal stimulation has been shown to cause a combination of decreased cyclic AMP and of increased calcium and cyclic GMP levels. In some of these cases, different subtypes of receptors are apparently involved, e.g., in rat hepatocytes, where α_1-adrenoceptors mediate phosphorylase activation by a calcium-dependent mechanism and α_2-adrenoceptors appear to mediate effects on cyclic AMP formation. In other cases of combined responses, the occurrence and involvement of two subtypes of receptors has not been established, e.g., for angiotensin II effects on rat hepatocytes and for ADP actions on human platelets.

In some other cases of hormonal effects, the nature of the intracellular signal(s) is far from being clear. For instance, insulin and some growth factors exert cellular changes after inter-action with membrane-bound receptors, but established intracellular signals such as cyclic nucleotides and calcium ions are apparently not or not solely involved in the cellular responses to these hormones.

ENZYMATIC BASIS OF HORMONE-INDUCED DECREASES IN CYCLIC AMP LEVELS

Decreased cellular levels of cyclic AMP can be the result of either decreased formation or increased degradation of the nucleo-tide. The latter mechanism has been shown to underly, at least partially, the effect of insulin on cyclic AMP levels in adipocytes and hepatocytes; insulin can stimulate the membrane-bound, high affinity form of cyclic AMP phosphodiesterase. For many other hormonal factors capable of lowering cyclic AMP levels, inhibition of adenylate cyclase has recently been established as the enzymatic basis. Receptor-mediated inhibition of the enzyme in cell-free membrane preparations has been demonstrated with α-adrenergic (via α_2-adrenoceptors) and muscarinic cholinergic agonists in several tissues, with some prostaglandins, adenosine and nicotinic acid in adipocytes, with opiates in neuroblastoma x glioma hybrid cells, with ADP in human platelets and with angiotensin II in rat liver[5,7].

Inhibitory effects of hormones and neurotransmitters on adenylate cyclase share some common features:

1. α-adrenergic and cholinergic agonists and other hormonal factors reduce adenylate cyclase activity without apparent time lag. Antagonists, so far available, reverse the inhibitory effects without apparent lag phase. Hormone-induced inhibition of adenylate cyclase is closely correlated with data obtained in receptor binding experiments and in studies on overall cellular responses with regard to the potency order of hormonal agonists and antagonists.

2. Hormone-induced inhibition of adenylate cyclase is observed on
    basal as well as on hormone-stimulated enzyme activities. In
    the absence of a stimulatory hormone, inhibition by maximally
    effective inhibitory hormone concentration varies between 50
    and 80%. Stimulatory hormones usually lower the % inhibition,
    but increase the total decrease in cyclic AMP formation. Less
    inhibition is generally seen when adenylate cyclase is stimulated
    by fluoride.

3. The presence of GTP is generally required for inhibitory hormonal
    effects on adenylate cyclase as it has been shown to be necessary
    for the enzyme stimulation by stimulatory hormones[8,9]. In
    several systems, half-maximally effective GTP concentrations
    were found to be 5- to 10-fold higher for inhibitory hormonal
    effects than those required for stimulation by hormones. Stable
    GTP analogues, which are only slowly hydrolyzed and cause
    persistent adenylate cyclase activation and potentiation of
    hormone-induced stimulation, prevented or reversed inhibitory
    hormonal effects on the enzyme. In contrast, after treatment
    with GTP plus cholera toxin, a condition causing a similar
    persistent activation, pronounced inhibitory hormonal effects
    have been observed. As shown in at least two systems, i.e.,
    rabbit myocardium and human platelets, the hormones appeared to
    counteract the GTP-induced activation seen after cholera toxin
    treatment.

4. It has been shown in several cellular systems that besides GTP
    a monovalent cation is required for effective inhibitory coupling
    of α-adrenergic, cholinergic and other receptors to adenylate
    cyclase. Na^+ appears to be most potent, followed by Li^+ and K^+
    Sodium concentrations required for inhibitory hormonal effects
    were between 30 and 200mM. Whereas the requirement for a mono-
    valent cation has been established for hamster and rat fat cells,
    rabbit myocardium and neuroblastoma x glioma hybrid cells, this
    requirement is not clear for human platelets, whose adenylate
    cyclase is inhibited by α-adrenergic agonists and ADP. In some
    systems, such as hamster and rat adipocytes, myocardium and
    liver, monovalent cations activated adenylate cyclase in the
    presence of GTP, and hormonal factors appeared to counteract
    this activation.

5. Inhibitory hormonal effects on adenylate cyclase activity are
    additionally affected by divalent cations such as manganese and
    magnesium. Whereas inhibitory effects were most pronounced at
    low concentrations of free Mn^{2+} or Mg^{2+}, hormone-induced
    inhibition was reduced or abolished when the Me^{2+} concentration
    was high.

Table 3.   Comparison of Stimulatory and Inhibitory Hormonal Effects
          on Adenylate Cyclase

Adenylate cyclase	stimulation	inhibition
onset / reversal	fast	fast
forms affected	basal, inhibited	basal, stimulated
maximal effects	several-fold	50 to 80 %
GTP	required	required (higher conc.)
cholera toxin + GTP	synergistic	permit inhibition
stable GTP analogues	synergistic	prevent inhibition
optimal Na^+ conc.	low	high (30 to 200 mM)
optimal Me^{2+} conc.	high	low

H·R-Binding*	stimulatory horm.	inhibitory hormones
modulation by GTP	+	+
modulation by Na^+	-	+
modulation by Me^{2+}	+	+

*H·R-Binding, hormone-receptor-binding.

The mechanism by which α-adrenergic, cholinergic and other
inhibitory receptors are coupled to adenylate cyclase is far from
being clear. As both stimulatory and inhibitory hormonal factors
exert their effects on adenylate cyclase by GTP-dependent mechanisms
and guanine nucleotides affect agonist binding to either group of
receptors (see Table 3), it is possible that one common GTP-controlled
membrane component is involved in both transducing mechanisms. The
finding that hormone-induced inhibition is prevented or reversed by
slowly hydrolyzable GTP analogues may indicate that inhibitory
hormonal effects on adenylate cyclase may involve increased inactivation of this enzyme by stimulation of the specific GTPase, which
catalyzes the degradation of the GTP bound to the regulatory trans-
sucing component[10]. On the other hand, the finding with the stable
GTP analogues may indicate the involvement of a GTP-dependent
phosphorylation process in hormone-induced inhibition.

Rodbell and associates[8] have proposed that different GTP-
dependent control components are involved in stimulatory and
inhibitory effect on adenylate cyclase. So far only one GTP-binding
protein with a molecular weight of about 43,000 has been isolated[11].
As this protein, which is ADP-ribosylated by cholera toxin, appears
to exist as a subunit of a higher molecular weight complex within
the membrane[12,13], it is possible that this GTP-binding protein is
associated with other regulatory protein subunits to form different

either stimulatory or inhibitory transducing components within the
membrane.

Sodium ions, which have been shown to affect binding of
α-adrenergic, cholinergic and other agonists to their receptors in
a manner similar to that observed with GTP, appear to affect
receptor-cyclase-coupling by an effect on the GTP-controlled transducing component. The exact mechanism of this control by monovalent
cations, however, and the side of the membrane, from which the
transduction process is affected by monovalent cations, is unclear.

HORMONAL CONTROL OF CYCLIC GMP FORMATION

Cholinergic agonists (via muscarinic receptors), α-adrenergic
and other stimuli can increase cyclic GMP levels in a variety of
cell types[14]. These effects of hormonal factors are restricted to
intact tissues or cells and appear to be the result of increased
cyclic GMP formation by the guanylate cyclase rather than decreased
nucleotide degradation. Hormonal effects on cyclic GMP formation
are thought to be the result of a chain of mediating cellular
reactions. As hormonal effects on cyclic GMP levels are generally
reduced or abolished when the extracellular Ca^+ concentration is
reduced, it is thought that a hormone-induced increase in the
cytoplasmic Ca^+ concentration is an early step in the coupling of
hormone receptors to guanylate cyclase [15]. Besides Ca^+, oxygen
has been shown to be required for hormone- and calcium-induced
elevations of tissue cyclic GMP levels at least in two tissues,
umbilical arteries[16] and kidney slices[17].

With none of the hormonal factors capable of increasing cyclic
GMP levels in intact cells, convincing and reproducible stimulatory
effects on cell-free guanylate cyclase preparations have been
observed[18]. The restriction of hormonal responses on cyclic GMP
formation to intact cells may be explained by hormone-induced changes
of the gradients of ionic concentrations, which are interrupted by
cell desintegration. The guanylate cyclase occurs in a soluble and
in a particulate form, the latter one being partially bound to plasma
membranes. Whereas the regulation of the particulate guanylate
cyclase activity is only poorly understood, some data on the
regulations have been obtained with crude or partially purified preparations of the soluble form of the enzyme. Potent stimulants of
this form of the enzyme are fatty acid peroxides such as prostaglandin G_2 and various hydroperoxyeicosatetraenoic acids (HPETEs)[19,20].
Additionally, reactive oxygen species such as the hydroxyl radical
can stimulate the enzyme[21].

On the basis of these findings with intact tissues and cell-
free guanylate cyclase preparations, it has been proposed that
hormone-induced stimulation of cyclic GMP formation involve a
Ca^{2+}-induced increase in the release and peroxidation of arachidonate

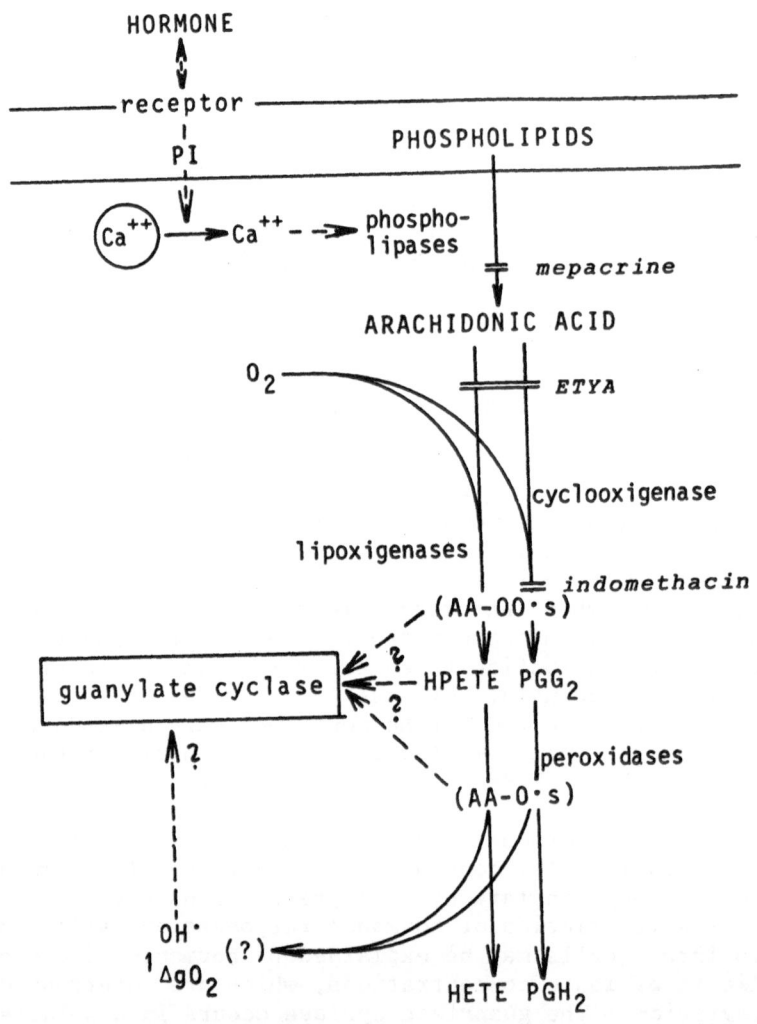

Fig. 2.   Proposed coupling of hormone receptors to soluble guanylate
cyclase. PI, phosphatidylinositol; ETYA, eicosatetraynoic
acid; AA, arachidonic acid; HPETEs, hydroperoxyeicosatetraenoic acids; PGG$_2$, prostaglandin G$_2$; PGH$_2$, prostaglandin H$_2$;
HETEs, hydroxyeicosatetraenoic acids; OH$^\cdot$, hydroxyl radical;
$^1\Delta$O$_2$, singlet oxygen. For further explanations see text.

and possibly other unsaturated fatty acids from membrane phospholipids (see Fig. 2). This concept has been strengthened by the
findings that mepacrine, a phospholipase inhibitor, and eicosa-
tetranoic acid (ETYA), an arachidonic acid analogue, preventing
fatty acid peroxidation by cyclooxygenase and lipoxygenases, can

prevent or reduce hormone- and Ca^{2+}-induced elevations of tissue cyclic GMP[22]. Inhibition of cyclooxygenase by indomethacin has resulted in a decrease in hormone effect on cyclic GMP levels in human platelets[23] but not in the rat ductus deferens. Whether fatty acid peroxides (prostaglandin G_2 and hydroperoxyeicosatetraenoic acids), radical fatty acid forms (occurring during peroxide formation and degradation by peroxidase) or hydroxyl radical (as a likely peroxidase product) are stimulants of guanylate cyclase <u>in vivo</u> needs to be elucidated. Moreover, it needs to be clarified whether additional, different receptor-guanylate cyclase-coupling mechanisms exist, especially for the membrane-bound form of guanylate cyclase.

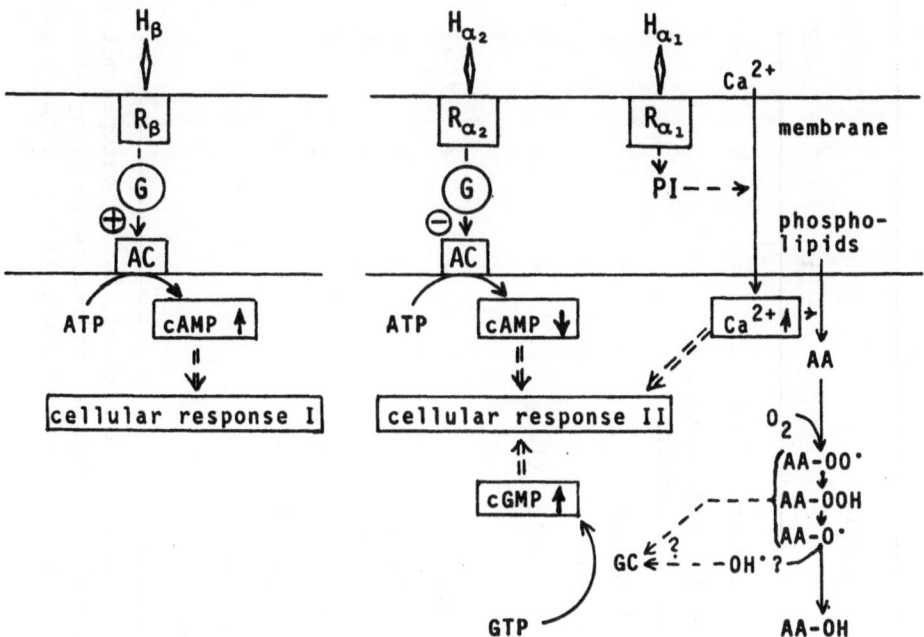

Fig. 3. Signal-generating systems coupled to membrane-bound hormone receptors. Hormonal agonists (H) can interact with receptors (R) that are coupled by a GTP-dependent transducing component (G) to adenylate cyclase (AC). Via different receptors, hormonal factors can cause either stimulation or inhibition of adenylate cyclase. Binding of agonists to other hormone receptors induces changes in phosphatidylinositol (PI) metabolism and stimulates influx of Ca^{2+} from inactive sites into the cytoplasm. Secondary formation of arachidonic acid (AA) peroxides is assumed to stimulate guanylate cyclase (GC). Adrenergic receptor subtypes (R_β, R_{α_1}, R_{α_2}) coupled to the three given signal-generating systems are shown as representative examples.

Table 4. Cellular Changes Induced by Hormonal Stimuli Acting via Membrane-Bound Receptors and Not Mediated by Increased Cyclic AMP Formation

Hormonal stimulus	Cell / Tissue	Cellular changes*				
		PI↑	Ca²⁺↑	cGMP↑	AC↓	GTP on H·R
adrenergic α₁	smooth muscle, liver	+	+	+		
α₂	platelets, adipocytes, liver, N x G hybrids§				+	+
cholinergic (muscarinic)	smooth muscle	+	+	+		
	myocardium			+	+	+
angiotensin II	smooth muscle, liver	+	+			
	liver				+	+
ADP	platelets	+	+	+	+	
histamine H₁	smooth muscle	+	+	+		
serotonin	smooth muscle	+	+	+		
vasopressin	smooth muscle, liver	+	+			
opiates	N x G hybrids§				+	+
adenosine	adipocytes				+	
prostaglandin E₂	adipocytes				+	

*PI↑, phosphatidylinositol metabolism increased; Ca²⁺↑, cytoplasmic calcium increased; cGMP↑, cyclic GMP level increased; AC↓, adenylate cyclase inhibition; GTP on H·R, modulation of hormone-receptor-binding by GTP. §N x G hybrids, neuroblastoma x glioma hybrid cells.

CONCLUSIONS

Membrane-bound receptors for hormones and neurotransmitters can be coupled to different other membrane components generating intracellular signals, which lead to the cellular response to the stimulus (Figure 3). For various extracellular signals, the occurrence of subclasses of cellular receptors has been established, which are coupled to different intracellular signal systems. Whereas β-adrenoceptors are generally linked in a stimulatory manner to adenylate cyclase, α-adrenoceptors can be coupled to two different signal-generating systems (Table 4). α-Adrenoceptors of the α_2-subtype have been shown in some cell types to be linked in an inhibitory manner to adenylate cyclase. In contrast, α_1-adrenoceptors generally appear to be coupled to changes in phosphatidylinositol metabolism, calcium distribution and cyclic GMP metabolism. When both of the latter types of signal systems are stimulated in one cell type, e.g., by an α-adrenergic agonist in rat hepatocytes, different receptor subtypes appear to be involved in the two responses. Whether one receptor can be coupled to both signal-generating systems in one cell, is not yet clear. Findings with human platelets may suggest this possibility. In this cell type, ADP and α-adrenergic stimuli cause adenylate cyclase inhibition and changes in calcium distribution accompanied by an effect on phosphatidylinositol metabolism (clearly shown for ADP) and increases in cyclic GMP levels. The picture is even more complicated in the rat myocardium, where the effect of muscarinic cholinergic agonists involves decreased cyclic AMP and increased cyclic GMP levels; the latter effect is apparently independent of increased phosphatidylinositol turnover and calcium influx. More sophisticated studies on receptor properties may reveal further features discriminating receptor subclasses, eventually further supporting the concept that one receptor can, at least under one given condition, only be coupled to one signal-generating system.

ACKNOWLEDGEMENTS

The authors' studies reported herein were supported by grants from the Deutsche Forschungsgemeinschaft.

REFERENCES

1. Robison, G.A., Butcher, R.W., and Sutherland, E.W. (1971) Cyclic AMP. Academic Press, New York and London.
2. Williams, L.T., and Lefkowitz, R.J. (1978) Receptor Binding Studies in Adrenergic Pharmacology. Raven Press, New York.
3. Michell, R.H. (1975) Biochim. Biophys. Acta 415:81-147.
4. Jones, L.M., and Michell, R.H. (1978) Biochem. Soc. Transact. 6:673-688.

5. Jakobs, K.H. (1979) Mol. Cell. Endocrinol. 16:147-156.
6. Jakobs, K.H., Aktories, K., Lasch, P., and Schultz, G. (1980) in: Hormones and Cell Regulation, Vol. 4, ed. by J.E. Dumont and J. Nunez, North-Holland, Amsterdam, pp. 89-106.
7. Jakobs, K.H., Aktories, K., and Schultz, G. (1981) Adv. Cyclic Nucleotide Res. 14: in press.
8. Rodbell, M. (1980) Nature 284:17-22.
9. Abramowitz, J., Iyengar, R., and Birnbaumer, L. (1979) Mol. Cell. Endocrinol. 16:129-146.
10. Cassel, D., Levkovitz, H., and Selinger, Z. (1977) J. Cyclic Nucleotide Res. 3:393-406
11. Pfeuffer, T. (1977) J. Biol. Chem. 252:7224-7234.
12. Howlett, A.C., and Gilman, A.G. (1980) J. Biol. Chem. 255:2861-2866.
13. Kaslow, H.R., Johnson, G.L., Brothers, V.M., and Bourne, H.R. (1980) J. Biol. Chem. 255: 3736-3741.
14. Goldberg, N.D., and Haddox, M.K. (1977) Ann. Rev. Biochem. 46: 823-896.
15. Schultz, G., Hardman, J.G., Schultz, K., Baird, C.E., and Sutherland, E.W. (1973) Proc. Natl. Acad. Sci. USA 70:3889-3893.
16. Clyman, R.I., Blacksin, A.S., Manganiello, V.C., and Vaughan, M. (1975) Proc. Natl. Acad. Sci. USA 72:3883-3887.
17. DeRubertis, F.R., and Craven, P.A. (1978) Metabolism 27:855-868.
18. Murad, F., Arnold, W.P., Mittal, C.K., and Braughler, J.M. (1979) Adv. Cyclic Nucleotide Res. 11:175-204.
19. Graff, G., Stephenson, J.H., Glass, D.B., Haddox, M.K., and Goldberg, N.D. (1978) J. Biol. Chem. 253:7662-7676.
20. Hidaka, H., and Asano, T. (1977) Proc. Natl. Acad. Sci. USA 74:3657-3661.
21. Mittal, C.K., and Murad, F. (1977) Proc. Natl. Acad. Sci. USA 74:4360-4364.
22. Spies, C., Schultz, K.-D., and Schultz, G. (1980) Naunyn-Schmiedeberg's Arch. Pharmacol. 311:71-77.
23. Glass, D.B., Gerrard, F.M., Townsend, D., Carr, D.W., White, J.G., and Goldberg, N.D. (1977) J. Cyclic Nucleotide Res. 3:37-44.

BIOCHEMICAL AND IMMUNOCHEMICAL STUDIES OF THE β-ADRENERGIC RECEPTOR

A. Donny Strosberg

Molecular Immunology
IRBM - CNRS and University of Paris VII
Place Jussieu 2 - Paris 75221

INTRODUCTION

Cell-cell interactions in multicellular organisms involve molecular recognition mechanisms in which soluble messengers such as hormones, neurotransmitters, antigens and mitogens and cell-bound receptors play a major role. Complex networks of interacting ligands and cells have been described in areas as diverse as endocrinology, immunology and neurobiology. Few receptor-ligands systems have however been studied at the molecular level and even fewer receptors have been isolated and characterized. The main difficulty rests with the low number of receptors per cell and the ignorance of which components are necessary to isolate active molecules. A striking exception is the acetylcholine receptor whose amino acid sequence determination is now under way[1,2].

Beyond the receptor-ligand interaction, one would like to study the transmembrane signalling and further metabolic changes induced by specific ligand binding. In this respect, the system of choice would be one where hormone binding would lead to adenylate cyclase activation and increased cyclic AMP production. The β-adrenergic receptor cyclase complex provides at the present time the best opportunity for such a study, and we will review here the progress of the investigations concerning this system.

We will successively consider the active components in the receptor-cyclase complex and then analyze progress on each of them in the various types of cells which have been used by the leading investigators in the field.

1. The β-adrenergic Receptor Cyclase Complex: Active Components

Effects of the catecholamine agonists (isoproterenol, norepine-phrine, epinephrine) on target cells were shown to be influenced by effectors as diverse as guanyl nucleotides, magnesium or manganese. Hormone sensitive adenylate cyclase is itself affected by fluoride, adenosine and also guanyl nucleotides. The various studies describing these receptor-cyclase modulations led to the formulation of a number of models which all involve the existence of independent receptor, hormone sensitive GTPase, guanyl nucleotide binding protein and adenylate cyclase catalytic units[3-6]. While several groups are now directing their major effort to the isolation of each of these components, others have attempted to describe their subtle inter-actions in the living cell.

2. The β-adrenergic Receptor

a. Binding properties. The availability, through the pioneering efforts of Lefkowitz, Auerbach, Gilman, Levitzki and their collabor-ators, of radio-labeled catecholamine antagonists such as [3H]-propranolol, [3H]-dihydroalprenolol or [^{125}I] hydroxy-benzylpindolol has permitted the characterization of specific binding sites on intact cells, purified membrane fractions, detergent solubilized membrane extracts and finally affinity-purified membrane fractions. Generally a single class of non-co-operative high affinity receptors could be defined with equilibrium dissociation constants (K_D) ranging from 1-10nM in cells as diverse as nucleated erythrocytes, cardiac or adipose tissue cells, lymphoma cells etc. Stereospecific binding was fast and reversible and corresponded well to cyclase activation. Based on the potency to stimulate adenylate cyclase, β-adrenergic receptors have been sub-classified as β_1 and β_2. At β_1 receptors (heart, adipose tissue, avian erythrocytes), norepinephrine is approximately equipotent with epinephrine, and 3-4-fold weaker than isoproterenol. At β_2 receptors (smooth muscle), norepinephrine is slightly weaker than epinephrine and about 100-fold weaker than isoproterenol. The ability to displace bound radiolabeled antagon-ists closely follows the biological effect, thus confirming the identity of binding sites and adrenergic receptors : densitization of target cells by prolonged exposure to adrenergic agonists was shown to correspond to an apparent decrease in the number of binding sites. The number of receptors varies from cell to cell but is fairly constant when normalized per unit of cell surface and averages from five hundred to two thousand molecules per cell.

b. Molecular properties of the β-adrenergic receptor. 1° Solu-bilization and purification. The solubilization by digitonin treat-ment of receptors from cardiac tissue cells and from turkey and frog erythrocytes encouraged various groups to attempt the molecular characterization of these elusive membrane components. Gel filtration and sucrose gradient fractionation yielded approximate molecular weights of 1.5×10^5 daltons for the frog[7], 2×10^5 daltons for the turkey erythrocyte[8] and only 0.7×10^5 daltons for the S49 lymphoma cell receptor[9]. Various affinity chromatography gels were prepared for performing the purification of the receptor. An alprenolol agarose gel effectively permitted the separation of an active but no more hormone sensitive-adenylate cyclase catalytic unit from a pharmacologically active β-adrenergic receptor[8,10]. This 20.000-fold purification of the iodinated receptor was followed by the visualization by autoradiography after sodium dodecysulfate electrophoresis of two major sub-units of 30 and 33.000 daltons, and two minor components of 52 and 170.000 daltons[3,11].

2⁰ Presence of one or several essential disulfide bonds. The reducing agent dithriothreitol has been shown to affect the binding of catecholamine agonists and antagonists on glioma cells and on membrane-bound or purified receptors from turkey erythrocytes[12]. On the latter, dithiothreitol inactivation was prevented by preliminary binding of β-adrenergic agonists or antagonists, suggesting that these drugs effectively protect one or several essential disulfide bonds. Protection proceeds either by direct shielding of such bonds located at the binding site or by induction of a conformation change of the receptor resulting in burying bonds distant from the binding site. Very similar effects were recently observed on whole turkey erythrocytes, thus suggesting that the essential disulfide bond(s) of the receptor are exposed at the cell surface.

3⁰ Presence of N-ethylmaleimide sensitive sites. Studies in our laboratory indicated that treatment of turkey erythrocyte membrane with the alkylation agent N-ethylmaleimide resulted in effects markedly different from those observed for dithiothreitol. Pre-treatment of plasma membranes with either N-ethylmaleimide or β-adrenergic agonists alone does not affect subsequent ligand binding to receptor sites. The simultaneous pre-treatment by the two types of compounds causes, however, the irreversible inactivation of nearly 50% of the sites[13,14]. A striking correlation was observed between the agonist character (i.e., maximal ability for adenylate cyclase stimulation) and the rate of alkylation by N-ethylmaleimide. Agonists are ineffective in potentiating the inactivation. These results suggest that binding of a β-adrenergic agonist but not of an antagonist induces a change in conformation of the receptor which leads to increased accessibility of an N-ethylmaleimide sensitive site (Fig. 1).

3. The GTP Regulatory Protein and Catecholamine Sensitive GTPase.

Catecholamine hormones stimulate adenylate cyclase only in the presence of guanine nucleotides and it is likely that a GTP regulatory protein transduces the hormonal signal to the enzyme[5,9,15]. The fact that the activation of cyclase is short-lived in the

presence of GTP and permanent in the presence of the non-hydrolyzable
analog guanyl-5'-yl imido diphosphate (Gpp(NH)p) further suggest
that the regulatory component(s) possess(es) GTPase activity. GTP-
binding proteins from solubilized pigeon erythrocyte membranes were
separated from adenylate cyclase by affinity chromatography. Eluted
regulatory protein restored GPP(NH)p and F-stimulated adenylate
cyclase activity to a preparation deprived of guanine nucleotide
binding proteins. A GTP-binding protein with a molecular weight
of 42.000 is covalently modified in the presence of $[^{32}P]$ NAD$^+$ and
cholera toxin by ADP-ribosylation. This protein confers an enhanced
GTP-stimulated activity on adenylate cyclase solubilized from non-
treatment membranes[16]. Cholera toxin also inhibits the catecholamine-
stimulated GTPase activity in turkey erythrocyte membranes[17], again
suggesting a close linkage between the GTP regulatory protein and
a hormone sensitive GTPase. Two recent series of experiments
illustrate the relationship between the GTP regulatory protein, the
hormone sensitive GTPase and the β-adrenergic receptor:

(a) Protection by GTP of the receptor against inactivation by NEM :
    NEM inactivates the catecholamine-sensitive GTPase but does not,
    by itself, affect the receptor. The simultaneous presence of NEM
    and catecholamine agonists inactivates about half of the total
    number of receptors and this inactivation is prevented by GTP
    or Gpp(NH)p[13,18].
(b) Solubilization of the hormone sensitive GTPase : Recent investi-
    gations have indicated that pre-treatment of receptor-containing
    membranes with agonists[19] plus GMP, Magnesium or EDTA[20,21,22]
    apparently increases the agonist binding affinity of the receptor.
    After pre-treatment with GMP plus agonist a hormone sensitive
    GTPase was solubilized by digitonin[23]. It is thus likely that
    in fact a receptor-GTPase complex remains associated in solution.

Reconstitution of an Active Receptor-Cyclase Complex

The group of Schramm has shown in a series of elegant experiments
that it is possible to reconstitute a hormone-responsive receptor-
cyclase complex by fusion of two types of cells, one deprived of its
original receptor the other of its adenylate cyclase enzyme[24] : a
catecholamine-sensitive cyclase activity was obtained by fusion of
Friend erythroleukemia cells lacking β-adrenergic receptors with N-
ethylmaleimide treated turkey erythrocytes without residual cyclase.
The receptor and enzyme thus behave as distinct components capable
of individual migration in the plasma membrane. Further experiments
demonstrated that it was the receptor that confers the hormone sensitivity to the adenylate cyclase moiety. More recently it was shown
that the β-adrenergic receptor determines the affinity but not the
intrinsic activity of the receptor cyclase system; a hybrid cell
which possesses the β_1 adrenoreceptor of a turkey and the cyclase of
a frog erythrocyte displays β_1-like relative drug affinities but β_2-
like relative drug intrinsic activities[25].

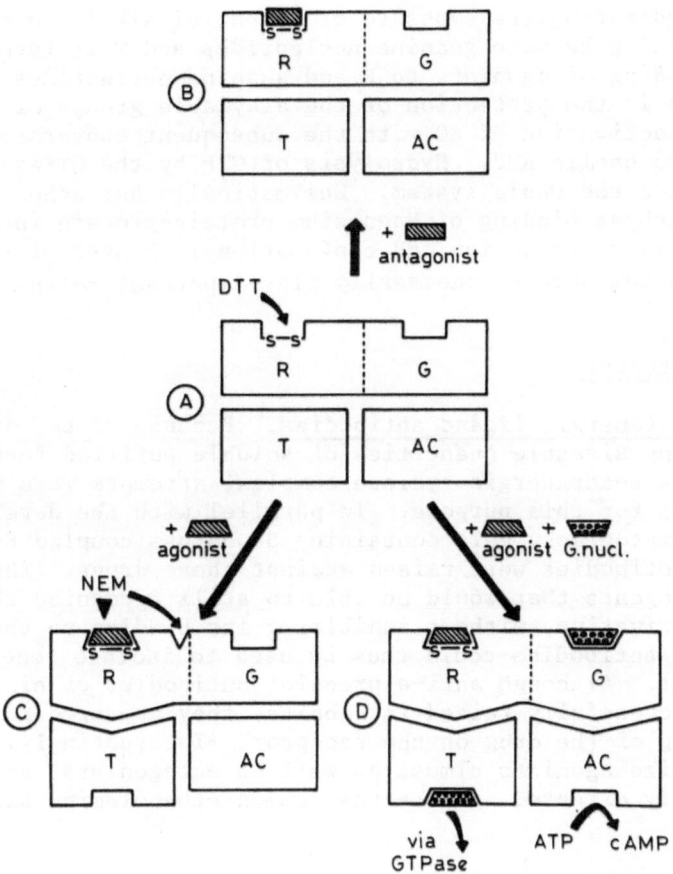

Fig. 1.   Schematic representation of the β-adrenergic receptor-
           cyclase complex of the turkey erythrocyte membrane and its
           interactions with agonists, antagonists guanine nucleotides
           (G nucl.), N-ethylmaleimide (NEM or dithiothreitol (DTT))
           (modified from reference 3). The system is composed of
           β-adrenergic receptor (R), one or several transmitter com-
           ponents (T,G or both) which may have catecholamine sensitive
           GTPase activity (T) and the adenylate cyclase catalytic unit
           (AC). Whether the GTPase (T) and/or the guanyl nucleotide
           regulatory site(s) are actually part of the receptor (R) is
           not yet clear. The scope of the model is to illustrate the
           molecular phenomena associated with β-adrenergic stimulation
           of the system. Stoichiometry, mobility and vectoral location
           of the different compounds are not considered. (A) No bound
           ligands : R can be inactivated by the reducing agent dithio-
           threitol (DTT), indicating the exposure of essential disul-
           fide bonds. (B) β-adrenergic antagonists bound to R : disul-
           fide bonds of R are shielded, but R can be inactivated by

NEM, indicating the exposure of essential alkylable groups.
Interaction between guanine nucleotides and T is favoured.
(D) Binding of agonists to R and guanine nucleotides to G
results in the protection of the alkylable groups of R and
in the activation of AC with the subsequent conversions of
ATP into cyclic AMP. Hydrolysis of GTP by the GTPase (T)
regulates the whole system. Enzymatically but other mechanisms such as binding of Magnesium protein-protein interactions, re-distribution and confomational changes of the
various components necessarily play important roles.

Immunochemical Studies

1. Anti β-adrenergic ligand antibodies. Because of the difficulty of obtaining sizeable quantities of soluble purified functional
components of the β-adrenergic-cyclase complex, attempts were made
to use antibodies for this purpose. In parallel with the development
of affinity chromatography gels containing Sepharose coupled β-adrenergic ligands, antibodies were raised against these drugs. The hope
was to obtain reagents that would be able to still recognize the
catecholamine derivative, without inhibiting its binding to the
receptor. These antibodies could thus be used to isolate receptor-
hormone complexes. Although anti-alprenolol antibodies of high
affinity were successfully raised in rabbits, they appeared to inhibit the binding of the drug on the receptor. Interestingly, they
seemed to recognize agonists almost as well as antagonists, and therefore were probably directed against the common ethanolamine side-chain
of the drugs.

2. Anti anti β-adrenergic ligand antibodies. The anti-alprenolol
antibodies were of no use in the isolation of the receptor-hormone
complex but turned out to be invaluable tools in further investigations. A recent observation[26] had indicated that antibodies raised
against anti-insulin antibodies were apparently able to bind themselves to the receptor, probably by "mimicking" the hormone. We
therefore re-injected the anti-alprenolol antibodies into rabbits
and also analyzed the sera for binding activity on the receptor.
After excluding the possibility that somehow, catecholamine agonists
or antagonists were present, we indeed could show binding of purified
anti-anti-alprenolol antibodies to cell bound, membrane bound or even
purified β-adrenergic receptors[27]. More strikingly, this binding
appeared to cause adenylate cyclase stimulation in the absence of
hormone[27]. In its presence, increased stimulation was observed.
The anti-anti-alprenol antibodies thus appear to mimic catecholamine
hormones both in terms of binding to the receptor and in terms of
activation of the adenylate cyclase catalytic unit. These antibodies
undoubtedly constitute useful tools for the analysis of the interactions between the various components of the system.

   3. Anti β-adrenergic receptor antibodies. The availability of purified β-adrenergic receptors obtained by affinity chromatography prompted us to raise antibodies against these elusive membrane proteins. Initial attempts by immunizing rabbits were unsuccessful. Immunization of mice resulted in the obtention of antibodies that hemagglutinated turkey erythrocytes from which the receptors were purified[28]. Although these antibodies did not appear to inhibit the binding of β-adrenergic ligands, they were shown to specifically recognize the receptor : they immunoprecipitated radiolabeled receptor, they bound receptor-hormone complexes and finally they stained, by immunofluorescence, S49 Wild Type lymphoma cells known to bear receptors, but did not stain receptor deficient mutant cells derived from the Wild Type line[28]. The anti-receptor antibodies again were shown to stimulate the turkey erythrocyte adenylate cyclase enzyme[28]. It thus seems that antibodies may interact with the different components of the β-adrenergic receptor cyclase system both by mimicking the hormone and by intervening in the hormone transmitted signals. Whether the re-distribution pattern in the membrane is modified by the antibodies constiutes but one of the several fascinating questions which will now be examined.

ACKNOWLEDGEMENTS

   The work described here was supported by grants from the NFWO, FGWO, IWONL and Janssen Pharmaceutica (Belgium) and by CNRS and INSERM (France).

REFERENCES

1. J. Giraudat and J. P. Changeux, Trends in Pharmacol. Sci. 198-202 (1980).
2. A. Devillers-Thiery, J. P. Changeux, P. Paroutaud and A. D. Strosberg, FEBS Lett. 104:99-105 (1979).
3. A. D. Strosberg, G. Vauquelin, O. Durieu-Trautmann, C. Delavier-Klutchko, S. Bottari and C. Andrè, Trends in Biochem. Sci. 5:11-14 (1980).
4. A. Levitzki and E. Helmreich, FEBS Lett. 101:213-219 (1979).
5. M. Rodbell, Nature 284:17-22 (1980).
6. R. J. Lefkowitz, Ann. Int. Med. 91:450-458 (1979).
7. L. E. Limbird and R.J. Lefkowitz, J. Biol. Chem. 252:799-802 (1977).
8. G. Vauquelin, P. Geynet, J. Hanoune and A. D. Strosberg, Eur. J. Biochem. 98:543-556 (1979).
9. T. Haga, K. Haga and A. G. Gilman, J. Biol. Chem. 252:5776-5782 (1977).
10. G. Vauquelin, P. Geynet, J. Hanoune and A. D. Strosberg, Proc. Natl, Acad, Sci. U.S.A. 74:3710-3714 (1977).

11. O. Durieu-Trautmann, C. Delavier-Klutchko, G. Vauquelin and
    A. D. Strosberg, J. Supramol. Struct., in press.
12. G. Vauquelin, S. Bottari, L. Kanarek and A.D. Strosberg, J. Biol.
    Chem. 254:4462-4449 (1979).
13. S. Bottari, G. Vauquelin, O. Durieu-Trautmann, C. Delavier-
    Klutchko and A. D. Strosberg, Biochem. Biophys. Res. Commun.
    86:1311-1318 (1979).
14. G. Vauquelin, S. Bottari and A. D. Strosberg, Mol. Pharmacol.
    17:163-171 (1979).
15. E. M. Ross, A. C. Howlett, M. Ferguson and A. D. Gilman J. Biol.
    Chem. 253:6401-6412 (1978).
16. D. Cassel and T. Pfeuffer, Proc. Natl. Acad. Sci. 75:2669-2673
    (1978).
17. D. Cassel and Z. Selinger, Proc. Natl. Acad. Sci. 74:3307-3311
    (1977).
18. G. Vauquelin, S. Bottari, C. André, B. Jacobsson and
    A. D. Strosberg, Proc. Natl. Acad. Sci. 77:3801-3805 (1980).
19. J. M. Stadel, A. Delean, and R. J. Lefkowitz, J. Biol. Chem.
    255:1436-1441 (1980).
20. P. M. Lad, T. B. Nielsen, M. S. Preston and M. Rodbell, J. Biol.
    Chem. 255:988-995 (1980).
21. M. E. Maguire, E. M. Ross and A. G. Gilman, Adv. Cyclic Nucleotide
    Res. 8:1-83 (1977).
22. S. J. Bird and M. E. Maguire, J. Biol. Chem. 253:8826-8834 (1978).
23. C. Delavier-Klutchko, O. Durieu-Traumann, P. O. Couraud, C. André
    and A. D. Strosberg, FEBS Lett. 117:341-343 (1980).
24. J. Orly and M. Schramm, Proc. Natl. Acad. Sci. U.S.A. 73:4410-4414
    (1976).
25. L. J. Pike, L. E. Limbird and R. J. Lefkowitz, Nature 280:502-504
    (1979).
26. K. Sege and P. A. Peterson, Proc. Natl. Acad. Sci. U.S.A. 75:2443-
    2447 (1978).
27. A. B. Schreiber, P. O. Couraud, C. André, B. Vray and
    A. D. Strosberg, Proc. Natl. Acad. Sci. U.S.A., in press.
28. P. O. Couraud, C. Delavier-Klutchko, O. Durieu-Trautmann and
    A. D. Strosberg, Abst. Proc. Fourth. Internat. Conf. Cyclic
    Nucleotides, Brussels.

CYCLIC NUCLEOTIDE PHOSPHODIESTERASES

Samuel J. Strada

Department of Pharmacology
The University of Texas Medical School at Houston
Houston, Texas   77030

INTRODUCTION

Discovery and Early Studies

Once adenosine 3',5'-monophosphate (cyclic AMP) was established
as the heat stable factor that mediated the activation of liver
phosphorylase by epinephrine and glucagon[1], the search began to
identify an enzyme to inactivate the factor. A number of general
phosphatases and diesterases failed to affect it. However, an enzyme
capable of destroying the biological activity of cyclic AMP was
detected in extracts of heart, brain and liver[2]. The enzyme was
later identified in a great many tissues[3] and shown to form adenosine
5'-phosphate as the product of hydrolysis. Partially purified enzyme
activity from dog heart was activated by magnesium ion and inhibited
by the methylxanthine caffeine. The possible complexity of this
activity, as a system, was indicated by its presence in soluble as
well as particulate fractions. Approximately 60% of the phosphodiesterase activity in homogenates of beef heart was found to be
associated with particulate fractions after low-speed centrifugation.
The apparent K_m of the partially purified soluble enzyme activity
was around 100μM. It was inhibited, in a competitive manner, by
theophylline>theobromine = caffeine. The activity was stimulated
by imidazole which shifted the pH optima for enzyme activity from
around 8 (in tris buffer) to approximately 7.4. Subsequent attempts
to purify phosphodiesterase from other tissue sources, e.g., rabbit
brain[4], dog heart[5] and rat brain[6] failed to yield enzyme preparations
with specific activities greater than those reported for the beef
heart enzyme[3].

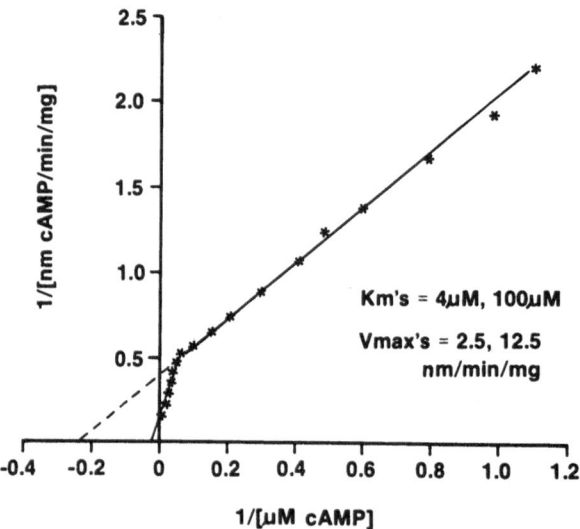

Fig. 1.   Example of the anomalous kinetic behavior of cyclic AMP
          phosphodiesterase activity of rat kidney homogenates.
          Enzyme Activity was measured by the method of Thompson et
          al.[17]   From Van Inwegen et al.[22].

     A puzzling feature of these early studies was that the Km for
cyclic AMP appeared to be much higher, by several orders of magnitude, than the concentration of cyclic AMP in most cells. However,
in 1968 Brooker et al.[7] reported the presence of a second phosphodiesterase activity in crude brain fractions with a Km around
1×10^{-6}M. The demonstration of this "higher affinity" enzyme activity depended on the utilization of a more sensitive isotopic
method of assay, and suggested an additional complexity in the enzyme
system. Since the apparent Km of this latter enzyme is closer to
the concentration of cyclic AMP found in most cells, it has been
suggested[8] that this form of the enzyme may be physiologically more
important insofar as the metabolism of cyclic AMP is concerned. However, the anomalous kinetic data which led to the conclusion, an
example of which is shown in Figure 1, could also have been interpreted in terms of one enzyme exhibiting negative cooperativity[9].

Later Developments

     More definitive evidence for the existence of multiple molecular
forms of cyclic nucleotide phosphodiesterase was provided by the
physical separation of enzyme forms using techniques such as molecular sieze and ion-exchange chromatography. The first example of form
separation was achieved by Agarose A 0.5 gel chromatography[10] of
several rat tissues. The multiple forms which have been identified

Table 1. Differential Properties of Multiple Forms of Cyclic
          Nucleotide Phosphodiesterase

1. Kinetic parameters	7. Sensitivities to physiological regulators (hormones, proteins, lipids, ions)
2. Substrate affinities	
3. Molecular size	8. Sensitivites to pharmacologic effects (drugs and other chemicals)
4. Charge distribution	9. Variable tissue distributions
5. Electrophoretic mobilities	10. Cellular specificity
6. Isoelectic points	11. Different genetic regulations

now in most tissues by several separation techniques differ in substrate specificity and affinity, subcellular localization, molecular
weights, sensitivity to pharmacological agents and to endogenous
inhibitory and stimulatory factors (see Table 1).

The presence of these multiple activities, which vary in
quantity, type and stability from tissue to tissue[11,12] has precluded
extrapolation of experimental data from one laboratory to another,
especially when different methods of enzyme isolation and assay
conditions are used. The biochemical and physiological relationships
of the enzyme forms have been difficult to elucidate until recently,
principally because the enzyme systems have been isolated from
heterogeneous tissues, such as heart, brain and liver, and are
usually studied in an unpure state. However, at least two distinct
enzyme forms[13,14] have recently been purified to apparent homogeneity
and their biochemical and pharmacological properties investigated in
detail. Additional studies conducted primarily with homogenous cell
preparations, have shown that multiple mechanisms operate within
the cell to regulate the activities and other properties of the
individual forms.

ANALYSIS OF CYCLIC NUCLEOTIDE PHOSPHODIESTERASES

Enzymatic Assay

The general principles and essential details for determining
cyclic AMP or cyclic GMP hydrolysis have been reviewed[15]. The
"batch" assay of Thompson and Appleman[16], or modifications of it,
has provided a relatively simple, sensitive, and reproducible method
of activity analysis and represents the most commonly used procedures. Briefly, tritiated substrates are converted by PDE and
the 5'-nucleotidase of cobra venom to ^{3}H-adenosine or ^{3}H-guanosine;

the nucleoside products are separated from unreacted substrate by addition of an anion-exchange resin directly to the reaction vessel.

This method has been used to measure PDE activity in a variety of crude and partially purified enzyme preparations. Care must be taken to ensure direct stoichiometry, a sometimes troublesome difficulty in tissue homogenates. These problems (which are especially pronounced in muscle preparations, for example) are minimized by short incubation times and, theoretically, should not arise if the side reaction products (for example, those produced by 5'-AMP deaminase) are also converted to nucleosides or other uncharged or basic compounds that cannot bind to anion-exchange resins. The limits of the methods will then be determined solely by the specific activities of substrate isotopes.

There have been reports in the literature criticizing the "batch" assay method and its modifications because of some non-specific nucleoside binding (adenosine, guanosine, inosine) to commercial anion-exchange resins (e.g., Bio-Rad Dowex or AG 1-X8)[17]. These reports have correctly pointed out a potential for underestimating total activity of the enzyme. Reports vary widely as to the extent of binding; different batches have been found to bind adenosine, guanosine, and/or inosine. The binding seems to be of a non-specific nature; it is not reversed by high concentrations of nucleoside or by pre-washing of resin with a variety of agents including alcohols, detergents, boiling water, or several acids and bases. We recently published a modification of the assay procedure which advantageously does not raise blank values[17]. We have found that inclusion of methanol during the binding step largely eliminates it. Perhaps more importantly, it does not decrease nucleotide binding as do other published procedures for alleviating nucleoside binding. In addition, ethanol can be used to alleviate adenosine binding to Dowex 2-X8[18].

Various published methods for PDE assay have been compared[17]. Nonradioactive PDE assay methods that measure phosphate release, changes in ultraviolet absorbance, or changes in pH are usually too insensitive and/or susceptible to interfering substances to be of general utility. Interference is also a problem with those assays which employ multiple additional enzymatic reactions such as the firefly luciferin-luciferase method. By and large, the very best assays use radioisotopes along with some physiocochemical means for separating substrates from end-products. Thin-layer chromatography has been used. An aditional complication is that because crude enzyme preparations contain endogenous phosphatase, nucleotidase, and deaminase activities, multiple products are often created in PDE reaction mixtures. To accurately assess phosphodiesterase activity all degradation products should be measured.

Resolution of Forms

Procedures most exploited to resolve major phosphodiesterase forms have been: ion-exchange chromatography, preparative electrophoresis, isoelectric focusing, gel filtration, and sucrose-gradient ultracentrifugation.

Since the initial studies in rat liver by Russell et al.[19], DEAE-cellulose anion-exchange chromatography has been used to separate multiple enzyme forms from a variety of tissue sources. As a classical biochemical approach to separate multiple enzyme forms based on net protein charge, this technique is simple and straight-forward; when experimental detail is adhered to closely, it provides a good method to compare data from different laboratories. In this procedure, detailed and catalytically distinct enzyme forms can be obtained relatively free of cross-contamination and in quantities adequate for kinetic analysis.

A DEAE-cellulose chromatographic profile of the cyclic nucleotide PDE's from rat cardiac tissue[20] is shown in Fig. 2. This profile is analogous to that reported by Russell et al.[19] for rat liver, but is not identical to those found in, for example, platelets[21] or rat kidney[22]. In tissues where good separations can be obtained, the differences in substrate specificity, affinity, and cooperativity can be readily compared for each of the enzyme peaks.

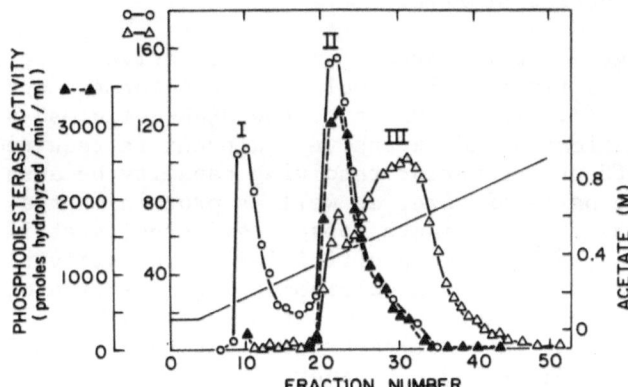

Fig. 2. Separation of cyclic nucleotide phosphodiesterases of rat heart by DEAE-cellulose anion-exchange chromatography. The profiles were obtained by assaying PDE activity at concentrations of cAMP of 1μM (Δ) or 100μM (▲) and of cGMP of 1μM (o). The symbols I, II, III represent the order of enzyme elution from the column. From Terasaki and Appleman[20].

Briefly, fraction I is relatively cyclic GMP-specific and shows Michaelis-Menten kinetic behavior for cyclic GMP hydrolysis. This fraction is activated by calcium ion in the presence of heat-stable activator protein. Fraction II shows positive cooperativity for cyclic AMP hydrolysis, and the cyclic AMP PDE activity of this fraction is activated by μM concentrations of cGMP[20]. Fractions I and II have high apparent Km's for cyclic AMP (~50μM) which distinguish them from fraction III, which has a low Km for cyclic AMP (~1μM). In addition, fraction III is more specific for cyclic AMP than for cyclic GMP as substrate. This form may have originated in the particulate fraction of the cell and appears to share some of the properties of the lower molecular agarose forms reported earlier by Thompson and Appleman[10].

Users of this method must be aware of several conditions which affect the separation of cyclic nucleotide phosphodiesterases. Inherent in the complex enzymology of phosphodiesterases are reported activity and/or form changes due to ionic strength, pH, medium hydrophobicity, proteolysis, cations, "aging" of homogenates, and sulfhydryl reagents. Therefore, as with all preparation procedures, seemingly minor changes in the protocols can give rise to markedly different results. It has been reported that this procedure can reveal latent activity[19], and the possibility exists that an accurate profile of activities of the original tissue extract may not be obtained. Enzyme forms separated by DEAE-cellulose should therefore be monitored after isolation for the occurence of differential activity changes during storage.

Acrylamide gel electrophoresis on a preparative scale has been used to separate multiple cyclic nucleotide PDE forms, most notably in brain tissue[23,24]. In contrasts to the DEAE-cellulose procedure, relatively sophisticated and expensive equipment is required, but this technique offers a superior resolving capacity because of its separation on the basis of size, as well as protein charge. The PDE forms separated by this procedure possess different stabilities, kinetic properties, substrate specificities and sensitivities to inhibitors and to an endogenous calcium-dependent activator protein[25]. The procedure has been used extensively by Weiss and his colleagues to differentiate activator-sensitive and insensitive forms of rat brain PDE's[23-26]. However, it is not clear if these forms are distinct isozymes or anomalies of tissue extraction and/or electrophoretic fields. The extent to which such factors as proteolysis membrane perturbations, subcellular redistribution, phosphorylation-dephosphorylation, ionic fluxes and lipid alterations contribute to the appearance or disappearance of the various forms is not known. The relationship of these electrophoretic-separable forms to the lower molecular weight enzyme forms isolated by gel filtration or ultracentrifugation, or to the different charge enzymes separated by ion-exchange chromatography or isolectrofocusing is also not established.

Distribution of Forms

The resolution of "soluble" activities (i.e., activity not found in the 105,000g pellet after 1-2 hours of centrifugation) has been achieved by a variety of techniques. The most commonly used procedures are gel filtration, DEAE-cellulose anion-exchange chromatography, isolectric focusing, polyacrylamide gel electrophoresis, starch gel electrophoresis and linear sucrose gradient centrifugation. A potential drawback of the latter technique is the length of time required for each centrifugation (12-24 hours) if the stimulated or unstimulated enzyme forms are not stable or if there is a prominent "aging" effect which occurs in some tissues.

Differential centrifugation methods can be used to separate soluble from particulate activities, but the technique of discontinuous sucrose-gradient centrifugation may be more effective for this purpose. To study hormone-induced changes, activities must be compared to a high recovery, quantitative basis, a criterion fulfilled by this technique. In addition, sufficient quantities of both soluble and particulate material are provided to enable a kinetic analysis of the enzyme fractions. This method is also useful as a screening procedure when little background is available on the enzyme in a given tissue. When differential centrifugation alone is used, a major problem encountered is an excess of soluble low-affinity cyclic nucleotide phosphodiesterase activity relative to particulate activity and the consequent contamination of the particulate fraction, which frequently results even with extensive washing. Our own studies have employed discontinuous sucrose-gradient fractionation procedures for various purposes to study phosphodiestrase activities. These have included various effects of hormonal pertubations and endocrinipathies on particulate activity from adipose and liver tissues (see below), and the redistribution of particulate enzyme during cell growth in culture[27].

In summary, a variety of conditions, such as pH, sulfhydryl reagents, temperature, and hydrophobic mediums, markedly affect the results obtained with techniques used to separate non-membrane-bound forms of cyclic nucleotide phosphodiesterase. Where they have been studied, particulate cyclic nucleotide phosphodiesterase activities have shown complex kinetic behavior, hydrolysis of cyclic AMP and sometimes cyclic GMP as well, regulation of cyclic AMP hydrolysis by cyclic GMP, and a susceptibility to many of the factors that influence soluble enzyme activities.

REGULATION OF PHOSPHODIESTERASES

Non-Genetic Mechanisms

Most mammalian tissues can be shown to contain at least three distinct phosphodiesterase activities, which may be controlled separately. For the most part, this hypothesis is based on findings in

Table 2. Prominent Phosphodiesterase Forms

---

1. High affinity (low Km), lower molecular weight, negatively co-
   operative, cyclic AMP substrate specific
2. Low affinity (high Km), higher molecular weight, Michaelis-
   Menten behavior, cyclic nucleotide substrate-non-specific,
   higher affinity for cyclic GMP, Ca^{++} and modulator sensitive
3. Low affinity (high Km), cyclic AMP substrate-specific site,
   sensitive to cyclic GMP modulation

---

crude or partially purified preparations. The general characteristics of these three activities based on several criteria are listed
in Table 2. Briefly, one phosphodiesterase form has a high affinity
for cyclic AMP and is regulated by the cyclic AMP substrate concentration in a negatively cooperative manner. A second form has a
higher molecular weight than the first and a greater affinity for
cyclic GMP than for cyclic AMP. This form may be regulated by calcium
and a dissociable protein factor referred to earlier as an "activator"
or "modulator" or more recently as calmodulin. A third form is a
low-affinity cyclic AMP phosphodiesterase activity that can be regulated by cyclic GMP. The regulation of cyclic nucleotide phosphodiesterase activity probably involves a number of different control
mechanisms. In recent years, several hormones (a partial list is
shown in Table 3) have been shown to alter cyclic nucleotide PDE
activities.

Hormonal modulation of PDE could involve any or all of known
enzyme control mechanisms acting alone or in concert. However, the
cellular location of the higher affinity, cyclic AMP specific enzyme
form is perhaps the cell membrane, and therefore, may be subject to
more limited controls. It was hypothesized earlier that the membrane-
bound form of the enzyme is involved in the mechanism of action of
some hormones[8]. Technically, this poses a difficult problem, because
in addition to being associated with membranes, this form of the
enzyme represents only a small portion of the total activity. Some
cells have no detectable membrane-bound phosphodiesterase activity,
e.g., human lymphocytes but, nevertheless, have enzymes that are
activated by agents which do not traverse the cell membrane, e.g.,
mitogenis lectins. On the other hand, other hormones, e.g., steroids,
which affect cyclic nucleotide phosphodiesterase activity do not have
the cell membrane as their commonly accepted locus of primary action.
Most of these agents decrease "soluble" enzyme activities, and the
cytosolic portion of the cells is thought to be the location of their
receptors. In every instance shown in Table 3, the intact cell is
a necessary prerequisite to demonstrate a hormonal effect on phosphodiesterase activity. This has posed the difficult experimental
problem of differentiating a direct from an indirect effect on the
enzyme.

Table 3. Hormones Affecting Cyclic Nucleotide Phosphodiesterase
          Activities

Hormone	Tissue	Preparation	Substrate	Effect
Insulin	Fat Cell	Particulate	cAMP	increase
	Liver	Particulate	cAMP	increase
	Muscle	Whole tissue	cAMP	increase
	Cell Culture	Homogenate	cAMP, cGMP	increase
Glucagon	Liver Cell	Particulate	cAMP	increase
Growth Hormone	Liver	Particulate	cAMP	increase
Catecholamines	Fat Cell	Particulate	cAMP	increase
	Pineal	Homogenate	cAMP	increase
	Cell Culture	Soluble	cAMP	increase
ACTH	Fat Cell	Particulate	cAMP	increase
	Adrenal Cortical Carcinoma	Homogenate	cGMP	increase
Prostaglandins	Cell Culture	Homogenate	cAMP	increase
	Cell Culture	Homogenate	cAMP	decrease
Cholecystokinin	Gall Bladder	Homogenate	cAMP	increase
Thyroxine	Fat Cell	Particulate	cAMP	decrease
	Thyroid	Soluble	cAMP, CGMP	decrease
	Thyroid	Particulate	cAMP	redistribute
	Bone	Soluble	cAMP	decrease
Glucocorticoids	Liver	Homogenate	cAMP, CGMP	decrease
	Muscle	Homogenate	cAMP, cGMP	decrease
	Heart	Homogenate	cAMP	decrease
	Lung	Homogenate	cAMP	decrease
	Hepatoma	Soluble	cAMP, cGMP	decrease
	Lymphocyte	Sonicate	cAMP	decrease
	Testis	Soluble	cAMP	decrease
Aldosterone	Bladder	Homogenate	cAMP	decrease
Estrogen	Uterus	Soluble	cAMP	decrease
Gonadotropins	Testes	Soluble	cGMP	decrease

Effects of insulin on the enzyme have been studied in the
greatest detail, but the role of phosphodiesterase in the biological
actions of insulin is unknown. In most instances, effects of insulin
are exerted on high affinity membrane-bound forms of phosphodiesterase, and do not involve protein synthesis, at least in liver
and fat cells. The mechanism by which insulin activates cyclic AMP
phosphodiesterase has not been elucidated[27]. However, some recent
experiments with insulin-treated hepatocytes suggest a role for a
phosphorylation-dephosphorylation mechanism[28]. More direct support
of this interesting hypothesis will require an actual demonstration
of a correlation between phosphate incorporation into membrane-bound
PDE and enzymatic activity changes in PDE in response to insulin
treatment. A common finding in all previous studies is the persistent
activation of the enzyme caused by insulin seen after the initial
activity increase. Kono et al.[29] have hypothesized that this is not
an effect of insulin per se but, rather, a consequence of sulfhydryl
oxidation, since basal activity is also affected. Studies in the
cellular slime mold Dictyostelium discoideum indicate specialized

roles for sulfhydryl interactions and redistribution between membrane-
bound and soluble cyclic nucleotide phosphodiesterases in the regulation of growth and differentiation in these organisms[30]. An insu-
lin-induced cyclic AMP phosphodiesterase activity increase has not
been localized completely to one membrane fraction.

Ideally, the study of the influence of hormones or other treatments on cyclic nucleotide phosphodiesterase activities requires
detection of activity in the intact cell or tissue, a situation which
is difficult to achieve. Conventionally, investigators have resorted
to testing enzyme activity or enzyme protein immediately upon breaking the cell in hopes of achieving an approximation of the phosphodiesterase complement of the intact cell. Estimations of activities
of higher affinity enzymes can be obtained in crude homogenates by
assaying at a substrate concentration well below the Km of the lower
affinity enzyme. However, in all known cases, in order to study
hormone stimulation, it has been necessary to assess the activity
of one or the other enzyme forms and/or analyze particulate versus
soluble activity. Thus, until immunological criteria are established,
it is necessary to fractionate phosphodiesterase activities in order
to study hormone regulation.

One problem of analyzing high-affinity enzyme activity becomes
more apparent when one appreciates that only a small portion of the
total cyclic AMP hydrolysis of an homogenate is particulate in nature.
In addition, usually only a small percentage of the total soluble
activity is accounted for by the high-affinity enzyme. The amounts
of high-affinity cyclic AMP enzyme and cyclic GMP hydrolytic capacity
are highly variable from tissue to tissue. Lower-affinity cyclic AMP
phosphodiesterase and cyclic GMP phosphodiesterase often show a
similar distribution in tissues, but this coincidence has not been
substantiated in all tissues. Additionally, the activities of each
enzyme should always be optimized for the particular homogenization
conditions and procedures employed.

Cell-cell interactions, serum factors, and growth conditions
of cultured cells can also markedly alter the kinetic parameters and
forms of cyclic nucleotide phosphodiesterase by mechanisms independent of protein synthesis[31]. However, the effects of insulin on
cyclic nucleotide phosphodiesterases in cultured BHK-21 cells and
the activation of lymphocyte PDE's by PHA and other mitogens are
sensitive to inhibitors of macromolecular synthesis[32,33].

Numerous studies with phosphodiesterases from different sources
have demonstrated a variety of effects of cyclic nucleotides themselves on enzyme activity. Whether or not the two cyclic nucleotides
interact under physiological circumstances to modify the hydrolysis
of each other in positive or negative directions is uncertain, however, some experimental evidence suggests that they may indeed do so.
Relatively low concentrations of cyclic GMP have been shown to stimu-

late the rate of cyclic AMP hydrolysis by crude phosphodiesterase
preparations from heart, thymic lymphocytes, fat cells, and liver.
Cyclic AMP has not been shown to stimulate cyclic GMP hydrolysis
in any system, but the two cyclic nucleotides do competitively inhibit the hydrolysis of each other where they serve as substrates
for the same enzyme. In certain tissues, e.g., uterus, cyclic GMP
can apparently inhibit cyclic AMP hydrolysis non-competitively.

Genetic Control

In vivo studies have shown modulation of enzyme activities
in intact tissues, particularly during development. For example,
specific activities of cyclic nucleotide phosphodiesterase fluctuate
during organ development. Such changes occur in (a) rat uterus
where low K_m cyclic AMP phosphodiesterase shows an abrupt decrease
in specific activity 25-30 days postnatally[34], (b) rat testis where
development shows a decrease in cyclic GMP phosphodiesterase activity
20-26 days[35], (c) rat cerebral cortex where there is fivefold increase in total cyclic AMP phosphodiesterase activity between 1-20
days of age[36], (d) rat pineal gland with a 20-fold increase in low K_m
cyclic AMP phosphodiesterase activity 1-20 days postnatally, and
(e) rat cerebellum[24] and testis[37] where the amount of endogenous
protein activator of phosphodiesterase increases during development.
The change in low K_m cyclic AMP phosphodiesterase activity in the
uterus may involve estrogenic effects[34]. The enzyme changes in testis
may also be hormonally mediated[39]. Activity changes during development also reflect the appearance of specific cell types and physiological responding systems (e.g., specific synaptic pathways). The
mechanisms responsible for these enzyme activity changes and their
functional significance are generally not known.

Cyclic nucleotides themselves also affect their own hydrolytic
rates and the hydrolysis of other cyclic nucleotides via a genetic
mechanism. One response to a prolonged elevation of cellular cyclic
AMP may be the induction of more hydrolytic enzyme[40] (see below).

Direct attempts have been made to correlate rates of cyclic AMP
formation and degradation during growth stages. Parallel increases
in the specific activities of adenylyl cyclase and cyclic AMP phosphodiesterase were seen as the density of cultured fibroblasts increased[41]. Contact-inhibited cells increased their cyclic AMP content as the cells reached confluency; the elevated level corresponded
to an increase in the ratio of total adenylyl cyclase to phosphodiesterase activity at higher cell densities. Cyclic AMP levels
were not elevated at high cell densities in virally transformed,
non-contact-inhibited cells, nor was there a change in the ratio
of cyclase to phosphodiesterase activity.

The mechanisms responsible for changes in enzyme activities may
be related, in some instances, to intracellular levels of cyclic AMP.

For example, increased levels of cyclic AMP have been shown to induce
de novo synthesis of cyclic AMP phosphodiesterase in cultures of
fibroblasts, neuroblastoma, glioma, lymphoma, and 3T3-Li cells.

Studies of the phosphodiesterase enzyme system of mouse lymphoma
cells led to speculation that activation of protein kinase was essential for the induction of phosphodiesterase. The parent lymphosarcoma
cell line is normally killed by dibutyryl cyclic AMP, whereas mutant
cells are resistant to the cytotoxic effects of nucleotide. Protein
and messenger RNA synthesis are inhibited by the cyclic AMP derivative
in the sensitive cells but not in the resistant line. Dibutyryl
cyclic AMP or prostaglandin E_1 "induce" phosphodiesterase activity
in the parent cell but not in the resistant mutant line[42,43]. A defect found in the cells not showing phosphodiesterase induction is
a deficiency in the cyclic AMP binding sub-unit of protein kinase
activity. Although these findings are intriguing, attempts to relate
changes in cyclic nucleotide levels to altered rates of synthesis
and degradation and ultimately to growth properties of these and
other cultured cell lines have met with limitations[44].

What is clear is that the amount, type, and number of forms
of cyclic AMP and cyclic GMP phosphodiestrase of cultured fibroblasts
are dependent upon the density of the cells and the phase of growth
cycle[45]. The results suggest important differences between mechanisms regulating growth characteristics and modulation of cyclic
nucleotide phosphodiesterases of cultured cells on a short-term
(nongenetic) and long-term (genetic) basis. They attest further to
the utility of investigating experimaental conditions that influence
cell growth for their effects on specific forms and properties of
phosphodiesterase. The acute-control system causes short-term
changes in cyclic nucleotide phosphodiesterase activities and could
be responsible for rapid oscillatory changes in cyclic nucleotide
levels. The longer-term control may participate in the recovery
and maintenance of cyclic nucleotide levels in response to cellular
stimuli such as cyclic AMP elevations and/or GMP reductions.

The kinetic and other data described above suggest that culture
conditions can alter forms, quantities, and substrate affinities
of enzyme activites of normal and transformed cultured cells. Enzyme
parameters are influenced by cell-cell contact, cell density, serum
levels, cyclic nucleotide concentrations, and viral transformation.
Many of these parameters are also modified by organ development and
hormonal modulation. Eukaryotic cells seem to control cyclic nucleotide levels through a variety of mechanisms influencing phosphodiesterase catalysis. Mechanisms responsible for the changes in
phosphodiesterase characteristics remain primarily speculative for
most of the experimental conditions discussed and further experiment-
ation is needed to relate these changes to cellular function.

ACKNOWLEDGEMENT

The experimental work from our laboratory that was described in this article was supported by a grant (GM21361) from U.S.P.H.S. The author gratefully acknowledges the many intellectual and experimental contributions made by my friend and colleague, Dr. W. Joseph Thompson, during the course of collaborative efforts to study the biological role of cyclic nucleotide phosphodiesterases.

REFERENCES

1. T. W. Rall and E. W. Sutherland, J. Biol. Chem. 232:105 (1958).
2. E. W. Sutherland and T. W. Rall, J. Biol. Chem. 232:1077 (1958).
3. R. W. Butcher and E. W. Sutherland, J. Biol. Chem. 237:1244 (1962).
4. G. I. Drummond and S. Perrott-Yee, J. Biol Chem. 236:1126 (1962).
5. K. G. Nair, Biochemistery 5:150 (1966).
6. W. Y. Cheung, Biochemistery 6:1079 (1966).
7. G. Brooker, L. J. Thomas and M. M. Appleman, Biochemistry 7:4177 (1968).
8. W. J. Thompson and M. M. Appleman, Biochemistry 7:4177 (1968).
9. T. R. Russell, W. J. Thompson, F. W. Schneider and M. M. Appleman, Proc. Natl. Acad. Sci. U.S.A. 69:1791 (1972).
10. W. J. Appleman and M. M. Appleman, J. Biol. Chem. 246:3145 (1971).
11. J. N. Wells and J. G. Hardman, Adv. Cyclic Nucl. Res. 8:119 (1977).
12. S. J. Strada and W. J. Thompson, Adv. Cyclic Nucl. Res. 9:265 (1978).
13. W. J. Thompson, P. M. Epstein and S. J. Strada, Biochemistry 18:5228 (1979).
14. R. K. Sharma, T. H. Wang, E. Wirch and J. H. Wang, J. Biol. Chem. 255:5916 (1980).
15. W. J. Thompson, G. Brooker and M. M. Appleman, Methods Enzymol 38:205 (1974).
16. W. J. Thompson and M. M. Appleman, Biochemistry 10:311 (1979).
17. W. J. Thompson, W. J. Terasaki, P. M. Epstein and S. J. Strada, Adv. Cyclic Nucl. Res. 10:69 (1979).
18. J. Londesborough, Anal. Biochem. 71:623 (1976).
19. T. R. Russell, W. L. Terasaki and M. M. Appleman, J. Biol. Chem. 248:1334 (1973).
20. W. L. Terasaki and M. M. Appleman, Metabolism 24:311 (1975).
21. H. Hidaka and T. Asano, Biochim. Biophys. Acta 429:485 (1976).
22. R. G. Van Inwegen, W. J. Pledger S. J. Strada and W. J. Thompson, Arch. Biochem. Biophys. 175:700 (1976).
23. P. Uzunov and B. Weiss, Biochim. Biophys. Acta 284:220 (1972).
24. S. J. Strada, P. Uzunov and B. Weiss, J. Neurochem. 23:1097 (1974).
25. B. Weiss, R. Fertel, R. Figlin and P. Uzunov, Mol. Pharmacol. 10:615 (1974).

26. B. Weiss, Adv. Cyclic Nucl. Res. 5:195 (1975).
27. W. J. Thompson and S. J. Strada, in: "Receptors and Hormone Action", Vol.3, L. Birnbaumer and B. W. O'Malley, eds., 553 Academic Press, New York (1978).
28. R. J. Marchmont and M. D. Houslay, Nature 286:904 (1980).
29. T. Kono, F. W. Robinson and J. A. Sarber, J. Biol. Chem. 250: 7826 (1975).
30. A. Robertson and J. Grutsch, Life Sci. 15:1031 (1976).
31. S. J. Strada and W. J. Pledger, in: "Cyclic Nucleotides in Disease", B. Weiss, ed., 3, University Park Press, Baltimore, Md. (1975).
32. W. J. Pledger, W. J. Thompson, P. M. Epstein and S. J. Strada, J. Cell. Physol. 100:497 (1979).
33. P. M. Epstein, J. S. Mills, E. M. Hersh, S. J. Strada and W. J. Thompson, Cancer Res. 40:379 (1980).
34. G. M. Stancel, W. J. Thompson and S. J. Strada, Molec. Cell. Endo. 3:283 (1975).
35. J. J. Heindel, M. I. Hintz, E. Steinberger and S. J. Strada, Endo. Res. Commun. 4:311 (1977).
36. B. Weiss and S. J. Strada in: "Fetal Pharmacology", L. Boureus, ed., 205, Raven Press, New York (1973).
37. J. A. Smoake, S. Y. Song and W. Y. Cheung, Biochim. Biophys. Acta 341:402 (1974).
38. E. A. Gardner, W. J. Thompson, S. J. Strada and G. M. Stancel, Biochemistry 17:2995 (1978).
39. E. G. Beale, J. R. Dedman and A. R. Means, Endocrinology 100:1621 (1977).
40. M. Vaughan, in: "Eukaryotic Cell Function and Growth: Regulation by Intracellular Cyclic Nucleotides", J. E. Dumont, B. L. Brown and N. J. Marshall, eds., 195, Plenum Press, New York (1976).
41. J. A. Nesbitt, W. B. Anderson, Z. Miller, I. Pastan, T. R. Russell and D. Gospodarowicz, J. Biol. Chem. 251:2344 (1976).
42. H. R. Bourne, G. M. Tomkins and S. Dion, Science 181:952 (1973).
43. H. R. Bourne, P. Coffino, K. L. Thomkins, and Y. Weinstein, Adv. Cyclic Nucl. Res. 5:771 (1975).
44. F. J. Chlapowski, L. A. Kelly and R. W. Butcher, Adv. Cyclic Nucl. Res. 5:245 (1975).
45. W. J. Pledger, R. M. Gardner, P. M. Epstein, W. J. Thompson, S. J. Strada, and L. Wlodyka, Exptl. Cell Res. 118:389 (1979).

PROSTAGLANDINS AND CYCLIC NUCLEOTIDES

S. Nicosia and R. Paoletti

Inst. of Pharmacology and Pharmacognosy
University of Milan
20129 Milan, Italy.

The literature on the interactions of prostaglandins (PGs) and cyclic nucleotides is growing at an ever increasing rate; despite this fact, it is still very difficult to integrate the different and often conflicting reports into a unified picture.

PG-cyclic nucleotide interactions have been reviewed recently[1]; here we have selected to discuss a few aspects only of this topic that seem to us particularly relevant to understand the current status of the research. The references given should be regarded as examples of the problem discussed rather than as an exhaustive bibliography record.

Are PGs Hormones?

Many of the actions of PGs are mediated by a variation in the level of intracellular cyclic nucleotides[1] and in this respect PGs resemble a number of hormones. However, PGs cannot be considered as classical hormones; in fact, they are synthetized in almost all mammmalian cells[2] with the exception of erythrocytes, and act on most tissues of different origin. Therefore, there is no organ specificity for the production or effect of PGs, as there is for hormones. Moreover, PGs do not survive circulation, since they are metabolized more than 90% by a single passage through the lungs[3]. PGI_2 is a notable exception because it is synthetized by lung and metabolized by this organ to a lesser extent than other PGs[4]; not even PGI however, can be considered a hormone, since its circulating concentration is too low to play a role in either platelet aggregation[5] or vascular tone[6].

Not being "classical" hormones, PGs could be short-range hormones, that is intercellular messengers; most of the data available
on PG-cyclic nucleotide interactions are compatible with this hypothesis.

In addition, PGs could act as intracellular messengers, i.e.,
as modulators of the action of other hormones within the cell where
they are produced; examples of this role of PGs will be presented.

PGE_1, PGE_2 and PGI_2

PGE_1 and PGE_2 can affect cAMP levels in a number of tissues[1],
generally through an action on adenylate cyclase, even if an effect
on phosphodiesterase has been suggested in some instances[7,8]. It
has often been found that PGE_1 is more effective than PGE_2 in modifying
cAMP levels, even in those tissues where PGE_1 concentration and turnover are extremely low. Such an apparent paradox can be explained
by the recent discovery that PGE_1 displaces bound $[^3H]$-PGI_2 in platelets[9]. It is likely that such a non-specific interaction of PGE_1
with PGI_2 receptor can occur also in other tissues. Indeed, PGI_2
receptors coupled to stimulation of cAMP accumulation have been
demonstrated in platelets[10,11], fibroblasts[12], anterior pituitary[13],
myometrium and endometrium[14], renal cortex and medulla[15] and granulosa
cells[16]. In arteries[17-29] and in fat cells[21,22] PGI_2 exhibits both
inhibitory and stimulatory effects on cAMP production, depending
on the dose and on the presence of phosphodiesterase inhibitors.

While PGE_1 and PGE_2 enhance adenylate cyclase activity in most
tissues, it is well known that they inhibit epinephrine-stimulated
cAMP accumulation in isolated rat fat cells[23]. Until recently,
it had been possible to reveal an action of PGE on cAMP system only
in intact adipocytes; however, Aktories et al. have shown that PGE_1
is active also in fat cell membranes, provided that GTP and Na^+ are
present[24].

On the contrary, it is still disputed[25,26] whether PGs are
active in broken cell preparations from central nervous system
tissue, while their activity on cAMP accumulation has been demonstrated in cerebral cortex slices[27], in neuroblastoma[28,29] and
neuroblastoma x glioma[30] cell lines.

Endoperoxides (PGG_2, PGH_2) and Thromboxane A_2

PGG_2 and PGH_2 have been shown to inhibit PGE_1-stimulated cAMP
accumulation in platelets[31] and epinephrine-stimulated adenylate
cyclase in rat adipocyte ghosts[32], respectively. However, the
possible relevance of such findings should be carefully considered

in view of the fact that endoperoxide preparation usually involves
the addition of p-hydroxy-mercuric-benzoate[33]; such a compound,
which copurifies with the endoperoxides[34], is a well known thiol
reagent and inhibitor of adenylate cyclase activity. Pure endoperoxides possess very little, if any, inhibiting activity[34]. In
addition, inhibition of adenylate cyclase in platelets can be ascribed at least in part to Thromboxane A_2 (TXA_2) which is synthetized from endoperoxides, and is an inhibitor of cAMP formation
in these cells[35].

$PGF_{2\alpha}$

Stimulation of adenylate cyclase by $PGF_{2\alpha}$ has been claimed
to occur in a number of tissues[1], but the high concentrations
required suggest that it is a non-specific effect. Actually, $PGF_{2\alpha}$
can bind to PGE receptor, even if with much lower affinity[36], and
therefore $PGF_{2\alpha}$ may mimick PGE effects in some instances.

On the other hand, $PGF_{2\alpha}$ can inhibit stimulated cAMP formation
in tissues where other PGs have no or little stimulatory effect[37,38],
and this is probably a genuine effect.

Desensitization

PGs share with hormones and neurotransmitters the property
of inducing densitization. For instance, pre-treatment of frog
erythrocytes with PGE_1 reduces the response of adenylate cyclase
to subsequent challenge with PGE_1, but not with other agents such
as isoproterenol or NaF[36].

In other systems, however, PGE_1 is not a specific desensitizing
agent; in lung fibroblasts, for instance, it has been demonstrated[39]
that PGE_1 pre-treatment decrease the sensitivity of adenylate cyclase
to GTP, GPP(NH)P, PGE_1 + GTP and epinephrine + GTP by the same
percentage. It has been suggested in this case that the mechanism
of PGE-induced refractoriness could be at the level of the GTP-
binding protein instead than at the level of the receptor[39].

Recently, a possible role of arachidonic acid and/or PGs in
the onset of desensitization to other hormones has been suggested:
the use of phospholipase A_2 inhibitors, which abolish arachidonic
acid release and therefore PG formation, inhibits also isoproterenol
induced refractoriness in C_6 astrocytoma cells[40].

Inhibition versus Stimulation

It has been proposed that some PGs can interact with a stimulatory receptor, whereas others would interact with an inhibitory
one[41].

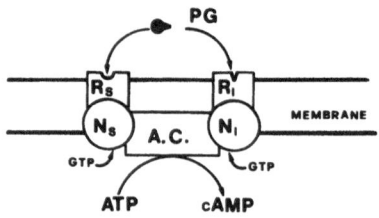

Fig. 1.   Proposed model for biphasic action of PGs in the same cell.
R_S= stimulatory receptor, R_I= inhibitory receptor; A.C.=
catalytic moiety of adenylate cyclase; N_S= nucleotide
binding protein necessary for stimulation; ^SN_I= nucleotide
binding protein necessary for inhibition.

We propose here that the same PG can actually bind to two
different receptors, stimulatory and inhibitory, respectively and
that both receptors might be present in the same cell and coupled
to the same cyclase (Fig. 1).

Evidence for the existence of two types of receptors for the
same PG stems from the biphasic nature of the dose-response curves
for a given PG in arteries[17-20], fat cells[21,22,42] and liver[43].   Inhibitory receptors (R_I) show higher affinity than stimulatory ones
(R_S), except in the liver where the reverse applies; the different
affinity of the receptors for the same PG and the actual PG concentration would regulate the interaction with either receptors.   Both
stimulation and inhibition of adenylate cyclase by PGs requires
GTP[24,43], and therefore the model includes the presence of different
guanyl nucleotide binding protein (N_S and N_I).   However, it should
be emphasized that, while the existence of these different receptors
is clearly proven, their coupling to the same cyclase and even their
presence within the same cell has not been yet demonstrated.

EFFECTS ON GUANYLATE CYCLASE

Less is known about the regulation of guanylate cyclase than
about adenylate cyclase.   For instance, it is not known why those
hormones that stimulate cGMP formation in intact cells, generally
have no effects on either soluble or membrane bound guanylate cyclase
in broken cell preparations[44].

Some of the arachidonic acid metabolites, (namely the endoperoxides PGG_2 and PGH_2[1,45] and related compounds such as fatty
acids[44,45] seem to be an exception; in fact these substances can
activate guanylate cyclase preparations in cell-free systems.

It should be noted, however, that such activation is probably
related to the well documented sensitivity of the enzyme to free
radicals and redox agents[44], and is not a hormonal effect.

This hypothesis is suggested by the following observations: endo-peroxides and fatty acid hydroperoxides, that can generate radicals, stimulate soluble splenic guanylate cyclase[45], while stable PGs and saturated or unsaturated fatty acids are not effective at con-centrations up to 10^{-5}M. On the other hand, platelet guanylate cyclase can be stimulated by unsaturated fatty acids at 10^{-4}M, but the addition of lipoxides clearly potentiate the stimulation[52]; lipoxidase catalizes the formation of fatty acid peroxides, which in turn can generate free radicals.

Therefore, the endoperoxides and hydroperoxides probably act directly at the level of the catalytic site, affecting the redox state of the enzyme, and their action might not be mediated through a receptor. It should be noted, however, that such compounds are formed in cells as intermediates in arachidonic acid metabolism, and that a physiological role cannot be excluded.

In particular, hydroperoxy acids are synthetized by lipo-oxy-genases while the endoperoxides are the products of cyclo-oxygenase, and the involvement of both lipo-oxygenase and cyclo-oxygenase path-ways in the control of cGMP formation has been suggested also by other authors[47].

Stable PGs have been reported to stimulate cGMP accumulation in various preparation: $PGF_{2\alpha}$ is active in vascular tissue, uterus, liver, fibroblasts and central nervous system[1]; PGI_2 and PGE_1[46] are effective in a neuroblastoma cell line, and an action for PGEs has been reported also in a few other instances[1].

PGs AS MODULATORS OF HORMONE ACTION

So far we have discussed the effects of different PGs on cyclic nucleotide levels but the interactions of PG and nucleotide systems are more complex and perhaps not fully understood. In fact, it has been shown that not only cAMP levels can be affected by PGs, as discussed earlier, but also cAMP can in turn modify PG production.

For instance, cAMP increases PG formation in a number of differ-ent tissues[1], probably by acting at the level of arachidonic acid release from phospholipids. The mutual interactions of PGs and cAMP in such tissues seem to suggest the existency of a sort of positive feed-back system, the significance of which is so far unknown.

On the contrary, the picture is completely different in adipocytes, where PGEs are inhibitors of adenylate cyclase[23,24], except at high concentrations. In these cells a regulatory system based on negative fee-back can be envisioned[48]; a hormone-induced increase in intracellular cAMP would trigger a stimulation of PG synthesis, and PGs would in turn inhibit further cAMP production.

In this respect PGs can be said to act as modulators of hormone action.

A different pattern of PG-cAMP interactions has been suggested in platelets. In this cells, cAMP inhibits PG and Thromboxane B_2 formation[49,54]; the different groups, however, do not agree on whether the site of action of cAMP is the phospholipase or the cyclo-oxygenase. In either case, exogenous PGI_2, which elevates cAMP levels, would cause a decrease in pro-aggregatory TXA_2 formation, thus reinforcing its own direct anti-aggregatory effect.

The complexity of the interactions between PGs and cAMP suggests once again that PGs are not classical hormones, but that they probably play a unique role in the regulation of cell metabolism.

ACKNOWLEDGEMENTS

This work was partially supported by a grant from the Italian National Research Council (CNR No 9.02396.65).

REFERENCES

1. B. Samuelsson, M. Goldyn, E. Grantrom, M. Hamberg, S. Hammarstrom and C. Malmsten, Ann. Rev. Biochem. 47:997 (1978).
2. S. Bergstrom, L. A. Carlson and J. R. Weeks, Pharmacol. Rev. 20:1 (1968).
3. S. H. Ferreira and J. R. Vane, Nature, 216:868 (1967).
4. R. J. Gryglewski, R. Korbut and A. Ocetkiewicz, Nature, 273:765 (1978).
5. M. L. Streer, D. E. MacIntyre, L. Levine and E. W. Saltzman, Nature, 283:194 (1980).
6. J. B. Smith, M. L. Ogletree, A. M. Lefer and K. C. Nicolaou, Nature, 274:64 (1978).
7. M. A. Amer, N. R. Marquis, in: "Prostaglandins in cellular biology", P. W. Ramwell and B. B. Pharriss, eds., Plenum Press, N.Y., p.93 (1972).
8. G. M. Nemeck, K. P. Ray and R. W. Butcher, J. Biol. Chem. 254:598 (1979).
9. A. M. Siegl, J. B. Smith, M. J. Silver, J. C. Nicolaou and D. Ahern, J. Clin. Invest. 63:215 (1979).
10. J. E. Tateson, S. Moncada and J. R. Vane, Prostaglandins, 13:389 (1977).
11. R. R. Gorman, S. Bunting and O. V. Miller, Prostaglandins, 13:377 (1977).
12. R. R. Gorman, R. D. Hamilton and N. K. Hopkins, J. Biol. Chem. 254:1671 (1979).
13. M. Toth, M. Todd and F. Hertelendy, Prostaglandins, 17:105 (1979).

14. M. F. Vesin, L. DoKhac and S. Harbon, Mol. Pharmacol. 16:823 (1979).
15. C. A. Herman, T. V. Zenser and B. B. Davis, Biochim. Biophys. Acta 582:496 (1979).
16. A. K. Goff, J. Zamecnik, M. Ali and D. T. Armstrong, Prostaglandins 15:875 (1978).
17. K. Schror and P. Rosen, Naunyn-Schmiedeberg's Arch. Pharmacol. 306:101 (1979).
18. A. Dembinska-Kiec, W. Rucker and P. S. Schonhofer, ibid 308:107 (1979).
19. W. R. Kukovetz, S. Holzmann, A. Wurm and G. Poch, J. Cycl. Nucleot. Res. 5:469 (1979).
20. O. V. Miller, J.W. Aiken, D. P. Hemker, R. J. Shebuski and R. R. Gorman, Prostaglandins 18:915 (1979).
21. B. B. Fredholm, P. Hjemdahl and S. Hammerstrom, Biochem. Pharmacol. 29:661 (1980).
22. G. Basile and A. Nicosia, to be published.
23. R. W. Butcher and C. E. Baird, J. Biol. Chem. 243:1713 (1968).
24. K. Aktories, G. Schultz and K. H. Jakobs, FEBS Letters 107:100 (1979).
25. H. O. J. Collier and A. C. Roy, Nature 248:24 (1974).
26. G. P. Tell, G. W. Pasternak and P. Cuatrecasas, FEBS Letters 51:242 (1975).
27. F. Berti, M. Trabucchi, V. Bernareggi and R. Fumagalli, in: "Advances in the Biosciences", S. Bergstrom and s. Bernhard, eds., Pergamon Press, p.475 (1973).
28. B. Hamprecht and J. Schultz, FEBS Letters 34:85 (1973).
29. A. J. Blume and C.J. Foster, J. Biol. Chem. 251:3399 (1976).
30. M. Brandt, C. Buchen and B. Hamprecht, J. Neurochem. 34:643 (1980).
31. O. V. Miller and R. R. Gorman, J. Cyclic Nucleot. Res. 2:79 (1976).
32. R. R. Gorman, M. Hamberg and B. Samuelsson, J. Biol. Chem. 250:6460 (1975).
33. M. Hamberg, J. Svensson, T. Wakabayashi and B. Samuelsson, Proc. Natl. Acas. Sci. U.S.A. 71:345 (1974).
34. S. Nicosia, unpublished observations.
35. O. V. Miller, R. A. Johnson and R. R. Gorman, Prostaglandins 13:599 (1977).
36. R. J. Lefkowitz, D. Mullikin, C. L. Wood, T. B. Gore and C. Mukherjee, J. Biol. Chem. 252:5295 (1977).
37. D. G. Gardner, E. M. Brown, R. Windeck and G. D. Aurbach, Endocrinology 104:1 (1979).
38. S. Champion, B. Haye and C. Jacquemin, FEBS Letters 46:289 (1974).
39. R. B. Clark and R. W. Butcher, J. Biol. Chem. 254:9373 (1979).
40. P. Mallorga, J.F. Tallman, R. C. Henneberry, F. Hirata, W. T. Strittmatter and J. Axelrod, Proc. Natl. Acad. Sci. U.S.A. 77:1341 (1980).
41. H. S. Kantor, P. Tao and H. C. Kiefer, Proc. Natl. Acad. Sci. U.S.A. 71:1317 (1974).

42. H. Kather and B. Simon, J. Clin. Invest. 64:609 (1979).
43. V. Thomasi and E. Ferretti, Molec. Cell Endocrinol. 2:221 (1975).
44. F. Murad, W. P. Arnold, C. K. Mittal and J. M. Braughler,
    Adv. Cyclic Nucleot. Res. 11:175 (1979).
45. G. Graff, J. H. Stephenson, D.B. Glass, M. K. Haddox and
    N. D. Goldberg, J. Biol. Chem. 253:7662 (1978).
46. R. Ortman, FEBS Letters 90:348 (1978).
47. D. Y. Gruetter and L. J. Ignarro, Prostaglandins 18:541 (1979).
48. C. Dalton and W. C. Hope, Prostaglandins 6:227 (1974).
49. C. Malmsten, E. Granstrom and B. Samuelsson, Biochem. Biophys.
    Res. Comm. 68:569 (1976).
50. M. Minkes, N. Stranford, M. Y. Chi, G. J. Roth, A. Raz,
    P. Needleman and P. W. Majerus, J. Clin. Invest. 59:449
    (1977).
51. J. A. Lindgren, H.E. Claesson, H. Kindahl and S. Hammarstrom,
    in: "Advances in Prostaglandin and Thromboxane Research",
    B. Samuelsson, P. W. Ramwell and R. Paoletti, eds., Raven
    Press, New York, Vol.6, p.275 (1980).
52. H. Hidaka and T. Asano, Proc. Natl. Acad. Sci. U.S.A. 74:3657
    (1977).

CYCLIC NUCLEOTIDE METABOLISM IN WHOLE CELLS

R. Barber and R. W. Butcher
Graduate School of Biomedical Sciences
The University of Texas Health Science Center at Houston
P. O. Box 20334, Astrodome Station
Houston, Texas   77025

INTRODUCTION

While we have learned a great deal about the enzymes of cyclic AMP metabolism, the control of intracellular cyclic AMP levels is only dimly perceived. This is partly because of the way in which cyclic AMP was discovered: this discovery was in fact the logical outcome of a beautiful series of experiments by Sutherland, Rall and their co-workers, designed to show that a hormone (epinephrine) could work in cell-free systems. That hormones could do so was not a popular notion at the time, so they attempted to relate, at least in qualitative fashion, events occurring in homogenates or subcellular fractions to those known to occur in the intact tissue[1]. How well they succeeded is evidenced by this NATO course.

In addition, efforts to understand the fine control of intracellular cyclic AMP metabolism by hormones, drugs, and other factors has been confounded by a variety of factors which are not often appreciated. Amongst these are the following:

1. The range of cellular cyclic AMP concentrations which are rate-limiting upon a particular biological process is very narrow[2]. Unfortunately, such changes in cyclic AMP concentrations are often so small as to make accurate study impossible or at least very discouraging. As an expedient, most studies, including those presented here, have dealt with supermaximal concentrations of hormones or other agonists or antagonists. Thus, the levels of cellular cyclic AMP, which we know the most about, are far in excess of those which might be expected to be found in the in vivo situation.

2. There are inherent problems in studying regulatory enzymes in
   cell-free preparations. Not only have membranous architecture
   and other aspects of intracellular topography been destroyed,
   but also the availabilities of cofactors, substrates, end-product
   inhibitors, and other effectors, have been changed. Given that
   there are few (if any) manipulations which can be made with intact
   cells which do not in some way alter cellular cyclic AMP levels,
   one is confronted with the disquieting feeling that a kind of
   biological Heisenberg uncertainty principle may obtain.

KINETICS OF CYCLIC AMP METABOLISM

There are at least four components to the overall process called
cyclic AMP metabolism. To have a firm quantitative grasp on the process, we need to know the rates of synthesis and hydrolysis during
the stimulus, how rapidly cyclic AMP escapes from the intracellular
to the extracellular space, and, how rapidly and to what extent it
is bound to other cellular constituents.

Given a detailed knowledge of the rates and extents of the processes listed above, we can then determine how a cell meets its
requirements for the control of a particular intermediate of hormone
action (in this case, cyclic AMP), not only in terms of the kinetic
parameters of the controlling enzymes (which will of course include
the time course of desensitization (vide infra)), but the temporal
and relative importance of the processes. An additional regard will
be an understanding of the energy burden placed on the cell by the
system.

APPROACHES TO THE STUDY OF INTRACELLULAR cAMP METABOLISM

In principle, four approaches are available:

The classic biochemical technique of homogenization and assay.
This approach is tried and true, and helpful, but is seriously limited
in the context of determining cellular cAMP metabolism because of:
a. changes in the parameters of the enzymes upon cell rupture.
b. changes in the topological arrangement of the enzymes.

This last point is especially important for membrane bound
enzymes, where the broken cell preparation is usually a collection
of vesicles. In which case, the dominating factors in measured reaction rates might be more dependent on the history and the physical
properties of the vesicles rather than on the enzymes themselves.
c. changes in escape, by definition, cannot be measured in cell free
systems.

The approach of measuring endogenous cyclic AMP levels at various points in time. In this type of experiment, one measures cellular cyclic AMP levels at some point in time making use of assays of endogenous concentrations of the compound such as radioimmunoassay or the Gilman binding assay. Unfortunately, this type of experiment is limited in that one can only measure accumulation at some point in time.

The mathematical modeling approach. Here a series of assumptions are made concerning the kinetic properties of the enzymes (or processes) involved in cAMP accumulation. These assumptions are cast into mathematical form and developed in order to obtain an expression relating intracellular cAMP accumulation to time. The expression is fitted to experimental data so as to obtain numerical values for the enzyme parameters.

The radioactive tracer approach. This is another classic technique well proven in both physiology and biochemistry. It involves the addition of an appropriate radiolabelled compound to the hormone-stimulated cells and following the progress of the label with time through the various metabolic pools.

APPLICATIONS

All of the approaches listed above have, with greater or lesser degrees of success, been used[3]. In our own experiments, we have made use of all and have finally settled on the combination of all four approaches as being the only feasible method of determining cyclic AMP turnover available at this time.

Qualitatively, the system was solved by Sutherland and his co-workers nearly two decades ago. An additional ingredient prerequisite to a detailed kinetic ananlysis was first noticed by Rall and his co-workers in brain slices - the phenomenon of desensitization or refractoriness of the cAMP accumulation system[4]. The existence of this phenomenon meant a priori that one or more of the controlling factors in accumulation, i.e., synthesis, hydrolysis or escape, changed with time.

Cell free systems. These have been widely used to study cAMP metabolism[3,5-9]. The basic weakness with this approach has been mentioned already. It sometimes requires the implicit assumption that enzyme activities determined for vesicle preparations are little altered from those that existed within the intact cell. Sometimes the assumption need not be that rigorous, but still demands that changes brought about by various treatments to whole cells are reflected ultimately in broken cell measurements.

Intracellular Accumulation. Measurements of intracellular accumulation represent paradoxically the most direct approach experimentally and the most equivocal theoretically. Frequently, statements are made concerning the significance of differences in accumulation at only one or two time points, and this is a very sloppy practice cAMP accumulation at any time point is the resultant of the activities of the adenylate cyclase, the phosphodiesterase (or phosphodiesterases) and of escape all since the time of stimulation (or even prior to that). Any particular level at a particular time could, in principle, have been achieved in an infinite number of ways. For instance, suppose that after five minutes of treatment A, two units of cAMP have accumulated, and after treatment B, one unit has accumulated. This could be (and frequently is) interpreted by saying that the rate of synthesis might have been unchanged while the rates of removal by hydrolysis and escape were increased. Moreover even if the investigator were sure that all of the difference in accumulation could be attributed to changes in hydrolysis, it would not be possible to say from these time points alone to what extent the hydrolytic parameters had been changed.

Figure 1 illustrates one theoretical time course of accumulation to two units at five minutes and three time courses reaching one unit at the same time. The upper curve was calculated by assuming a steady rate of synthesis of 0.5083 units/min and a first order hydrolysis constant of $0.1min^{-1}$. The lower solid curve was obtained by simply halving the rate of synthesis so as to achieve half of the synthesis at five minutes (or any other time). The dashed curve was obtained with an unchanged rate of synthesis but with the hydrolysis constant increased to $0.456min^{-1}$. The final curve (in dots and dashes) just above it has the same initial rate of syntheis (0.5083 units/min as the first curve and the same hydrolysis constant, but in this case we supposed that the cyclase is desensitizing in first order fashion with a rate constant of $0.2875min^{-1}$. Thus, knowledge that a drug or treatment lowered the 5min accumulation to one half (by itself) cannot show what aspects of the accumulation were changed. The one section of an accumulation time course where interpretation might be simplified occurs at very short times after the onset of stimulation (times of the order of a few seconds). Here, the initial rate of accumulation is equal to the rate of synthesis provided that: 1) the rate of desensitization is not very large, 2) the response time of the cells to the hormone is not very slow, 3) there is no zero order component of hydrolysis, and 4) the rate constants for the higher order components of hydrolysis are not large. Conditions 1), 2) and 4) were indicated by determining accumulation at a series of short times and showing that the initial accumulation curve was linear and that it extrapolated back through the origin. Condition 3) was indicated by showing that the dose response curve extrapolated back through the origin. If there were a significant zero order component of hydrolysis (or escape) then there should have been a significant level of hormone that would produce no increase in accumulation. For WI-38 cells, we have been unable to detect this.

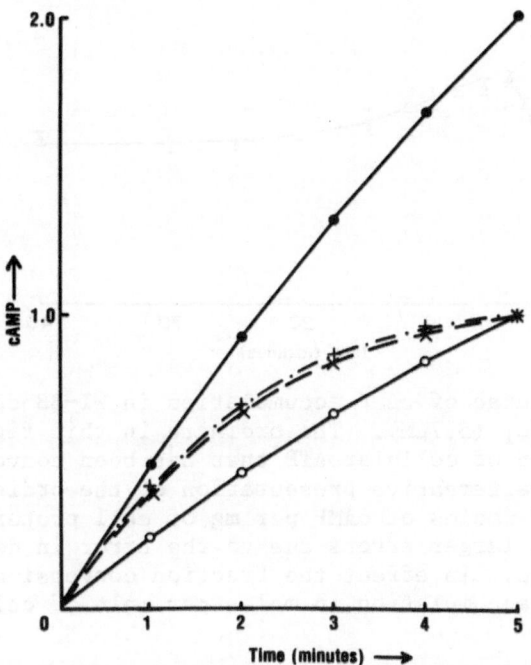

Fig. 1.   Theoretical time courses to illustrate how changes in
          accumulation of cAMP at a single time point can be achieved
          in a variety of ways. An explanation of the figure is
          given in the text.

     In order to be able to interpret accumulation data in terms of
the individual processes, complete time courses and some sort of
model are necessary to assist in their interpretation. Few investigators have attempted this.

     Tracer techniques. It is possible to label cells with [³H]-
adenine and show rapid conversion to ATP[10-12]. In the presence of
hormone, cAMP is formed and can be expressed as a percentage conversion of ATP (or rather, cellular adenine nucleotide). If one makes
use of a fixed labelling period, this technique allows the same sorts
of experiments on the kinetics of cAMP accumulation as have been made
using endogenous methods (such as the Gilman assay). On the other
hand, in the field of intracellular cAMP kinetics, pulse tracer techniques have been largely neglected, though (with appropriate safe-
guards) they offer the best opportunity of a direct approach.

     Our studies. In our studies with cultured fibroblasts (princi-
pally the untransformed line WI-38), we have made use of all three
methods outlined above. Our approach was to use mathematical techniques to extract information pertaining to the enzymes from experi-

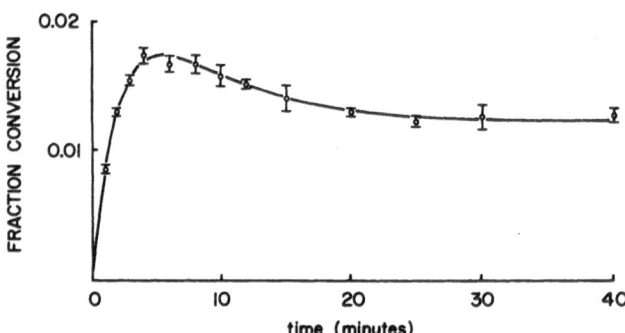

Fig. 2. The time course of cAMP accumulation in WI-38 cells stimu-
         lated by PGE_1 (5.7μM). The ordinate in this figure gives
         the fraction of cellular ATP that has been converted to
         cAMP. The alternative presentation of the ordinate values
         in terms of pmoles of cAMP per mg of cell protein gives
         values with larger errors due to the error in determining
         cell protein. In effect the fraction conversion used here
         gives cAMP accumulation in moles per mole of cellular ATP.

mental time courses and then to use broken cell and tracer studies
for confirmation. This is, therefore, the order in which the events
will be described now.

Figure 2 shows a typical time course for cAMP accumulation in
WI-38 cells stimulated with 5.7μM PGE_1.

The fact that the time course passed through a maximum was
sufficient by itself to demonstrate desensitization (it being known
that the hormone had not been destroyed nor the cells killed). If
there were no desensitization and if removal of the cyclic nucleotide
(by hydrolysis and escape) were first order, the mathematical analysis would be easy. The rate equation describing accumulation would
then be given by:

$$c = \frac{k_s}{k_e} (1 - e^{-k_e t})$$

This is illustrated in figure 3. A priori, there was some
reason to believe that desensitization was a phenomenon that concerned chiefly the synthetic apparatus since it was possible to
observe hormone specific desensitization in many systems[13,14], which
was difficult to explain in terms of changes in the phosphodiesterase. Also there was some evidence that removal of cAMP from these
cells was a first order phenomenon. This involved experiments in

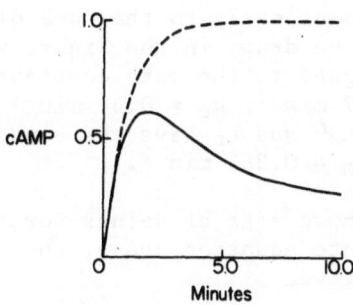

Fig. 3.   Theoretical constructs of cAMP accumulations with time.
(———) Pattern typical for most cells. (----) Pattern
expected with uniform synthesis and first order elimination.

which stimulating hormone was suddenly withdrawn. The subsequent
decay of cAMP within the cells was accurately first order. Finally,
it was clear (in view of the elevated steady state) that complete
desensitization did not occur within the times of our experiments[15].

The scheme devised to fit all of the above facts was as follows:

$$\text{unstimulated cyclase} \xrightarrow{\text{hormone}} \text{cyclase: stimulated and active} \underset{k_n}{\overset{k_m}{\rightleftharpoons}} \text{cyclase: stimulated and inactive}$$

If sufficient hormone is added to stimulate all of the cyclase,
the above scheme can be expressed as a differential equation and
solved to give:

$$k_s = k_s^{\,o} \left[\frac{k_n + k_m e^{-(k_m + k_n)t}}{k_m + k_n} \right]$$

where $k_s^{\,o}$ is the initial rate of synthesis and k_m and k_n are (respectively) the rate constants for inactivation and reactivation of the
cyclase. Thus, the rate of synthesis is expressed as a function of
time. The introduction of the further postulate that rate of removal
of cAMP was first order allowed the derivation of an equation that
expresses cAMP accumulation as a function of time:

$$c = k_s^{\,o} \left[\frac{k_n (1 - e^{-k_e t})}{k_e(k_m + k_n)} + \frac{k_m (e^{-k_m + k_n)t} - e^{-k_e t})}{(k_e - k_m - k_n)(k_m + k_n)} \right]$$

This can be fitted numerically to the data displayed in Figure 2 to give the continuous curve drawn in the figure when the following numerical values are assigned to the rate constants: $k_s{}^0 = 0.0105$ conversion/min, $k_n = 0.077$ min^{-1}, $k_e = 0.44$ min^{-1} and $k_m = 0.073$ min^{-1}, or alternatively, $k_s{}^0$ and k_n have the same values as above, but $k_e = 0.15$ min^{-1} and $k_m = 0.363$ min^{-1}.

When either of the above sets of values for the kinetic parameters are substituted into equation above, the relationship between accumulation and time is given by:

$$C = 0.01225 + 0.01762\, e^{-.15t} - 0.02987\, e^{-.44t}$$

Only the two sets of parameters quoted above are capable of giving the above equation, which is plotted as the continuous line in Figure 2.

Thus, this technique of curve fitting of a model equation to accumulation data cannot give a unique estimate for either the fractional turnover constant (k_e) or for the rate and extent of desensitization. However, it can limit each of those parameters to just two possible values. For instance, we have stated above that the fractional turnover constant for cAMP in WI-38 cells must either be 0.44 min^{-1} or 0.15 min^{-1}. It cannot be 0.05, 0.3 or 0.6 min^{-1}. Using these latter numerical values, a fit of the above equation to the experimental data is impossible, irrespective of adjustments to the other parameters.

To determine which of these sets of parameters corresponded to reality, it was essential to determine k_m and k_d independently[16]. Once either of those parameters was measured a firm prediction could then be made as to the value of the other.

Figure 4 shows the incorporation of tritium label into the cAMP in WI-38 cells stimulated with 5.7μM PGE$_1$. [8-^{3}H]Adenine was added to the system at various times after stimulation of the cells by PGE as indicated on the abscissa. The incubations were terminated 2 minutes after the addition of [8-^{3}H]Adenine. Assuming that the incorporation of [8-^{3}H]Adenine into [8-^{3}H]ATP was the same at all times, the decrease in the rate of synthesis of [8-^{3}H]cAMP began immediately after addition of the hormone. This process of desensitization had a $t_{1/2}$ of 4.5 min. The total amount of desensitization during the acute phase of stimulation of these cells corresponded to a loss of 50% of the activity of the adenylate cyclase system. In terms of the model this indicated numerical values for k_m and k_n of about 0.075 min^{-1} each, in agreement with the first set of kinetic parameters quoted above. Broken cell data on adenylate cyclase activities was in substantial agreement with these figures[17].

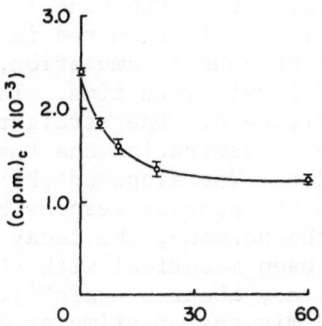

Fig. 4. Accumulation of adenylate cyclase into the cAMP of WI-38
         cells during 2 minute exposures to $[8-^3H]$ adenine.

Using this same experimental principle, the rate of recovery
of adenylate cyclase after removal of the stimulating hormone was
determined. In this case, the relative incorporation of radioactivity into cAMP over a three minute period was used. In Figure
5, the fraction of the adenylate cyclase activity that has recovered
is plotted against the time between washout and readdition of the
hormone, i.e., the time allowed for recovery. The continuous line
is the theoretical curve for recovery, assuming that the recovery
was a first-order process with a rate constant of 0.077 min^{-1} (i.e.,
the value previously assigned to k_n). While the errors are rather
large, the data are consistent with the idea that recovery was a
first-order process with a rate constant of 0.077 min^{-1}.

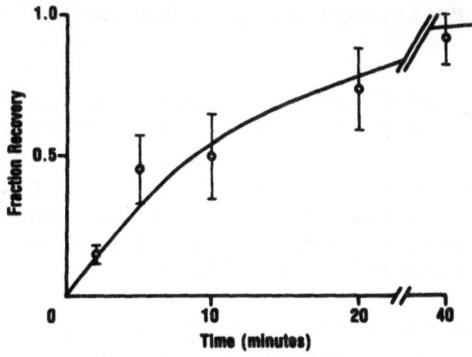

Fig. 5. Recovery of adenylate cyclase from the desensitized state
         after removal of the hormone. The ordinate is defined as
         the fraction (measured cyclase activity-desensitized cyclase
         activity)/(naive cyclase activity-desensitized cyclase
         activity).

In prospect, the determination of the fractional turnover constant would seem to be the easiest of the kinetic parameters to determine independently. All that is required in principle is to stimulate cells to a convenient cAMP accumulation, remove the hormone and follow the decay of cAMP levels with time. The results of such an experiment are shown in Figure 6. The straight-line relationship between the logarithm of the concentration and the time demonstrated that the decay was first order. The slope of the line gave the rate constant from the decay. If the process responsible for the removal of cAMP were unaffected by the hormone, the decay constant found by this experiment should have been identical with the fractional turnover estimated previously (i.e., about 0.4 min^{-1}). In fact, the values of 0.91 min^{-1} after 2 minutes of stimulation and 1.08 min^{-1} after 40 minutes of stimulation were much too high (Fig. 6).

A value for the fractional turnover constant of around 1.0 min^{-1} was not only incompatible with the model for desensitization discussed here, it was incompatible with any model that did not assume that the rate of synthesis of cAMP increased for several minutes after the onset of stimulation. The data presented in Figure 4 shows that such an increase did not occur. Thus, the conclusion that the rate of hydrolysis in the absence of agonist was 2.5 times greater than that during stimulation was unavoidable. In other words, directly or indirectly, the hormone had a profound effect on the activity of phosphodiesterase in intact cells. Direct measurements of phosphodiesterase activity in broken cell systems from naive and desensitized cells have corroborated this view[18].

While the above data were largely compatible with the model, there was a real need for a direct experiment to confirm or deny the veracity of our parameters. That is, we needed to determine k_d directly in the presence of hormone. To be able to do so, it was necessary to develop an isotope pulse technique[16,19,20]. We argued as follows:

ATP was the only known immediate precursor for cAMP within the cell, so any changes in the isotopic composition of the ATP should ultimately be reflected by similar changes in the cAMP. As a radioactive isotope is continuously added to the ATP, it will eventually be found in the cAMP, but with a delay dependent on the rate of cAMP turnover. A slower turnover would cause a longer delay in the accumulation of the radioactivity into the cAMP and vice versa.

The formal mathematical presentations of this idea, when cAMP accumulation is at a steady-state level, are straightforward and do not require unusual assumptions concerning the nature or kinetics of cAMP synthesis or removal and are published elsewhere. The bulk of the mathematical manipulation was obligatory only to obtain equations in which the fractional turnover constant (k_e) was expressed in terms of experimentally measurable quantities.

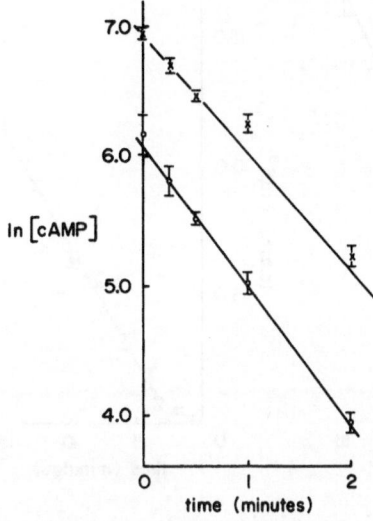

Fig. 6.   Decay of cAMP levels in WI-38 cells after removal of PGE_1.
(X) PGE_1 removed after 2 minutes of stimulation.   (O) PGE_1
removed after 40 minutes of stimulation.

The final form of the equation used to achieve this is:

$$\frac{(cpm)_c}{\int_o^t (cpm)_c dt} = \frac{k_e}{[ATP]} \cdot \frac{\int_o^t (cpm)_a dt}{\int_o^t (cpm)_c dt}$$

where $(cpm)_c$ and $(cpm)_a$ are, respectively, the counts for minute in
the cellular cAMP and ATP.   ATP   is the total cellular content of
ATP.

Figure 7 shows the incorporation of $[8\text{-}^3H]$ adenine into the ATP
of WI-38 cells. The cells had been stimulated with PGE_1 (5.7μM)
for 40 minutes previously. The zero time point on this graph marks
the addition of $[8\text{-}^3H]$ adenine to the incubation medium. The
incorporation of label into cAMP is given on the right. As expected
from the theoretical considerations discussed above, incorporation
into the cAMP pool was delayed relative to that into the ATP. There
was no perceptible delay at the shortest time studied (1 minute) for
the build-up of radioactivity in the ATP pool.

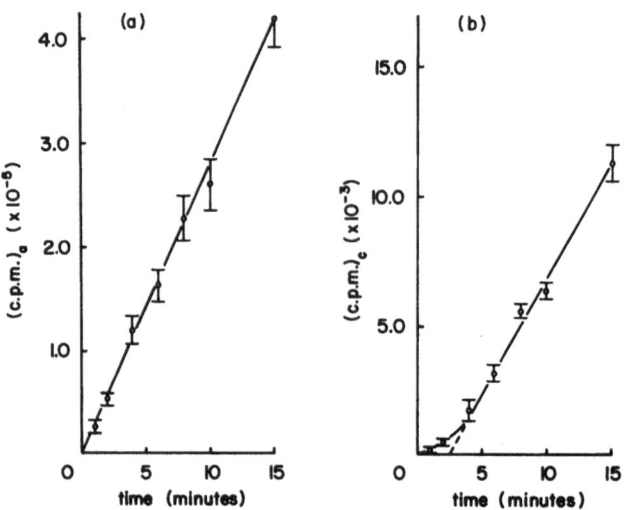

Fig. 7. Accumulation of radioactivity into the (a) ATP and (b) cAMP
of WI-38 cells as a result of incubation with $[8-^3H]$-
adenine. The labeled adenine was added to the incubation
medium at the zero points of the above graphs and was continuously present thereafter.

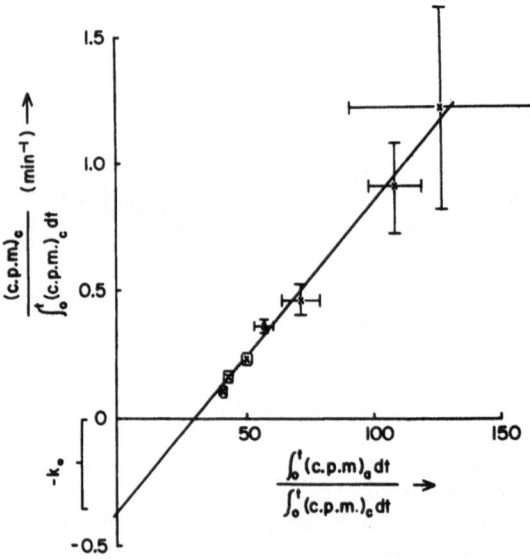

Fig. 8. Determination of the fractional turnover constant for cAMP
in WI-38 cells. An explanation of the ordinate and abscissa
is given in the text.

These data were used to generate the integral expressions quoted above, and in Figure 8 $(cpm)_c \Big/ \int_0^t (cpm)_c dt$ is plotted against $\int_0^t (cpm)_a dt \Big/ \int_0^t (cpm)_c dt$.

As anticipated by the derivation, the graph plotted as a straight line. k_e was equal to the negative intercept of the y-axis. The value found was 0.38 min^{-1}. This is good agreement with the value obtained from modeling of the time course and in contradiction to that obtained by washout of the hormone.

The kinetics of cAMP accumulation in turkey erythrocytes indicates a very different functional design from that in WI-38 cells. During 75 minutes of stimulation with isopropylnoradrenalin, we found neither significant desensitization nor hydrolysis. Thus, the cAMP accumulation system in the turkey cells is stripped down to the very simple version discussed earlier of uniform synthesis with first order removal (apparently almost only by escape) from the cells. This is illustrated in Figure 9, where it can be seen that the total

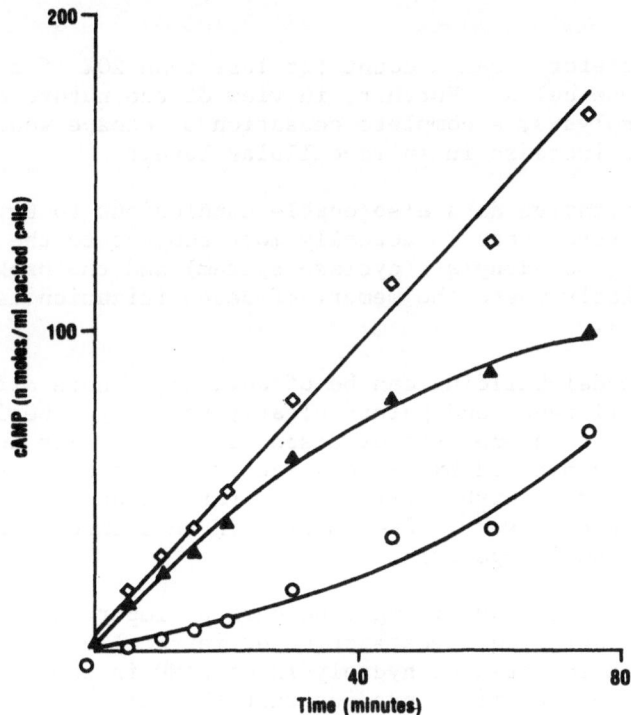

Fig. 9.   Accumulation of cAMP in turkey erythrocytes stimulated with 10 M isopropylnoradrenalin.   (O) Extracellular cAMP. (▲) Intracellular cAMP. (◇) Total cAMP.

accumulation is linear with time (implying no loss of cAMP to the
system and constant synthesis). Moreover, the intracellular accumulation is described by the simple equation of k_s/k_e $(1-e^{-k_e t})$ with
k_s equal to 2.36n moles for ml of packed cells and k_e equal to
0.017 min^{-1}. Parenthetically, Hoffman et al.[21] have presented data
claiming desensitization of turkey erythrocytes by catecholamines.
However, the time of the treatment by hormone was 5 hours and the
minimum time at which desensitization might be apparent is not clear.

cAMP TURNOVER IN CULTURED CELLS

     How may the information described above be helpful? In fact,
many previously qualitative questions can now be answered wholly or
at least in part. As an example, there is the question of the role
of escape. One thing which seems clear now is that escape is not
an important mechanism for control of cAMP levels in WL-38 cells.
We have been able to show that the rate of escape is proportional
to intracellular cAMP levels (in press) and that the rate constant
for the process is 0.068 min^{-1}. As mentioned above, the rate constant for total turnover in these cells is in the region of
0.4 min^{-1}.

     Escape, therefore, can account for less than 20% of the loss
of cAMP from these cells. Further, in view of the nature of the
kinetics of hydrolysis, a complete cessation of escape would result
in only a modest increase in intra-cellular levels.

     These quantitative data also enable conclusions to be drawn
concerning the system that is actually most subject to the time dependent changes (the adenylate cyclase system) and the broken cell
aspects show clearly where the memory of desensitization is stored
(in the membrane).

     Attempted model-building can be of advantage where differences
exist between cell types and points clearly to an hypothesis that
there exists more than one type of desensitization. For instance,
the time courses published by Nickols and Brooker for C6-2B cells
are quite inconsistent with a first order rate of desensitization
for the adenylate cyclase[22]. Presumably, in that case, a quite
different mechanism is operating.

     However, the most surprising (and perhaps important) result to
emerge from this detailed investigation of intracellular kinetics
is the difference in rates of hydrolysis of cAMP in the presence and
absence of hormone. We might predict that the value of k_e measured
after washout of the hormone is closer to the unstimulated state
than found during stimulation. But in any case, there seems little
doubt that at least in some cells hormones have profound effects on
the cellular PDEs and that these effects occur very rapidly.

Being able to determine cAMP turnover in intact cells with the discovery of the control of PDE activity by hormonal stimulation has posed a whole new set of questions. For example, what are the connec-ting elements between hormone stimulation and PDE modulation? And, what do these rapid changes in PDE activity require in terms of the topological position of the PDE within the cell? Before leaving the subject of hormone-sensitive PDEs, one might mention that it has a nice teleological appeal. In collaboration with the Dumont group in Brussels[23], we have shown that the fact of desensitization enables cells to meet the requirements of response and terminationn of response within preset times more economically with respect to energy use than would otherwise be the case. A system with a hormone-sensitive PDE would clearly be still more efficient, since the enzyme is least active when the cell is accumulating cAMP and most active when the cell is "attempting to get rid of it".

CONCLUSIONS

The range of problems approachable by turnover studies is potentially wide. For example, more precise knowledge on the sites of action of drugs or treatments is to be expected, since by determining which kinetic parameters are changed by the addition of a drug, one can determine which molecular apparatus is in fact suffering the effect. This approach has been used to determine whether the effect of carbachol on the accumulation of cAMP in WI-38 cells is primarily due to changes in the adenylate cylase or in the phosphodiesterase[16,24]. Figure 10 shows the accumulation of cAMP in cells treated with 5.7μM PGE$_1$ and 1.0μM carbachol (lower solid line). It is clear that the peak accumulation is shifted to shorter times for the lower curve and that, therefore, the carbachol exerted effects beyond a simple, uniform decrease in the rate of synthesis. It should also be noted that in the presence of carbachol, the steady state level of accumulation is reduced to about 15% of that for the cells which have not been treated by the drug. Since the steady state accumulation is given by the ratio of the rate of synthesis to the fractional turnover constant (k_s/k_e), this reduction could be due to a proportionate decrease in the rate of synthesis (an effect on the adenylate cyclase) or increase in the fractional turnover constant (an effect on the phosphodiesterase). This is resolved by determination of the fractional turnover constant by the tracer technique. Figure 11 shows the integral plots for cells subjected to that process in the presence or absence of carbachol. In fact, the carbachol caused a doubling of the turnover constant. Consequently, the reduction of accumulation of about 85% seen in the time course must be attributed to effects both on the adenylate cyclase (a reduction of activity to about one-third) and on the phosphodiesterase (a doubling of activity.

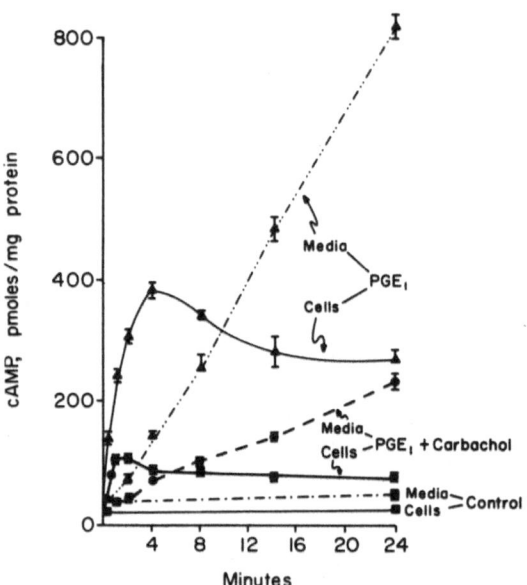

Fig. 10. Time course of cellular and medium cAMP changes in WI-38
          cells incubated with PGE$_1$ or PGE$_1$ and carbachol. The con-
          centrations of PGE$_1$ and carbachol were 5.7 and 1μM respec-
          tively.

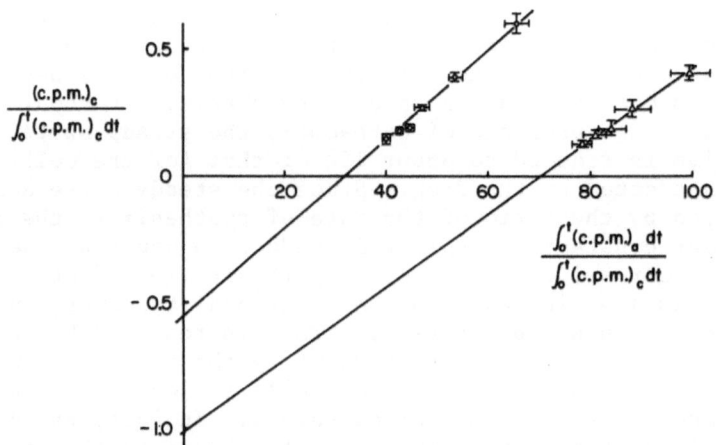

Fig. 11. Determination of the fractional turnover constant for cAMP
          in the presence (Δ) and absence (O) of carbachol. An ex-
          planation of the ordinate and abscissa is given in the
          text.

The cellular dynamics of cAMP, if precise enough, should have bearing on rather more fundametnal biochemical questions than the above exercise in drug action. For example, there are probably several types of desensitization. Beyond receptor internalization which does not correlate very well with at least the first phases of desensitization in many cells, there is little by way of physical events with which to associate those processes. However, knowing in accurate detail the rate of the desensitization process limits and confines the physical events to those which are compatible with the dynamics of the process. Much non-productive speculation can be avoided if a good phenomenological description is available.

We have discussed the dangers of drawing conclusions from insufficient or inappropriate data. Some discussions of the use of inaccurate data is also warranted. Though biochemistry is a branch of biology and seeks to answer questions of a biological nature, its techniques are essentially those of chemistry or physics. Those disciplines have achieved their present levels of development by a rigorous attention to quantitative detail so as to be able to distinguish between competing models of nature that may predict only slight differences in experimentally measurable parameters.

Questions of the type asked in the present communication require data of greater accuracy than those of a more simple nature. For example, the question "Does carbachol affect cAMP accumulation?", can be answered by two data points performed in triplicate in combination with a Student's t-test. The question "Which enzymes of the cAMP accumulation system does carbachol affect?", requires not only more data points, but also greater accuracy in those points. The present and future limits of the approaches used here probably rest with the accuracy of the data that are the arbiters of the competing models.

REFERENCES

1. E. W. Sutherland and T. W. Rall, The relation of adenosine-3'5'-phosphate and phosphorylase to the actions of catecholamines and other hormones, Pharmacol. Rev. 12:265 (1960).
2. G. A. Robison, R. W. Butcher and E. W. Sutherland, in: "Cyclic AMP," Academic Press, New York (1971).
3. W. L. Terasaki, G. Brooker, J. deVellis, P. Inglish, C.-Y. Hsu and R. D. Moyland, Involvement of cyclic AMP and protein synthesis in catecholamine refractoriness, in: "Advances in Cyclic Nucleotide Research," W. J. George and L. J. Ignarro, eds., Vol.9, pp.33-52, Raven Press, New York (1978).
4. S. Kakiuchi and T. W. Rall, The influence of chemical agents on the accumulation of adenosine-3'5'-phosphate in slices of rabbit cerebellum, Mol. Pharmacol. 4:367-378 (1968).
5. E. T. Brwoning, C. O. Brostrom and V. E. Groffi Jr., Altered adenosine cyclic 3',5'-monophosphate synthesis and degradation

by C-6 astrocytoma cells following prolonged exposure to norepinephrine, Mol.Pharmacol. 12:32 (1976).

6. V. Tom and H.-P. Bar, Adrenaline-induced desensitization of liver adenylate cyclase, Biochem. Pharmacol. 25:2103 (1976).

7. M. H. Makman, Properties of adenylate cyclase of lymphoid cells, Proc. Natl. Acad, Sci. U.S.A. 68:885 (1971).

8. J. Mickey, R. Tate and R. J. Lefkowitz, Subsensitivity of adenylate cyclase and decreased -adrenergic receptor binding after chronic exposure to (-)-isoproterenol in vitro, J. Biol. Chem. 250:5727 (1975).

9. B. Rapoport and R. J. Adams, Induction of refractoriness to thyrotropin stimulation in cultured thyroid cells, J. Biol. Chem. 251:6653 (1976).

10. J. F. Kuo and I. K. Dill, Antilypolytic action of valinomycin and nonactin in isolated adipose cells through inhibition of adenyl cyclase, Biochem. Biophys. Res. Commun. 32:333 (1968).

11. J. F. Kuo and E. C. DeRenzo, A comparison of the effects of lypolytic and antilypolytic agents on adenosine 3',5'-monophosphate levels in adipose cells as determined by prior labeling with adenine-8-^{14}C, J. Biol. Chem. 224:2252 (1969).

12. H. Shimizu, C. R. Greveling and J. W. Daly, A radioisotopic method for measuring the formation of enenosine 3',5'-cyclic monophosphate in incubated slices of brain, J. Neurochem. 16:1609 (1969).

13. Y.-F. Su, L. Cubeddu and J. P. Perkins, Regulation of adenosine 3',5'-monophosphate content of human astrocytoma cells: Desensitization to catecholamines and prostaglandins, J. Cyclic Nucl. Res. 2:257-270 (1976).

14. Y.-F. Su, G. L. Johnson, L. Cubeddu, B. H. Leichtling, R. Ortmann and J. P. Perkins, Regulation of adenosine 3',5'-monophosphate content of human astrocytoma cells: Mechanism of agonist-specific desensitization, J. Cyclic Nucleotide Res. 2:271-285 (1976).

15. R. Barber, R. B. Clark, L. A. Kelly and R. W. Butcher, A model of desensitization in intact cells, in: "Advances in Cyclic Nucleotide Research", W. J. George and L. J. Ignarro, eds., Vol.9 pp.507-516, Raven Press, New York (1978).

16. R. Barber, K. P. Ray and R. W. Butcher, Turnover of adenosine 3',5'-monophosphate in WI-38 cultured fibroblasts, Biochemistry 19:2560 (1980).

17. R. B. Clark and R. W. Butcher, Desensitization of adenylate cyclase in cultured fibroblasts with prostaglandin E$_1$ and epinephrine, J. Biol. Chem. 254:9373 (1979).

18. G. M. Nemecek, K. P. Roy and R. W. Butcher, Inhibition of cyclic nucleotide phosphodiesterase during exposure of WI-38 cells to prostaglandin E$_1$, J. Biol. Chem. 254:598-601 (1979).

19. R. Barber and R. W. Butcher, The turnover of cyclic AMP in cultured fibroblasts, J. Cyclic Nucleotide Res. 6:3 (1980).

20. R. Barber, K. P. Ray and R. W. Butcher, Temperature effects on cyclic AMP accumulation in cultured fibroblasts, J. Cyclic Nucleotide Res. 6:15 (1980).

21. B. B. Hoffman, D. Mullikin-Kilpatrick and R. J. Lefkowitz, Desensitization of beta-adrenergic stimulated adenylate cyclase in turkey erythrocytes, J. Cyclic Nucleotide Res. 5:355 (1979).
22. G. A. Nickols and G. Brooker, Temperature sensitivity of cyclic AMP production and catecholamine-induced refractoriness in a rat astrocytoma cell line, Proc. Natl. Acad. Sci. U.S.A. 75: 5520 (1978).
23. S. Swillens, E. Lefort, R. Barber, R. W. Butcher and J. E. Dumont, Consequences of hormone-induced desensitization of adenylate cyclase in intact cells, Biochem. J. 188:169 (1980).
24. R. W. Butcher, Decreased cAMP levels in himan diploid cells exposed to cholinergic stimuli, J. Cyclic Nucleotide Res. 4:411 (1978).

USE OF A GENETIC APPROACH TO STUDY CYCLIC AMP METABOLISM

Paul A. Insel

Division of Pharmacology, M-013
Department of Medicine
University of California, San Diego
La Jolla, California 92093

SUMMARY

In this review, I present a brief overview of a genetic approach for studying cyclic AMP metabolism and the application of this approach with murine S49 lymphoma cells. S49 clones having alterations in the pathway of cyclic AMP generation and function have proved useful in characterizing components involved in hormonal stimulation of adenylate cyclase, in activation of cAMP-dependent protein kinase, and in studying more distal cellular events mediated by cyclic AMP. The use of genetic methods in S49 and other cultured cell systems offers an important complementary approach to biochemical and pharmacological methods for studying the cyclic AMP pathway.

For the past several decades, genetic approaches have offered important tools in studies of prokaryotic and, to a lesser extent, eukaryotic organisms. The isolation of a relevant mutant has often been the first means of defining a basic biological phenomenon and in pinpointing biochemical mechanisms that mediate or result from such phenomena[1]. The underlying premise of the genetic approach is that the altered phenotype of a mutant cell results from alteration of a single peptide gene product. If that gene product can be unequivocally identified, it follows that: 1) the product is directly involved in expression of the particular phenotype that is altered and 2) when the change in the mutant gene product is sufficiently well-defined, strong inference can be made regarding the molecular interactions with other gene products in producing both mutant and normal phenotypes. Work with prokaryotes has established the conceptual bases and experimental methodology for genetic studies, and in recent years these genetic techniques have provided a powerful means to explore growth and function in eukaryotes.

Aside from the use of genetic approaches to ask "genetic questions" (e.g. defining chromosomal localization of genes coding for particular peptides, relative dominance or recessiveness of mutant phenotypes, effects of ploidy on gene expression, etc.), mutant clones isolated from wild-type (parental) cells have been useful in exploring a variety of important biochemical, cell biological, immunologic, and pharmacological questions. Experiments in cultured cell lines from which one can isolate spontaneous or induced mutant clones offer a number of advantages. Among them are the relative homogeneity of clonally-derived cells (thus distinguishing them from heterogeneous tissues), the accessibility of experimental material, the ease of technical manipulation, and the ability to identify the contribution of mutant gene products to particular cellular responses.

Genetic approaches for examining cyclic AMP metabolism have involved an increasingly large number of cell types[2], including murine neuroblastoma cells[3], murine macrophage-like cells[4], murine adrenocortical tumor Y-1 cells[5,6], Chinese hamster ovary cells[7,8], and murine S49 lymphoma cells. Of these, the S49 cell line has been the most extensively studied system. For this reason, and because results with S49 cells nicely demonstrate the usefulness of combining genetic and biochemical approaches for investigating cyclic nucleotide metabolism, I will focus the remainder of this article reviewing some of those results. Although I will limit my discussion to S49 variants involving the cyclic AMP pathway, it should be noted that mutant S49 clones have been isolated that have alterations in the pathways of glucocorticoid action and of purine and pyrimidine metabolism[9-12].

S49 cells originally arose as lymphoid tumors in a female Balb/c mouse injected with phages and Freund adjuvant; following explant, these cells have been grown in tissue culture for a number of years[13]. Characterization of S49 cells indicates that they have a stable pseudodiploid karyotype (40 telocentric chromosomes) and possess Thy-1 and TL surface antigens[13]. These and other properties, such as their sensitivity to killing by glucocorticoids, indicate that S49 cells are similar to immature T-lymphocytes. S49 cells can be maintained indefinitely in liquid suspension culture and can be grown in minimum essential medium plus serum or in defined, serum-free medium[14,15]. The cells (Figure 1) have membrane receptors for prostaglandin E_1 (PGE_1) and beta-adrenergic agents[16]. In addition, they contain an active adenylate cyclase, which catalyzes formation of cyclic AMP in response to beta-adrenergic agonists, cholera toxin, PGE_1, NaF, or guanyl nucleotides[16,17]. Endogenously formed cyclic AMP or exogenously added cyclic AMP supplied as N^6, 2'-dibutyryl cyclic AMP or as 8-Br cyclic AMP produces the following responses in S49 cells (Figure 1): 1) Induction of cyclic AMP phosphodiesterase (PDE)[18] 2) decrease in ornithine decarboxylase and S-adenosylmethionine decarboxylase (ODC, SAMDC) activities[19,20], 3) decrease in uptake of precursors of macromolecular synthesis, such as leucine, glucose, and uridine[21], and 4) growth arrest in the G_1 phase of the cell cycle[13]. These effects

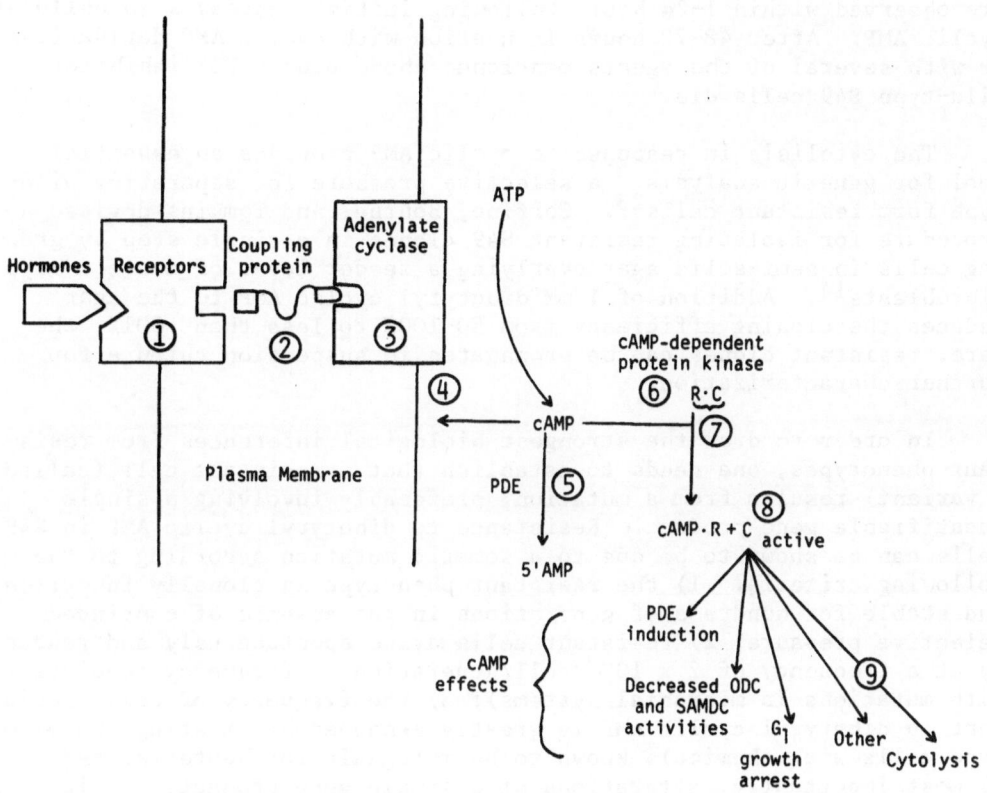

Site in diagram	Variant	Phenotype
①	β^d	Selective loss of β-adrenergic responses and receptors.
②	UNC	Receptors and cyclase present, but uncoupled.
③	" AC⁻ "	Deficient in adenylate cyclase activity (no functional N).
④	U200.95	Enhanced cyclic AMP extrusion.
⑤	K30A	Increased PDE activity.
⑥	K_a	Cyclic AMP-dependent kinase shows increased K_a for cAMP.
⑦	V_{max}	Decreased cAMP-dependent kinase per cell.
⑧	kin⁻	No detectable cAMP-dependent kinase.
⑨	D⁻	Kinase normal; cytolysis not induced by cAMP.

Fig. 1.

are observed within 1-24 hours following initial increases in cellular cyclic AMP. After 48-72 hours incubation with cyclic AMP derivatives or with several of the agents mentioned above plus a PDE inhibitor, wild-type S49 cells die.

The cytolisis in response to cyclic AMP provides an essential tool for genetic analysis - a selective pressure for separating wild-type form resistant cells[23]. Coffino, Bourne, and Tomkins devised a procedure for isolating resistant S49 clones in a single step by growing cells in semi-solid agar overlying a feeder layer of mouse embryo fibroblasts[14]. Addition of 1 mM dibutyryl cyclic AMP to the agar reduces the cloning efficiency from 50-100% to less than .001%; the rare, resistant clones can be propagated in suspension culture for further characterization.

In order to draw the strongest biological inferences from resistant phenotypes, one needs to establish that a resistant cell (called a variant) results from a mutation, preferably involving a single identifiable gene product. Resistance to dibutyryl cyclic AMP in S49 cells can be shown to be due to a somatic mutation according to the following criteria: 1) the resistant phenotype is clonally inherited and stable for hundreds of generations in the absence of continued selective pressure; 2) resistant cells arise spontaneously and randomly at a frequency of 2×10^{-7}/cell/generation (a frequency consistent with mutations in microbial systems); 3) the frequency of cells resistant to dibutyryl cyclic AMP is greatly enhanced by treating the wild-type cells with chemicals known to be mutagenic for bacteria; and 4) most importantly, alterations of a single gene product, cyclic AMP-dependent protein kinase, have been demonstrated for some of the cyclic AMP-resistant variants[14,24].

The vast majority of variant clones that are resistant to the cytolytic action of dibutyryl cyclic AMP are defective in cyclic AMP-dependent protein kinas (PK) activity. These variants fall into one of three general classes - those with absent cyclic AMP PK activity (Kin⁻ variants), those with decreased PK activity but with normal potency of the kinase for activation by cyclic AMP ("V_{max}" variants) and those with decreased molar potency of activation of the kinase by cyclic AMP ("K_a variants")[25]. The similarity between the patterns of the dose-response curves for activation of cytosolic PK and for responses produced by dibutyryl cyclic AMP in intact S49 cells offered strong inferential evidence that PK mediated those responses[25]. Such results demonstrate the utility of wild-type and variant S49 cells as a system to test whether particular biological responses occur through activation of PK. Other studies have indicated that the K_a class of PK variants results from lesions (apparently amino acid substitutions) in the regulatory subunits of PK[24] and that the Kin⁻ class of variants arise from regulatory mutations in expression of regulatory subunits of PK[26]. Thus, studies of these S49 mutants have contributed to understanding of the biochemistry of cyclic AMP PK and to defining the pivotal role of kinase in these cells. Other work using the cyclic AMP-resistant variants has shown that cyclic AMP extrusion and beta-adrenergic receptor mediated inhibition of Mg^{++} transport is independent of PK,[27,28] that generation of different classes of mutants can be used to test for mutagens[29] and that wild-type and mutant cells are useful in studies of regulation of cell growth (in vivo and in vitro)[30,31] and of protein synthesis and modification[32] by cyclic AMP.

Two other cyclic AMP-resistant variants (Figure 1) that do not appear to have lesions in PK are a cyclic AMP "deathless" clone, which has normal PK activity and all the other responses to dibutyryl cyclic AMP except cytolysis[33], and a variant that has enhanced cyclic AMP extrusion[27,34]. The "deathless" variant has proved useful in exploring regulation of cell cycle progression, PDE induction, and inhibition of ODC in S49 cells[19,30,35].

A second major class of S49 variants is comprised of cells that are resistant to cytolysis produced by agents that stimulate cyclic AMP synthesis (Figure 1). Theoretically, a clone will survive such selective pressure owing to any of a variety of lesions, ranging from membrane receptor to cyclic AMP PK. Assuming an equal efficiency of selection for each hypothetical variant, the procedure should preferentially yield whichever variant is present at the highest incidence in the population. When wild-type cells were cloned in medium containing the beta-adrenergic agonist isoproterenol and an inhibitor of PDE, the cloning efficiency was 0.15%. On initial characterization of these resistant variants (originally termed AC⁻ or cyc⁻) Bourne et al. found low (10-30% of wild-type) basal cyclic AMP content and adenylate cyclase activity in particulate preparations, and no response of the cells or particulates to any of the usual stimulants of cyclic AMP synthesis[16]. More recent data indicating that plasma membranes from those cells have Mn^{++}-stimulable adenylate cyclase activity[36,37] as well as results with heterokaryons and hybrid cells[37,38], in reconstitution experiments[36], and in labelling studies with cholera toxin and $[^{32}P]NAD^{40}$, all indicate that the lesion in these variants does not result from absent catalytic cyclase activity. Instead these cells have no functional nucleotide coupling ("N" or "G/F") activity[36, 39,40]. Other studies indicate that although cyc⁻ cells have a similar number of beta-adrenergic receptors as do wild-type cells[41,42], the receptors in cyc⁻ cells show neither regulation of agonist binding by guanyl nucleotides and Mg^{++} nor a receptor-mediated inhibition of Mg^{++} uptake as do wild-type cells[17,43].

The "cyc⁻" cells have proved to be of major importance as a "reagent" in reconstitution studies of the hormone-sensitive adenylate cyclase system. By virtue of their lack of N units, membranes from cyc⁻ cells have served as recipients of N donated from a variety of cell types. The pioneering efforts of Ross, Gilman, and co-workers in reconstituting adenylate cyclase activity using the S49 system has represented an exciting advance in the cyclase field[40,44]. Gilman's

laboratory has continued to exploit the cyc⁻ cells as an assay system
during characterization and purification of N[45],[46]. Bourne's laboratory has used these cells to characterize N in other cell types[47] and
to help define the defect in N which occurs in some patients with
pseudohypoparathyroidism.[48]

   Two other classes of variants that were selected by virtue of
their resistance to beta-adrenergic agonists are the beta[d] cells and
UNC cells (Figure 1). The former are cells which have decreased
numbers of beta-adrenergic receptors, but retain cyclase activity in
response to other effectors. Studies with beta[d] cells have helped to
indicate that all the beta-adrenergic receptors are necessary for a
maximal beta-adrenergic response in S49 cells (i.e., no "spare receptors").[49] The UNC cells have lost both beta-adrenergic and PGE$_1$-
stimulated adenylate cyclase activity, while retaining beta-adrenergic
receptors as well as cyclase activity in response to NaF, cholera
toxin and guanyl nucleotides[50]. Recent data indicate that the beta-
adrenergic receptors in intact UNC cells can be recoupled to generate
cyclic AMP if the cells are pre-incubated with cholera toxin[51] and
that the N proteins in UNC have more acidic isoelectric points on
two-dimensional gels[52].

   Other newer variants that have been recently isolated by virtue
of their resistance to cholera toxin-induced cytolysis include clones
called H21A, whose N's, although nonfunctional in reconstitution experiments, label as do those from wild-type cells with cholera toxin
and [32P]NAD and a clone called K30A, which has a fourfold increase
in PDE activity (H.R. Bourne, personal communication).

   The S49 variants having lesions in the membrane components of the
receptor-N-cyclase system have been used to study molecular mechanisms
regulating hormonal responsiveness of S49 cells. Performing pharmacological and biochemical experiments with these variants has offered
an additional aid in dissecting events modulating cellular cyclic AMP
synthesis. Examples include the regulation of beta-adrenergic receptors in intact cells, enhancement of cellular cyclic AMP accumulation
by inhibitors of microtubule assembly, and mechanisms mediating refractoriness to beta-adrenergic agonists[53-56]. Beta-adrenergic receptors in intact S49 cells have much lower affinity for agonists in
radioligand binding studies than in cyclic AMP accumulation experiments, and this lower affinity does not appear to result solely from
receptor interaction with N[42],[53]. By contrast, the loss in beta-adrenergic receptors observed in membranes from cells that have become
refractory to beta-adrenergic agonists appears to require receptor
interaction with N and may require activation of adenylate cyclase[54].
The enhancement of cyclic AMP accumulation by S49 cells produced
inhibitors of microtubule assembly results from an action distal to
the receptor binding site with agonists, perhaps through enhanced
coupling of membrane components of the cyclase system[55],[56].

   In conclusion, genetic approaches using S49 variant cells have

offered an important complement to more classical methods for studying
the generation and action of cyclic AMP. It seems highly probable
that the selection, characterization and use of mutants in this and
other cultured cell systems will continue to have a major impact in
furthering our understanding of cyclic nucleotide metabolism. Our
ability to develop new ways of applying selective pressure for or
against components in this metabolic pathway is likely to be the major
rate-limiting step to the more widespread application of these genetic
approaches. For example, variant S49 clones totally lacking in beta-
adrenergic receptors (beta-) or with the exception of deathless cells,
cells that have lost other individual responses to cyclic AMP ("G_1
arrest$^-$", "PDE induction$^-$", etc.) have not yet been isolated. The
isolation and characterization of such clones should offer new and
perhaps unexpected insights into the cyclic AMP pathway.

Work in the author's laboratory has been supported by grants
from the National Science Foundation, California Heart Association,
American Heart Association, and American Cancer Society Institutional
grant, and an Established Investigatorship of the American Heart
Association.

REFERENCES

1.   J. Cairns, G.S. Stent, and J.D. Watson, "Phage and the Origin
      of Molecular Biology", Cold Spring Harbor Lab., Cold Spring
      Harbor, New York (1966).
2.   M.E. Maguire, L.L. Brunton, R.A. Wiklund, H.J. Anderson, P.M.
      Van Arsdale, and A.G. Gilman, Res. Prog. Hormone Res. 33:633-
      667 (1976).
3.   R. Simantov, and L. Sachs, J. Biol. Chem. 250:3236-3242 (1975).
4.   N. Rosen, J. Piscitello, J. Schneck, R.J. Muschel, B.R. Bloom,
      and O.M. Rosen, J. Cell Physiol. 98:125-136 (1979).
5.   B.P. Schimmer, J. Tsao, and M. Knapp, Mol. Cell Endocrinol. 8:
      135-145 (1977).
6.   P.A. Rae, H.S. Gutman, J. Tsao, and B. Schimmer, Proc. Nat. Acad.
      Sci., U.S.A. 76:1896-1900 (1979).
7.   D. Evain, M. Gottesman, I. Pastan, and W.B. Anderson, J. Biol.
      Chem. 254:6931 (.979).
8.   M. M. Gottesman, A. LeCam, M. Bukowski, and I. Pastan, Somatic
      Cell Genetics 6:45-61 (1980).
9.   K.R. Yamamoto, U. Gehring, M. Stampfer, and C.H. Sibley, Rec.
      Prog. in Hormone Res. 32:3-32 (1976).
10.   M. Pfahl, R. Kelleher, and S. Bourgeois, Molec. and Cell Endo-
      crinol. 10:193-207 (1978).
11.   B. Ullman, L.J. Gudas, S. M. Clift, and D.W. Martin, Proc. Nat.
      Acad. Sci., U.S.A. 76:1074-1078(1979).
12.   B.B. Levinson, B. Ullman, and D.W. Martin, J. Biol. Chem. 254:
      4396-4401 (1979).
13.   K. Horibata, and A.W. Harris, Exp. Cell. Res. 60:61-77 (1970).

14. P. Coffino, H.R. Bourne, U. Friedrich, P.A. Insel, K.L. Melmon, and G.M. Tomkins, Rec. Prog. Hormone Res. 32:669-682 (1976).
15. F.J. Darfler, H. Murikami, and P.A. Insel, Proc. Nat. Acad. Sci. U.S.A. 77: in press, (1980).
16. H. R. Bourne, P. Coffino, and G. M. Tomkins, Science 187:750-752 (1975).
17. E.M. Ross, M.E. Maguire, T.W. Sturgill, R.L. Biltonen, and A.G. Gilman, J. Biol. Chem. 252:5761-5775 (1977).
18. H.R. Bourne, G.M. Tomkins, and J. Dion, Science 181:952-954 (1973).
19. P.A. Insel, and J. Fenno, Proc. Nat. Acad. Sci., U.S.A. 75:862-865 (1978).
20. J.M. Honeysett, and P.A. Insel, Fed. Proc. 39:869 (1980).
21. V. Daniel, H.R. Bourne, and G.M. Tomkins, Nature New Biol. 244:67-169 (1973).
22. P. Coffino, J. Gray, and G.M. Tomkins, Proc. Nat. Acad. Sci. U.S.A. 72:878-882 (1975).
23. V. Daniel, G. Litwack, and G.M. Tomkins, Proc. Nat. Acad. Sci. U.S.A. 70:76--9 (1973).
24. R.A. Steinberg, P.H. O'Farrell, U. Friedrich, and P. Coffino, Cell, 10:381-391.
25. P.A. Insel, H.R. Bourne, P. Coffino, and G.M. Tomkins, Science 190:896-898 (1975).
26. R.A. Steinberg, T. Van Daalen Wetters, and P. Coffino Cell 15:1351-1361 (1978).
27. L.L. Brunton, Adv. Cyclic Nucleotide Res. 14: in press (1981).
28. M. Maguire, and J. Erdos, J. Biol. Chem. 255:1030-1035 (1980).
29. U. Friedrich, and P. Coffino, Proc. Nat. Acad. Sci. U.S.A. 74:679-683 (1978).
30. P. Coffino, and W. Gray, Cancer Res. 38:4285-4288 (1978).
31. J. Hochman, A. Katz and Y. Weinstein, Europ. J. Cancer 15:11-16 (1979).
32. R.A. Steinberg, and P. Coffino, Cell 18:719-733 (1979).
33. I. Lemaire, and P. Coffino, Cell 11:149-155 (1977).
34. R.A. Steinberg, M.G. Steinberg, and T. Van Daalen Wetters, J. Cell Physiol. 100:579-588 (1979).
35. M. Kaiser, H.R. Bourne, P.A. Insel, and P. Coffino, J. Cell Physiol. 101:369-374 (1979).
36. E.M. Ross, A.C. Howlett, K.M. Ferguson, and A.G. Gilman, J. Biol. Chem. 253:6401-6412 (1978).
37. J. Naya-Vigne, G.L. Johnson, H.R. Bourne, ans P. Coffino, Nature 272:720-722 (1978).
38. J.D. Schwarzmeier, and A.G. Gilman, J. Cyclic Nucleotide Res. 3:227-238 (1977).
39. G.L. Johnson, H.R. Kaslow, and H.R. Bourne, Proc. Nat. Acad. Sci. U.S.A. 75:3113-3117 (1978).
40. G.L. Johnson, H.R. Kaslow, and H.R. Bourne, J. Biol. Chem. 253:7120-7123 (1978).
41. P.A. Insel, M.E. Maguire, A.G. Gilman, H.R. Bourne, P.Coffino, and K.L. Melmon, Mol. Pharmacol. 12:1062-1069 (1976).

42. P.A. Insel, and M. Sanda, J. Biol. Chem. 254:6554-6559 (1979).
43. S.J. Bird, and M.E. Maguire, J. Biol. Chem. 253:8826-8834 (1978).
44. E.M. Ross, T. Haga, A. Howlett, J. Schwarzmeier, L. Schleifler, and A.G. Gilman, Adv. Cyclic Nucleotide Res. 9:53-68 (1978).
45. A.C. Howlett, P.C. Sternweis, B.A. Macik, P.M. VanArsdale and A.G. Gilman, J. Biol. Chem. 254:2287-2295 (1979).
46. J. Northup, P.C. Sternweis, M.D. Smigel, L.S. Schleifer, and A.G. Gilman, Fed. Proc. 39:516 (1980).
47. H.R. Kaslow, Z. Farfel, G.L. Johnson, and H.R. Bourne, Mol. Pharmacol. 15:432-483 (1979).
48. Z. Farfel, A.S. Brickman, H.R. Kaslow, V. Brothers, and H.R. Bourne, New Engl. J. Med., 303:237-242 (1980).
49. G.L. Johnson, H.R. Bourne, M.K. Gleason, P.Coffino, P.A. Insel, and K.L. Melmon, Mol. Pharmacol. 15:16-27 (1979).
50. T. Haga, E.M. Ross, H.J. Anderson, and A.G. Gilman, Proc. Nat. Acad. Sci. U.S.A. 74:2016-2020.
51. P.A. Insel, F.J. Darfler, and M.S. Kennedy, manuscript submitted (1980).
52. L. Schleifler, J.C. Garrison, P.C. Sternweis, J.K. Northup, and A.G. Gilman, J. Biol. Chem. 255:2691-2694 (1980).
53. P.A. Insel, and L.M. Stoolman, Mol. Pharmacol. 14:549-561 (1978).
54. M. Shear, P.A. Insel, K.L. Melmon, and P. Coffino, J. Biol. Chem. 251:7572-7576 (1976).
55. P.A. Insel, and M.S. Kennedy, Nature, 273:471-473 (1978).
56. M.S. Kennedy, and P.A. Insel, Mol. Pharmacol. 15:215-223 (1979).

SENSITIVITY AND SPECIFICITY OF PROSTANOIC DERIVATIVES

RADIOIMMUNOASSAYS : NEW STRATEGY

Fernand Dray

Fra n° 8 Inserm, URIA Institut Pasteur

28 Rue du Dr. Roux - 75724 Paris Cedex 15 (France)

Prostanoic derivatives, as prostaglandin, thromboxane, prosta-cyclin and their metabolites, are generally present in biological media at very low concentrations (10^{-9} - 10^{-11} M). Therefore assays with high sensitivity often are required for their measurement. Two techniques are widely used : mass spectrometry combined with gas chromatography (GS-MS), which is particularly specific[1] and radio-immunoassay (RIA) which is particularly sensitive and convenient for large series[2,7]. In our laboratory, immunological methods have been developed using radioactive[8] or non-radioactive tracers[5,9]; iodinated derivatives (as PG-^{125}I-Histamine) have been shown to be more sensi-tive convenient[10-14]. In this paper some aspects of the use of iodin-ated tracers will be presented and a new strategy for increasing the specificity of RIA of prostanoic derivatives will be proposed.

PREPARATION AND SELECTION OF IMMUNE SERA

Preparation of the immunogen.

To act as immunogens, PGs are covalently linked to an antigenic carrier. The chemical action selected must enable an adequate number of molecules to be bound to the carrier, and maintain the structural integrity of the molecule (except for the group involved in the co-valent bond with the antigenic carrier). In our experiments, we used bovine serum albumin (BSA) as the carrier; the free carboxyl function of the hapten, activated by a carbodimide, forms a peptide bond with the free NH_2 groups of the anti genic carrier[8]. Under our conditions (shown in Table 1) in most experiments 15 to 20 moles of PG were bound per mole BSA.

Table 1. Preparation of prostaglandin immunogen

	REAGENT	QUANTITY (mg)	VOLUME (ml)	MEDIUM	REACTION TIME
STEP 1 (COOH ACTIVATION)	PG + (^{3}H)-PG	10	10	Sodium carbonate	
	+				
	E D C I *	10		pH 5.5	1 h.
STEP 2 (COUPLING)	+			Sodium carbonate	
	BOVINE SERUM ALBUMIN	20	10	pH 5.5	Overnight

* 1-Ethyl-3 (3-dimethyl-amino-propyl)-carbodiimide-HCl
- All reactions are carried out at room temperature
- Conjugate is dialysed against distilled water for 24 h. at +4°C
- Number of PG-residues per molecule of albumin is estimated on the basis of isotopic dilution.

Immunisation

Animals were administered the PG-BSA conjugate according to a method originally proposed by Vaitukaitis et al.[15]. Each rabbit recieved 30-40 intradermal injections of the conjugate (about 200 µg) dissolved in 1 ml physiological buffered saline and emulsified in an equal volume of Freund's complete adjuvant. Five animals were immunised in each series.

Booster injections, given two months after the primary injections, were carried out following the same schedule. Subsequent booster injections were given depending upon the development of antibody titre. These subsequent booster injections, using smaller amounts of antigen, were carried out when the antibody titre fell from a maximum value. Animals were bled every ten days after the fourth week of immunisation.

Study of binding parameters

The course of the immune response may be followed by measuring the titre of the antiserum, defined as the dilution giving 50% binding of the radioactive tracer. When the titre of the serum was high enough, its sensitivity, affinity and specificity were tested. The best bleedings were kept separately or pooled. They were stored

Table 2. Cross-reactions of prostaglandin antisera.

INHIBITORS	ANTISERA									
	E_1	E_2	$F_{1\alpha}$	$F_{2\alpha}$	D_2	$DHKE_2$	$DHKF_{2\alpha}$	TXB_2	$6KF_{1\alpha}$	$6,15DKF_{1\alpha}$
E_1	100	15	0.2	<0.1	<0.05	<0.1	<0.1	<0.1	5.6	<0.3
E_2	3.0	100	<0.1	0.1	<0.05	0.1	<0.1	<0.1	2.2	<0.3
$F_{1\alpha}$	0.1	<0.1	100	7.0	<0.05	<0.1	<0.1	<0.1	18.0	<0.3
$F_{2\alpha}$	0.3	0.8	29	100	0.04	<0.1	4.0	<0.1	11.0	<0.3
D_1	<1.0	<1.0	<0.3	<0.3	16	<0.1	<0.1	<0.1	1.5	<0.3
D_2	<1.0	<1.0	<0.3	<0.3	100	<0.1	<0.1	0.15	0.5	<0.2
$DHKE_2$	<0.5	<0.5	<0.5	<0.5	<0.05	100	7.0	<0.1	<0.1	1.4
$DHKF_{2\alpha}$	<0.1	<0.1	<0.1	<0.1	<0.05	<0.1	100	<0.1	<0.1	3.5
TXB_2	<0.1	<0.1	<0.1	<0.1	0.9	<0.1	<0.1	100	<0.1	<0.1
$6KF_{1\alpha}$	6.0	2.0	18	11	<0.05	<0.1	0.3	<0.1	100	0.1
$6,15DKF_{1\alpha}$	<0.3	<0.3	<0.3	<0.3	<0.1	1.4	3.5	<0.3	<0.1	100

Cross reactivities are calculated on the basis of quantity (pg) necessary for 50% tracer displacement. Abbrevations : $DHKE_2/F_{2\alpha}$ for 13,14-dihydro-15-Keto-$PGE_2/F_{2\alpha}$; $6KF_{1\alpha}$ for 6-Keto-$PGF_{1\alpha}$; $6,15 DKF_{1\alpha}$ for 6,15-diketo $PGF_{1\alpha}$

either at +4°C after addition of 0.02% sodium azide, or at -20°C
after dilution in an equal volume of glycerol. Table 2 shows the
specificity of certain selected antisera.

PREPARATION OF IODINATED

Preparation of PG derivatives

A PG "derivative" was synthesised as the substrate for iodination.
For our studies the PG molecule was coupled to histamine; in preliminary work tyramine or tyrosyl methyl ester was used and similar
binding parameters were obtained[10]. A peptide bound was formed between the free NH_2 group of histamine and the free COOH group of PG
(Table 3); the derivative then was purified and separated from the
reagents by thin-layer chromatography on silica gel in n-butanol/
acetic acid/water system (75:10:25 v/v). The addition of (^{3}H)-his-
tamine to the reaction mixture in a preliminary experiment allowed
localization and identification of the derivative formed (Figure 1).
After elution in methanol, the derivative was distributed in small
fractions and lyophilized.

Table 3. Preparation of prostaglandin-histamine derivative

	REAGENT	QUANTITY (μmol)	VOLUME (ml)	MEDIUM	REACTION TIME
STEP 1	PG	28.6	0.5	Ethanol-water 1:1, v/v	1 h.
(COOH ACTIVATION)	+				
	E D C I *	52			
STEP 2	+				
(COUPLING)	HISTAMINE (His)	90	0.5	Water	Overnight

* 1-Ethyl-3 (3-dimethyl-amino-propyl)-carbodiimide-HCl
- All reactions are carried out at room temperature.

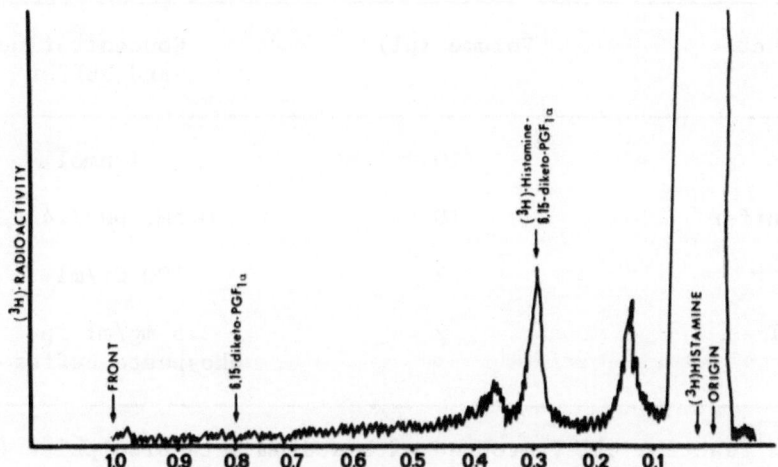

Fig. 1.   Purification of the coupled 6,15-diketo-PGF$_{1\alpha}$-(^{3}H)-histamine
by thin layer chromatography on silica gel (solvent system:
butanol/acetic acid/water 75:10:25).

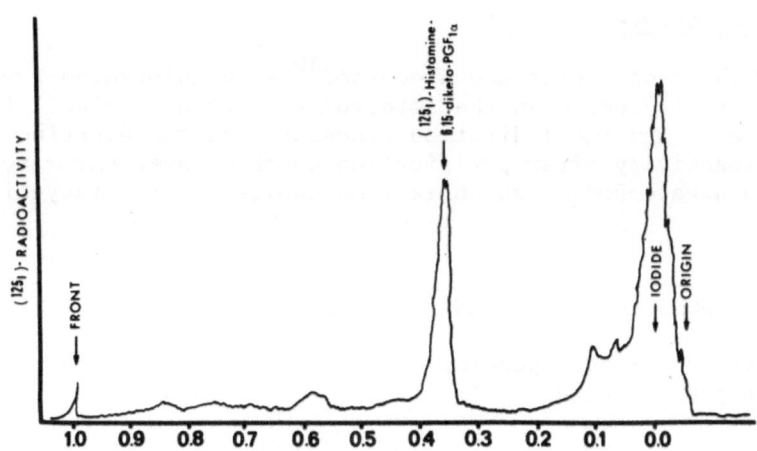

Fig. 2.   Autoradiochromatography of (^{125}I)-histamine-6,15-diketo-
PGF$_{1\alpha}$ on a silica gel plate (solvent system : chloroform/
methanol/water 80:20:2).

Table 4. Iodination procedure.

Product	Volume (μl)	Concentration and buffer
PG-His	10	1 nmol
Phosphate buffer	10	0.5M, pH 7.4
Na^{125}I	2	100 Ci/ml
Chloramine T	2	2.5 mg/ml in phosphate buffer

- 20 seconds reaction time; stopped with sodium meta-bisulphite (32μg)
- (^{125}I)-PG-His is purified by TLC (silica gel) in chloroform/water (60:-0:5, v/v).
- Radioactive spot is located by autoradiography and iodinated product is eluted from silica gel by ethanol and stored after distribution into small fractions at -20°C until use.

Iodination procedure

The methods of Hunter and Greenwood[16] with chloramine T was used to incorporate ^{125}iodide in the imidazol ring of histamine. Table 4 and Figure 2 show the iodination procedure and the distribution of the radioactivity after purification on thin-layer chromatography of the 6,15-diketo-PGF$_{1\alpha}$, which is a metabolite of prostacyclin.

THE RADIOIMMUNOLOGICAL REACTION

The method varies depending upon whether an iodinated or a tritiated tracer is used.

RIA using a tritiated tracer

To 5 ml polystyrene tubes were added successively 0.1 ml tritiated tracer (about 7,000 dpm), 0.1 ml PG standard or biological extract and 0.1 ml of a dilution of the antiserum such that the initial binding in the absence of standard or unknown PG is 40-50% of the total radioactivity. All the reagents were diluted in phosphate-buffered saline at pH 7.4, 0.1M, 0.9% NaCl, 0.1% gelatin. The tubes were incubated overnight at 4°C. Separation of the free fraction from that bound to the antiserum was carried out at 0°C by the addition of 1 ml charcoal-dextran. After incubation for 12 minutes in a melting ice bath, the tubes were centrifuged for 15 minutes at 2,000 g. The supernatant (bound fraction) was transferred to scintillation vials and counted in a liquid scintillation counter for 4 minutes. The standard curve was determined by plotting the log of the dose (pg/tube) on the abscissa against the value (%) of the ratio B/Bo (B being the amount of radioactivity bound to the antibody) on the ordinate. The results were then calculated by computer.

RIA using iodinated tracers

The technique is basically the same as for tritiated tracers, with the following modifications : 14,000 dpm iodinated tracer was added to each tube, and the buffer used contained 0.3% bovine gamma globulin instead of gelatin. Free and bound radioactivity were separated by adding 0.3 ml polyethylene glycol 6000 at 0°C to each tube (25 g/100 ml distilled water), mixing well and centrifuging at +4°C for 15 minutes at 2,000 g. The supernatant was decanted and the pellet (the bound fraction) was counted for 1 minute in a gamma counter.

Binding parameters of RIAs using iodinated tracers

These were tested for each antiserum in comparison with tritiated tracer. We always observed a higher dilution of the antiserum to bind 50% of tracer, usually no significant alteration of percentage of cross-reactions and very often an increase in sensitivity. This last point will be discussed later.

PURIFICATION OF BIOLOGICAL SAMPLES

Sampling of biological fluids (blood, urine, amniotic fluid, etc....) or tissues should be carried out according to a standard procedure adapted for each biological medium. Biosynthesis and metabolism of prostaglandins in the sample must be inhibited. In the case of urine, for example, each urination was collected in a clean vessel and immediately transferred to a bottle at 4°C containing meclofenamic acid. The total volume was measured after 24 hours and a fraction stored at -20°C.

Extraction

Two ml urine sample was pipetted into a centrifuge tube containing about 1,800 dpm of appropriate ^{3}H tracer for calculation of the extraction recovery. After acidification to pH 3.5 with citric acid, the PGs were extracted from the sample with three volumes of a mixture of cyclohexane and ethyl acetate (1:1, v/v). After 15 minutes vigorous shaking and centrifugation at 250 g, the organic phase was pipetted into a conical siliconised tube. The extraction was repeated and the two extracts pooled and evaporated to dryness under nitrogen.

CHROMATOGRAPHIC PURIFICATION

Silicic acid chromatography

Prostanoic derivatives were purified by passage over a silicic acid column : 500 mg silicic acid was equilibrated in 2 ml solvent 2 and washed first with 5 ml benzene/ethyl acetate/methanol (60:40:20, v/v) solvent 3, and then with 1.5 ml solvent 2. The extract was redissolved in 0.2 ml benzene/ethyl acetate/methanol (60:40:20, v/v) (solvent 1) and 0.5 ml benzene/ethylacetate (60:40, v/v) (solvent 2), the total volume of the extract was placed on the column and the compounds were eluted by three successive solvents:

- Elution 1 : 6 ml solvent 2 to extract the less polar lipids, pigments, PGA, PGB and 13,14-dihydro-15-keto-PGE_1 or PGE_2

- Elution 2 : 13 ml solvent 5 (benzene/ethyl acetate/methanol, 60:40:2, v/v) to extract PGE_1 and PGE_2 and also 13,14-dihydro-15-keto-$PGF_{1\alpha}$ and $PGF_{2\alpha}$

- Elution 3 : 4 ml solvent 3 to extract PGs F_α and 19-OH-PGE and PGF_α

Each eluate was evaporated under nitrogen and redissolved in an adequate quantity of buffer for radioimmunological assay. A portion was used to calculate recovery for the extraction and purification processes.

When thromboxane B_2, 6-keto-$PGF_{1\alpha}$ and 6,15-diketo-$PGF_{1\alpha}$ have to be measured elutions 2 and 3 were replaced by only one elution with 5 ml solvent 2.

High performance liquid chromatography (HPLC)

The eluate after silicic acid column was evaporated to dryness, then injected onto a column of μBondapack C_{18} (Waters) with elution conditions allowing the best separation of PGs and related compounds (Figure 3). The content of tubes was evaporated and RIA was carried out on each.

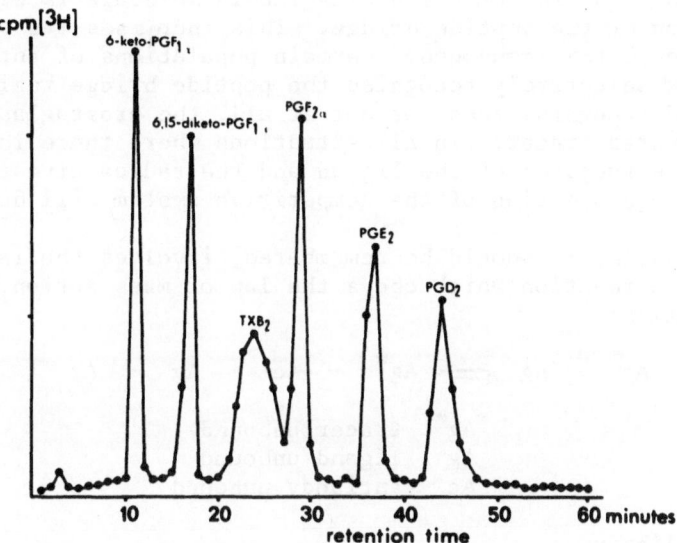

Fig. 3.   Separation of (^{3}H)-PG standard by rp-HPLC.

Column : Radial PAK A (8 x 100 mm); packing material : octa-
decylsilane; solvent system : water/acetonitril/acetic acid
70:30:0.1; flow rate : 1 ml/min. isocratic elution; linear
column pressure : 100 psi.

COMMENTS

PROBLEMS ARISING FROM THE USE OF IODINATED TRACERS PREPARED FROM
PG-HISTAMINE DERIVATIVES

Use of an iodinated derivative as a pose to a tritiated tracer
modifies, to a greater or lesser extent, the competitive properties
of the immunological system involved. There are two main factors
involved:

The specific activity of the iodinated tracer

The specific activity of the iodinated tracer is considerably
higher than that of a tritiated tracer, and, with appropriate purification, may be as high as that of sodium iodide (about 2,000 Ci/
mmole). In such cases, the weight of the tracer becomes very small
and is no longer a limiting factor in the sensitivity of the assay.

The structure of the iodinated derivative
_____

     The COOH function of the prostaglandin molecule is blocked in
the formation of the peptide bridge. This increases its similarity
of structure to the immunogen. Certain populations of antibodies,
however, may selectively recognise the peptide bridge region of the
molecule and recognise less, or not at all, the prostaglandin part
of the iodinated tracer. In all situations where there is a difference in the structures of the ligand and the radioactive tracer, the
thermodynamic properties of the competitive system will be modified.

     This system, it should be remembered, involves the isotopic
dilution of a reaction which obeys the law of mass action, according
to the equation:

$$Ag^* \; + \; Ac \; + \; Ag \; \rightleftharpoons \; Ag^* \; - \; Ac \; + \; Ag \; - \; Ac$$

Ag^* tracer unbound
Ag ligand unbound
Ac antibody unbound

and at equilibrium :

$$Ka^* \; = \frac{[Ag^* - Ac]}{[Ag^*]\,[\,Ac]} \qquad \text{and} \qquad Ka \; = \frac{[Ag - Ac]}{[Ag]\,[Ac]}$$

Ka^* and Ka = association constants for
each competitor expressed in L/M

     In the case of a tritiated tracer, where there is homology
between the structure of the ligand and the tracer, equations are
identical and $Ka^* = Ka$.

     Where in the case of an iodated tracer, theoretically there
are three possible results.

1) $^*Ka < Ka$: there is marked heterology of the two competitors
    (existance of a bridge, presence of iodine) and the iodinated
    tracer is not so well recognised by the antibody binding sites
    as is the unlabelled prostaglandin. In limiting cases the iodin-
    ated tracer does not participate in the competition.

2) $^*Ka > Ka$: the structure of the iodinated derivative resembles
    that of the immunogen more closely than that of the competitor,
    and may, in extreme cases, be so different from the competitor
    that it cannot be displaced. The competitive system can only
    function if the Ka can be elevated by modifying the structure
    of the ligand. Table 5 shows that the sensitivity is often
    improved when the carboxyl group of the ligand is esterified
    as methyl ester (13).

Table 5. Sensitivities of prostaglandins antisera using tritiated and iodinated tracers.

TRACER	ANTISERA							
	E_1	E_2	$F_{1\alpha}$	$F_{2\alpha}$	DHK E_2	DHK $F_{2\alpha}$	TXB_2	$6-K-F_{1\alpha}$
(^{3}H)-PG								
SENSITIVITY a	32	5	15	3.5	54	8	14	94
SPECIFIC RADIOACTIVITY (Ci/mmol)	90	117	79	178	66	85	125	20
FINAL DILUTION OF ANTISERUM	1/45.000	1/75.000	1/15.000	1/90.000	1/3.600	1/45.000	1/24.000	1/13.500
(^{125}I)-PG-His b								
SENSITIVITY	18	2.5	29	7	33	5	8	15
FINAL DILUTION	1/300.000	1/150.000	1/25.000	1/105.000	1/15.000	1/150.000	1/45.000	1/168.000

a Quantity of PG (pg) necessary to give 50% displacement of Bo : for all assays, the final dilution of the antiserum was adjusted to obtain 50-40% initial binding.

b Specific radioactivity of (^{125}I)-labelled PG-histamine tracers was estimated to 2000 Ci/mmol.

Table 6. Concentration of PG required (pmol/ml) to give 50% displacement of Bo using different tracers and inhibition.

TRACER/INHIBITOR	SYSTEM					
	PGE_1	PGE_2	$PGF_{1\alpha}$	$PGF_{2\alpha}$	DHK PGE_2	DHK $PGF_{2\alpha}$
(^{125}I)-labeled-PG-His/PG	0.17	0.025	0.82	0.20	0.95	0.09
(^{125}I)-labeled-PG-His/PG-ME [a]	0.09	0.028	0.14	0.08	0.26	0.09

[a] Abbreviation : ME, methyl ester

3) Ka = Ka : in this, the best situation, the radioactivity is higher,
    the dilution of the antiserum is greater and the competitive sys-
    tem functions well at lower concentrations of ligand, which dimin-
    ishes the detection threshold (Table 6). In addition the cross
    reactions are only slightly modified when compared with the homo-
    logous system using tritiated tracer.

PROBLEMS CONCERNING THE VALIDITY OF PROSTAGLANDIN RADIOIMMUNOASSAY

        These problems are threefold :

1) The radioimmunoassay system itself, in which can be distinguished
permanent factors : the parameters of antibody binding (affinity,
specificity), the relative affinity of the competitors for the antibody sites, and the experimental conditions and separation of bound
and free, and contingent factors related to individual assays, material present in the biological fluids or introduced or concentrated
during the purification procedures.

2) The purification of samples prior to radioimmunoassay; the extent
of the purification depends on the circumstances : in the case of a
new prostanoic derivative or a new biological milieu, we systematically employ all the steps shown in the Figure 4.

        The assay of 6-keto-PGF$_{1\alpha}$ in urine is an example of the necessity
of carrying out all these procedures (Table 7, Figure 5). In other
cases, however, experience has shown that one or more of the steps
may be omitted, and in some cases, a direct assay of the solution
of biological sample may be carried out. In these cases, the results
obtained after omission of each step must be compared, to ensure that
simplification does not engender error. For instance, it has now
been shown that for PGE and PGF$_\alpha$ assay, a purification using HPLC is
not necessary (Figure 5), that often, with our antisera, the results
obtained in the assay of PGE$_2$ after a simple extraction procedure
are valid; that TXB$_2$ may be assayed directly in a diluted biological
sample such as serum.

3) The sampling technique. For bioassay, GC/MS and RIA, rigorous
and reproducible conditions of sampling, and subsequent treatment
of samples, must be established. These should be adapted to each
individual biological medium so as to reduce to a minimum the in
vitro transformation by synthesis or degradation, of the prostanoic
derivatives analysed.

        In this way, the quantitative analysis of these compounds is of
value and may be an invaluable tool for the research biologist and
also the clinician.

Table 7.   Urinary 6-Keto-PGF$_{1\alpha}$ of 3 healthy adult women (pg/ml)

RIA after	1	2	3
a) Silicilic acid column	2974	5003	3085
b) HPLC*			
. Immunoreactivity of all the fractions	657	1937	1157
. Immunoreactivity of the peak corresponding to 6-Keto-PGF$_{1\alpha}$ standard	284	620	313

* See the profile of Fig. 5.

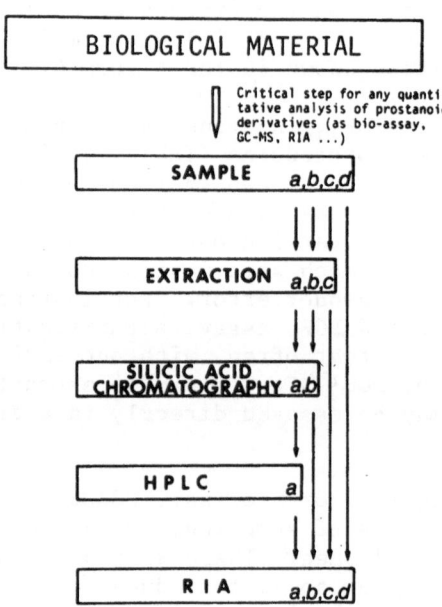

HPLC: High Performance Liquid Chromatography
RIA: Radioimmunoassay

Fig. 4.   Purification procedure : diagram "à la carte".

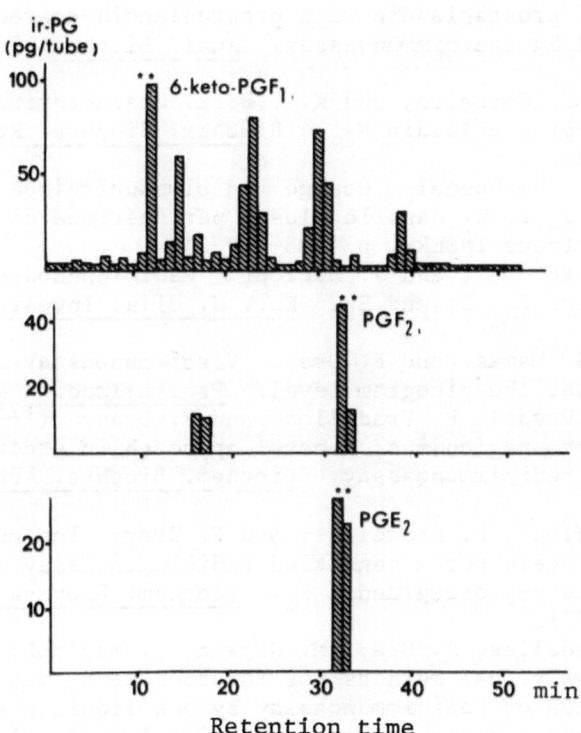

Fig. 5.   Immunoreactive (ir) materials detected after rp-HPLC of
          human urinary extracts obtained after extraction and sili-
          cilic acid column chromatography. Upper peak : each profile
          was found using 6-keto-PGF$_{1\alpha}$ or PGF$_{2\alpha}$ or PGE$_2$ antiserum.

REFERENCES

1.   K. Grenn, M. Hamberg, and B. Samuelsson. Quantitative analysis
     of prostaglanin and thromboxanes by mass spectrometric methods.
     In: "Advances in Prostaglandin and Thromboxane Research",
     B. Samuelsson and R. Paoletti Ed., Raven Press. N.Y. p.47
     (1976).
2.   L. Levine and H. Van Vanukis. Antigenic activity of prostaglan-
     dins. Biochem. Biophys. Res. Commun. 41:1171 (1970).
3.   B.V. Caldwell, S. Burstein, W.A. Brock, and L. Speroff. Radio-
     immunoassay of the F prostaglandins. J. Clin. Endocrinol.
     33:171 (1971)
4.   B.M. Jaffe, J.W. Smith, W.Y. Newton, and C.W. Parker. Radio-
     immunoassay for prostaglandins. Science 171:494 (1971)

5. F. Dray, E. Maron, S.A. Tillson, and M. Sela. Immunochemical
    detection of prostaglandin with prostaglandin-coated bacterio-
    phage T4 and by radioimmunoassay. Anal. Biochem. 50:399
    (1972)

6. K.T. Kirton, J.C. Cornette, and K.L. Barr. Characterization of
    antibody to prostaglandin $F_{2\alpha}$. Biochem. Biophys. Res. Commun.
    47:903 (1972)

7. F. Dray, and B. Charbonnel. Dosage radioimmunologique des pros-
    taglandines F_α et E_1 dans le plasma périphérique de l'homme
    normal. Colloque INSERM, p. 133 (1973)

8. F. Dray, B. Charbonnel, and J. Maclouf. Radioinnunoassay of
    prostaglandins F_α, E_1 and E_2. Eur. J. Clin. Invest. 5:311
    (1975).

9. J.M. Andrieu, S. Mamas, and F. Dray. Viroimmunoassay of prosta-
    glandin $F_{2\alpha}$ at the picogram level. Prostaglandins 6:15 (1974).

10. J. Maclouf, M. Pradel, P. Pradelles, and F. Dray. (^{125}I) deri-
    vatives of prostaglandins; a novel approach in prostaglandins
    analysis by radioimmunoassay. Biochem. Biophys. Acta 431:139
    (1976).

11. H. Sors, J. Maclouf, P. Pradelles, and F. Dray. The use of
    iodinated tracers for a sensitive radioimmunoassay of 13,14-
    dihydro-16-keto-prostaglandin F_α. Biochem. Biophys. Acta
    486:553 (1977).

12. H. Sors, P. Pradelles, F. Dray, M. Rigaud, J. Maclouf, and P.
    Bernard. Analytical methods for thromboxane B_2 measurement
    and validation of radioimmunoassay by gas liquid chromatog-
    raphy-mass spectrometry. Prostaglandins 16:277 (1978).

13. J. Maclouf, H. Sors, P. Pradelles, and F. Dray. Improved sensit-
    ivity of iodinated histamine-prostaglandin radio-imunoassay
    by prostaglandin methyl esters. Anal. Biochem. 87:169 (1978).

14. F. Dray, K. Gerozissis, B. Kouznetzova, S. Mama, P. Pradelles,
    and G. Trugnan. New approach to the RIA of prostaglandins
    and related compounds using iodinated tracers. In: "Advances
    in Prostaglandin and Thromboxane Research", B. Samuelsson,
    P.W. Ramwell and R. Paoletti Ed., Raven Press, N.Y. p. 167
    (1980).

15. J.L. Vaitukaitis, J.B. Robbins, E. Nieschlag, and G.T. Ross,
    A method for producing specific antisera with small doses
    of immunogen. J. Clin. Endocrinol. Metab. 33:988 (1971)

16. W. M. Hunter, and F.C. Greenwood. Preparation of iodine-131
    labelled human growth hormone of high specific activity.
    Nature (London) 194:495 (1962).

REGULATION OF CELLULAR FUNCTIONS BY PHOSPHORYLATION AND DE-
PHOSPHORYLATION OF PROTEINS
AN INTRODUCTION

F. Hofmann

Pharmakologisches Institut universität Heidelberg
Im Neuenheimer Feld 366
D-6900 Heidelberg
Germany

INTRODUCTION

During the last decade it has become apparent that activation
of specific protein kinases by intracellular generated signals -
such as calcium, cAMP, cGMP and other less well defined small molecules - and concomitant phosphorylation of key regulatory proteins
or enzymes is one important mechanism by which many hormones, neurotransmitters and autacoids (locally generated and acting hormones)
regulate cellular functions. From the results obtained so far it is
quite clear that protein phosphorylation is not the only regulatory
mechanism triggered by the binding of hormones to plasma membrane
receptors. This mechanism is only operative in such situations in
which the cellular response follows hormone binding within seconds
to minutes. This mechanism phosphorylation/dephosphorylation has not
been observed in very fast responses (milliseconds) - presumably since
the protein kinase catalyzed phophotransfer is too slow to modify
a significant number of protein molecules before the onset of such
fast cellular responses. In addition protein phosphorylation is not
primarily involved in slow cellular responses occurring within hours -
presumably since the covalent modification carried out by the protein
kinase is not stable for hours but easily reversed by the action of
one or several protein phosphatases. However, although the very fast
and very slow cellular responses are not mediated by phosphorylation
of some regulatory proteins, these responses may be modulated by the
phosphorylation or dephosphorylation of some proteins involved in
these responses.

Table 1. Intracellular Signal Transfer Systems

Signal	Signal Receptor Protein	Receptor-Regulated System
cAMP	Regulatory Subunit I and II of cAMP-dependent Protein Kinase	Catalytic Subunit of cAMP-dependent Protein Kinase (eukaryocytes)
	CRP (cyclic AMP receptor protein)	Gene Expression (E.coli)
cGMP	Regulatory Domain of cGMP-dependent Protein Kinase	Catalytic Domain of cGMP-dependent Protein Kinase
Calcium	Intestinal Calcium-Binding Protein	Calcium Transport
	Parvalbumin	Calcium Sequestration?
	Calsequestrin	Calcium Storage?
	Regulatory Myosin Light Chain (mollusca)	Contraction; Actomyosin ATPase
	Troponin-C (cardiac, skeletal muscle)	Contraction; Actomyosin ATPase
	Calmodulin	Many Systems

INTRACELLULAR SIGNAL TRANSFER

The first step after generation of an intracellular signal is the binding of the signal to a specific receptor protein. In general this binding process is followed by a second reaction, which initiates a change in the activity or function of some key regulatory protein. The intracellular signal receptor proteins for cAMP, cGMP and calcium and the second reactions controlled by these receptor proteins are given in Table 1. From this compilation it is quite clear that only a very limited number of receptor proteins (at most 3) are known for cAMP or cGMP. For cAMP this number decreases to two if only eukaryotic systems are considered and only one "second reaction" controlled by these two receptor proteins is known, namely the activity of a specific protein kinase termed catalytic subunit of cAMP-dependent protein kinase. Similar data are available for the cGMP controlled signal system. In contrast, quite a number of specific, intracellular calcium receptor proteins are known (Table 1). Not only the number of receptor proteins differentiates the calcium system from the cyclic nucleotide systems, but also the finding that one receptor protein may regulate quite distinct "second reactions" as exemplified by calmodulin (Table 2). It has been shown that one cell type may contain several of the enzymes regulated by one receptor protein. Thus, two or more receptor proteins controlling different "second reactions" may be present in one cell. In this case, expression of each reaction depends at least partially if not totally on the time during which the free concentration of calcium is increased and on the speed of the particular "second reaction"; for example, skeletal muscle contains troponin-C, which controls muscle contraction and acto-myosin ATPase activity, and calmodulin which receptor activates in the presence of calcium myosin light chain kinase and regulates thereby the phosphate content of myosin light chain-2. During a single twitch calcium activates only the troponin-C regulated system. No change in the phosphate content of myosin light chain-2 has been observed, presumably since the duration of the twitch-induced increase in calcium concentration is too short (msec.) to allow the phosphorylation of a significant number of myosin light chain-2 molecules. However, the myosin light chain kinase is activated, the phosphate content of myosin light chain-2 is increased, and the properties of the contractile system are changed, if a tetanus of a few seconds is used. The latter type of stimulation increases the calcium concentration for a longer period of time than a single twitch. Thus, the effect of calcium on cellular function is not only dependent on the magnitude of the increase in the intracellular concentration of calcium but also on the duration of the elicited increase. This consideration suggests that regulation of cellular functions by calcium depends on the following parameters: a) type of calcium receptor protein(s) present in the particular cell; b) type of"second reaction(s)" controlled by the receptor protein(s); c) duration of the signal increase. Although it is clear that calcium and cyclic nucleotides may regulate cellular function by quite different mechanisms, it is also quite significant

Table 2.   Calmodulin regulated systems

A. Protein Kinases

   myosin light chain kinase (several isozymes?)
   phosphorylase kinase

B. Other Enzymes

   phosphodiesterase (cAMP, cGMP)
   adenylate-cyclase (brain, adrenal and
     prokaryotic cells?)
   calcium ATPase (erythrocytes)
   calmodulin-binding protein I (calcineurin)
     and II
   guanylate cyclase (Tetrahyema pyriformis)
   NAD-kinase (plants)
   phospholipase A_2?

C. Influenced Enzyme System Unknown

   desaggregation of microtubules
   membrane phosphorylation

---

that each of these intracellular signals is capable to activate a
protein kinase. The reactions catalyzed by these protein kinases
are of intermediate speed. It is therefore possible, that phosphorylation and dephosphorylation of proteins is the common mechanism on
which the short term regulation (sec to min) of cellular functions
by hormones is based. The following pages will only give a short
introduction into the protein kinase field. The equally important
field of the protein phosphatases has not been covered since it is
felt, that this area is still too controversial to allow a simplified
introduction.

PROTEIN KINASES

    Protein kinases are a class of enzymes which catalyze the trans-
fer of the γ-phosphate of ATP to a serine or threonine residue of
certain proteins. This modification of a protein is reversed by
protein phosphatases which hydrolyze the phosphate bond. In many
cases these phosphorylation/dephosphorylation reactions are controlled by the activation of protein kinase by an intracellular signal
according to the following scheme:

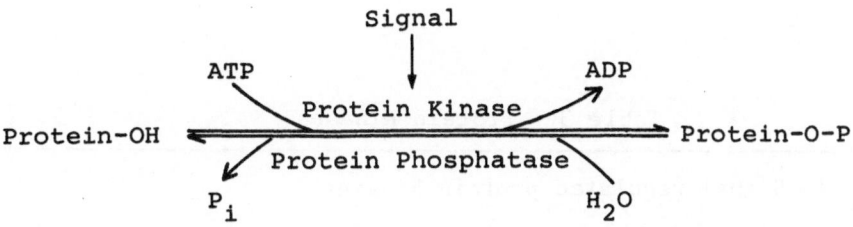

However a number of protein kinases have been found, which are not
regulated by intracellular signals. Therefore other control mechanisms must exist which prevent that the protein phosphorylation/
dephosphorylation reaction is working as an ATPase. More recently,
the phosphorylation of tyrosine residues has been detected in cells
which were transformed by certain viruses. Tyrosine phosphorylation
occurs also in non-transformed cells although at a much lower level
than in transformed cells; e.g., the kinase activated by epidermal
growth factor catalyzes the phosphorylation of tyrosine. It has
been suggested that the content of tyrosine phosphate of a cell may
be connected in some way with the growth properties of this cell.

The protein kinases involved in the reaction shown above are
listed in Table 3. These enzymes can be divided into two groups of
enzymes. Group I contains protein kinases which are known to be
regulated by intracellular signals (second messengers). Group II
contains those protein kinases which are not regulated by intracellular signals and for which the activation/inactivation mechanism
is unknown or non-existent. Most of the enzymes listed in group I
have a rather broad substrate specificity (exceptions are the calcium
(calmodulin)-regulated enzymes). In contrast to this, most of the
enzymes listed in group II have a rather narrow substrate specificity.

Type of changes induced by phosphorylation/dephosphorylation

Phosphorylation of a protein by anyone of the below listed
protein kinases may change following properties of an enzyme or other
protein:

a) Incorporation of phosphate may change the K_m or K_D for a
substrate, cofactor or small activator molecule. This type of reaction is found in many phosphorylated proteins.

b) Phosphorylation may change the affinity (K_D) for a regulator
protein for an enzyme. This may lead to activation or inhibition of
an enzyme; i.ex. phosphorylation of the regulatory subunit of type II
of cAMP-dependent protein kinase facilitates activation of the
enzyme by cAMP; phosphorylation of inhibitor I of protein phosphatase

Table 3. Protein Kinases

I Signal regulated protein kinases

a) cyclic nucleotide regulated
   cAMP-dependent protein kinase (I and II)
   cGMP-dependent protein kinase

b) calcium (calmodulin) regulated
   phosphorylase kinase
   myosin light chain kinase (3 isozymes)

c) dsRNA-regulated-protein kinase

d) calcium/diacylglyceride-regulated protein kinase

II Non regulated protein kinases

"casein" kinase I (uses only ATP)
"casein" kinase II (uses ATP and GTP)
pyruvate dehydrogenase kinase
rhodopsin-kinase
eIF$_2$-kinase
histone kinases
transformation (virus) induced protein kinase
EGF-kinase
others

---

dsRNA: double stranded RNA
EGF: epidermal growth factor

leads to inhibition of the protein phosphatase. Phosphorylation of a first protein may also change the affinity (K_D) for a small molecule to a second protein. This situation has been found with cardiac troponin, in which phosphorylation of the inhibitory subunit of troponin (TN-I) changes the calcium binding properties of troponin-C (TN-C).

c) Phosphorylation of several residues in one protein (multisite phosphorylation) may change the dephosphorylation rate of another site on the same protein. This type of reaction has been found in glycogen synthase and pyruvate dehydrogenase, in which the phosphorylation of the first and second site is important for the regulation of the enzyme activity and phosphorylation of the third site reduces the dephosphorylation rate of the first and/or second site.

In addition to these three types of changes introduced by phosphorylation/dephosphorylation other reactions have been observed occasionally. There are examples which indicate that phosphorylation increases or decreases the stability of a protein against denaturation. Although this type of reaction has not been explored very carefully, the occurrence of this type of change may suggest that phosphorylation/ dephosphorylation may also regulate the turnover of a given protein, by changing its susceptibility to proteolysis.

Enzymes and proteins affected by phosphorylation

Table 4 shows a list of enzymes and other proteins, the function of which may be changed by phosphorylation and dephosphorylation. As is evident from this list, which is by no means complete, many cellular functions may be regulated or modulated by phosphorylation of key regulatory enzymes and other proteins. Many of these key regulatory enzymes or other proteins are phosphorylated by different protein kinases allowing modulation of their activity by different signal systems. During the last years it has become evident that not only the key regulatory enzymes are phosphorylated but that the protein kinases are also phosphorylated either by an autocatalytic process or by an other protein kinase. In the few examples which have been studied extensively (cAMP-dependent protein kinase type II, smooth muscle myosin light chain kinase, phosphorylase kinase), phosphorylation of a protein kinase may drastically change the activitability of the enzyme by its own signal. For example, phosphorylation of phosphorylase kinase by cAMP-dependent protein kinase reduces the concentration of calcium required for activation of phosphorylase kinase from about 1 uM to 0.1 uM. Phosphorylation of smooth muscle myosin light chain kinase by cAMP-dependent protein kinase decreases the affinity of smooth muscle myosin light chain kinase for calmodulin by an order of magnitude. These findings suggest that signal-dependent protein kinases regulate cellular functions not only directly by phosphorylation of key regulatory enzymes but also by interfering with the transduction of an other signal system by phosphorylating the protein kinase regulated by the other signal.

Table 4. Examples of Enzymes and Other Proteins the Properties of
which can be Altered by Phosphorylation and Dephosphorylation (Early 1980)

Protein	In vivo-phosphorylation shown	Kinase-type Activity		change
		cAMP	others	
A) Carbohydrate and Lipid Metabolism				
glycogen phosphorylase	yes		+	↑
phosphorylase kinase	yes	+		↑
glycogen synthase	yes	+	+	↓
phosphofructokinase	yes		+	−
pyruvate kinase (liver)	yes	+		↓
pyruvate dehydrogenase	yes		+	↓
hormone-sensitive lipase	(yes)	+		↑
acetyl-CoA-carboxylase	(yes)		+	(↓)
glycerophosphat-acyl-transferase	no	?		
hydroxymethylglutaryl-CoA-reductase	(yes)		+	
B) Protein Synthesis				
histones	yes	+	+	−
ribosomal proteins (S6)	yes	+	+	−
initiationsfactors (eIF$_2$,eIF$_3$)	yes		+	1)↓
C) Contractile Systems				
troponin I (heart)	yes	+		2)↓
myosin light chain (smooth muscle)	yes		+	3)↑
myosin light chain (skeletal and heart. muscle)	yes		+	−
phospholambein (sarco-plasmatic reticulum, heart)	yes	+	+	4)↑
filamin	yes	+		−
desmin	yes		+	−
D) Others				
tyrosine hydroxylase	no	+		
protein I and II (brain)	yes	+	+	−
R$_{II}$ of cAMP-kinase	yes	+		5)↑
R$_I$ of cAMP-kinase	yes		+	6)↓
phosphatase inhibitor I	yes	+		7)↑

1) Phosphorylation of eIF$_2$ is thought to inhibit formation of the
   initiation complex.

2) Phosphorylation of troponin I decreases calcium sensitivity of
   myosin ATPase.

(continued)

Table 4. (continued)

3) Phosphorylation increases actomysin-ATPase and contraction.

4) Phosphorylation increases calcium uptake by sarcoplasmatic reticulum.

5) Phosphorylation of R_{II} changes sensitivity for activation.

6) Phosphorylation decreases amount of cAMP bound.

7) Only the phosphorylated inhibitor inhibits phosphatase.

Symbols: ↑ ,increase in activity;
      ↓ ,decrease in activity;
      – ,functional change unknown;
   (yes),in vivo phosphorylation not shown beyond doubt but
      very much likely

READING LIST

The reader who is interested in more details is referred to the following review articles which have appeared recently.

1. P. B. Chock, S. G. Rhee and E. R. Stadtman, (1980) "Interconvertible enzyme cascades in cellular regulation" Ann. Rev. Biochem. 49:813-843.
2. J. R. Knowles (1980) "Enzyme-catalyzed phosphoryltransfer reactions" Ann. Rev. Biochem. 49:877-919.
3. E. G. Krebs and J. A. Beavo (1979) "Phosphorylation-dephosphorylation of enzymes" Ann. Rev. Biochem. 48:923-959.
4. D. A. Walsh and R. H. Cooper (1979) "The physiological regulation and function of cAMP-dependent protein kinases" Biochem. Action of Hormones Vol. VI:1-75.
5. D. B. Glass and E. G. Krebs (1980) "Protein phosphorylation catalyzed by cyclic AMP-dependent and cyclic GMP-dependent protein kinases" Ann. Rev. Pharmacol. Toxicol. 20:363-388.
6. G. N. Gill and R. W. McCaine (1979) "Guanosine-3',5'-monophosphate-dependent protein kinase" Curr. Topic. Cell Reg. 15: 1-45.
7. Th. R. Soderling (1979) "Regulatory functions of protein multisite phosphorylation" Mol. Cell. Endocrinol. 16:157-180.
8. C. Baglioni (1979) "Interferon-induced enzymatic activities and their role in the antiviral state" Cell 17:255-264.
9. B. R. G. Williams and J. M. Kerr (1980) "The 2-5 A (pppA$^{2'}$p$^{5'}$ A$^{2'}$p$^{5'}$A) system in interferon-treated and control cells" TIBS p. 138-140.
10. R. H. Kretsinger (1980) "Structure and evolution of calcium-modulated proteins" CRC Crit. Rev. Biochem. 8:119-174.

11. R. H. Kretsinger (1979) "The informational role of calcium in
     the cytosol" Advances Cyclic Nucleot. Res. 11:1-26.
12. D. J. Wolff and C. O. Brostrom (1979) "Properties and functions
     of the calcium-dependent regulator protein" Advances in
     Cyclic Nucleot. Res. 11:27-88.
13. A. R. Means and J. R. Dedman (1980) "Calmodulin - an intracel-
     lular calcium receptor" Nature 285:73-76.
14. J. H. Wang and D. M. Waisman (1979) "Calmodulin and its role
     in the second-messenger system" Curr. Topic. Cell. Regul.
     15:47-107.
15. C. B. Klee, T. H. Crouch and P. G. Richman (1980) "Calmodulin"
     Ann. Rev. Biochem. 49:489-515.
16. J. T. Stull (1980) "Phosphorylation of contractile proteins in
     relation to muscle function" Advances in Cyclic Nucleot. Res.
     13:39-93.
17. R. S. Adelstein and E. Eisenberg (1980) "Regulation and kinetics
     of actin-myosin-ATP interaction" Ann. Rev. Biochem. 49:921-956.

REGULATION OF BIOCHEMICAL PROCESSES THROUGH PROTEIN PHOSPHORYLATION

AND DEPHOSPHORYLATION : SEVERAL IMPORTANT EXAMPLES

Francoise Lamy

Institut de Recherche Interdisciplinaire
Universite Libre de Bruxelles

INTRODUCTION

In 1969, Greengard has postulated that all the effects of cAMP are secondary to the phosphorylation of cellular proteins by cAMP activated protein kinases[1]. Much evidence has now accumulated in many laboratories which supports this concept (Fig.1). Recent studies indicate that a diverse group of regulatory agents, including but by no means limited to those agents acting through cAMP, may achieve certain of their biological actions through effects on the phosphorylation of specific proteins. These regulatory agents include hormones and neurotransmitters whose action involve cGMP, Ca^{2+} or still unknown mediators[2]. The protein kinases which are regulated by agents that do not act through cAMP are termed cAMP independent protein kinases. They include a cGMP-dependent protein kinase, Ca^{2+}-dependent protein kinases, a double-stranded RNA-dependent protein kinase, etc. Other effectors specific for other kinases will certainly be discovered.

The number of enzymes known to undergo phosphorylation-dephosphorylation has risen to more than twenty and many nonenzymic proteins must be added to this list[3]. The reactions involved in the phosphorylation and dephosphorylation of proteins are shown in equations (a) and (b):

$$\text{Protein} + \text{nATP} \underset{\text{protein kinase}}{\overset{\longrightarrow}{\rightleftharpoons}} \text{Protein} - \text{Pn} + \text{nADP} \qquad \text{(a)}$$

$$\text{Protein} - \text{Pn} + \text{nH}_2\text{O} \underset{\text{phosphoprotein phosphatase}}{\overset{\longrightarrow}{\rightleftharpoons}} \text{Protein} + \text{nPi} \qquad \text{(b)}$$

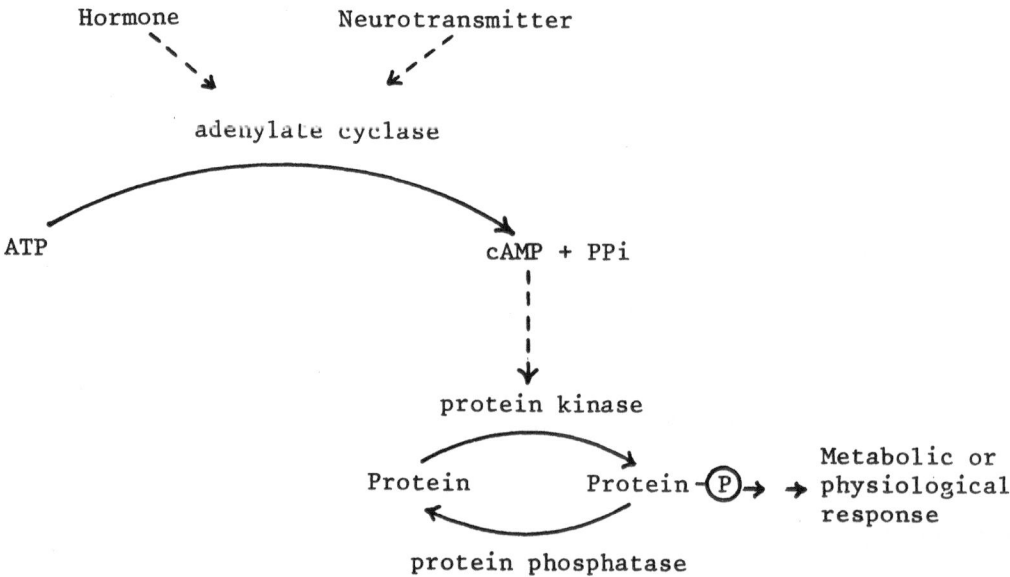

Fig. 1. Diagram representing the presumed role played by protein
         phosphorylation in mediating the biological effects of
         hormones and neurotransmitters acting through cyclic AMP.

The amino acid residue(s) to which the phosphoryl group is transferred
is usually a serine or a threonine but there are some examples where
protein kinases catalyse the transfer of the phosphoryl group from
ATP to histidine, lysine or tyrosine residues in proteins.

     In this chapter, we will describe several systems which are
regulated through reversible protein phosphorylation. More attention
will be paid to protein phosphorylations controlled by cAMP dependent
protein kinases as they have been more extensively studied.

     In 1973, Krebs proposed five criteria that must be satisfied
before an effect mediated by cAMP can be said to occur via the phosphorylation of a protein[4]:

1- Cell type involved contains a cAMP-dependent protein kinase
2- Protein substrate exists which bears a functional relationship to
   the process mediated by cAMP
3- Phosphorylation of the substrate alters its function in vitro
4- Protein substrate is modified in vivo in response to cAMP
5- A phosphoprotein phosphatase exists to reverse the process

     These criteria have been re-examined in 1977 by Nimmo and Cohen
in the light of the results obtained since then[5]. They made the
following remarks:

a) The first criterion is no longer relevant since the distribution of cAMP-dependent protein kinase in mammalian tissues is ubiquitous.
b) The rate at which a protein is phosphorylated _in vitro_ is important in view of the ability of cAMP-dependent protein kinase to catalyze slow phosphorylation of a number of proteins, particularly when they are no longer in their native conformations.
c) As several physiological substrates for cAMP-dependent protein kinases are subject to phosphorylation by at least one further protein kinase whose activity is unaffected by cAMP, it is insufficient to demonstrate that a protein which serves as a substrate for cAMP-dependent protein kinase _in vitro_ can be extracted from a tissue in a phosphorylated form or even that its phosphorylation is increased by administration of a hormone. As already mentioned, hormones can affect the phosphorylation of proteins by acting on cAMP-independent protein kinases. It is thus important to demonstrate that a protein becomes phosphorylated _in vivo_ in response to a hormonal stimulus at the same site(s) that is phosphorylated by cAMP-dependent protein kinases _in vitro_.
d) The fifth criterion is also no longer relevant: the existence of specific phosphatases[6] shows that there is not a single enzyme in mammalian cells which reverses all the phosphorylations catalyzed by cAMP-dependent protein kinase. A major problem appears not to be that of finding an enzyme which reverses a given phosphorylation but rather of deciding which of several such enzymes is the "relevant" phosphatase _in vivo_.

Nimmo and Cohen have redefined four new criteria that must be met before an effect mediated by cAMP can be said to occur through phosphorylation of a protein[5]:

1- A protein substrate for cAMP-dependent protein kinase should exist which bears a functional relationship to the process mediated by cAMP. The rate of phosphorylation of that protein, in its native state, should be adequate to account for the speed at which the process occurs _in vivo_ in response to cAMP.
2- The function of the protein should be shown to undergo a reversible alteration _in vitro_ by phosphorylation and dephosphorylation, catalyzed by cAMP-dependent protein kinase and a protein phosphatase.
3- A reversible change in the function of the protein should occur _in vivo_ in response to cAMP.
4- Phosphorylation of the protein should occur _in vivo_ in response to a hormone at the same site(s) phosphorylated by cAMP-dependent protein kinase _in vitro_.

These criteria have so far been rigorously met only for the activation of phosphorylase kinase by cAMP-dependent protein kinase. The regulation of this enzyme will be described in the next section relative to glycogen metabolism in skeletal muscle. Other systems

that are less well understood in molecular terms but in which control
through protein phosphorylation has been implicated will then be
discussed with respect to the criteria listed above. This review is
not exhaustive. Only the systems which have been the most extensively
studied will be considered.

Enzyme Regulation By Reversible Phosphorylation

Glycogen Metabolism in Skeletal Muscle

    Glycogen metabolism is the system where the regulation of key
enzymes through phsophorylation-dephosphorylation mechanisms has been
the most widely studied. The recognition over the past few years
that the control of many cellular processes by physiological stimuli
may be explainable in terms of changes in the state of phosphorylation
of key regulatory proteins makes it a model system of unique general
interest.

    It is well established that adrenaline stimulates glycogenolysis
in skeletal muscle through the sequence of reactions:

Adrenaline $\longrightarrow$ raised $\longrightarrow$ activated $\longrightarrow$ phosphorylated
                      cAMP   protein kinase   phosphorylase kinase

$\xrightarrow{Ca^{2+}}$ elevated
          phosphorylase a

Each of these events has been shown to take place in the skeletal
muscle in vivo as well as in vitro[6].

    Phosphorylase. The conversion of phosphorylase b to a catalyzed
by phosphorylase kinase involves the phosphorylation of a unique
serine residue on each polypeptide chain of the tetramer. This
serine residue is located 14 amino acids from the N-terminus of the
polypeptide chain which comprises 841 amino acids. The reconversion
to phosphorylase is catalyzed by phosphorylase phosphatase.

    Phosphorylase kinase. The activity of phosphorylase kinase is
dependent of Ca^{2+} and can be stimulated a further 20-40-fold at pH 6.8
by phosphorylation catalyzed by cAMP-dependent protein kinase. Phosphorylase kinase has a MW of 1,280,000. Its quaternary structure is
$(\alpha\beta\gamma\delta)_4$ where the MW of the α, β, γ and δ subunits are 145,000;
125,000; 45,000 and 17,000 respectively. The α- and β-subunits are
the components phosphorylated by cAMP-dependent protein kinase[6]. The
δ-subunit is identical to the calcium binding protein termed calmodulin and seems to be the component which confers Ca^{2+} sensitivity to
the phosphorylase kinase reaction[7].

When phosphorylase kinase is phosphorylated _in vitro_ in the presence of cAMP-dependent protein kinase, the phosphorylation of the enzyme reaches a plateau when 2 molecules of phosphate have been incorporated per ($\alpha\beta\gamma\delta$). The α and β subunits are each phosphorylated to the extent of one molecule per subunit. A serine residue is implicated in both cases but amino acid sequences at the phosphorylated sites are different. The only similarity is the presence of 2 adjacent basic amino acids just N-terminal to each phosphoserine. The β subunit is phosphorylated first, in paralell with a 40-fold rise in activity. The α subunit is phosphorylated more slowly after a short lag period, without any further effect on enzyme activity. Although the α subunit is phosphorylated slightly slower than the β subunit, its rate of phosphorylation is similar to that of other physiological substrates for cAMP-dependent protein kinase.

The dephosphorylation of the phosphorylated α- and β- subunits is catalyzed by two distinct protein phosphatases termed α-phosphorylase kinase phosphatase and β-phosphorylase kinase phosphatase respectively. Although phosphorylation of the α-subunit does not effect activity directly, it appears to alter the conformation of the enzyme in such a way that the dephosphorylation of the β-subunit by its specific phosphatase is enhanced at least 50-fold. This shows that the reversible activation of the enzyme correlates with the reversible phosphorylation of the β-subunit, and that phosphorylation of the α-subunit plays a role in controlling the reversal of phosphorylase kinase activation. Regulation of phosphorylase kinase activity proceeds as follows: phosphorylase kinase is first activated by phosphorylation of the β-subunits. When 2 β-subunits have been phosphorylated per mole of enzyme $(\alpha\beta\gamma\delta)_4$, the phosphorylation of the α-subunits begins. The transition to the form which is an effective substrate for β-phosphorylase kinase phosphatase does not take place until 2 α-subunits have been phosphorylated per mole of enzyme. There is thus a lag period during which the kinase and phosphatase reactions do not compete. After this lag period, enzyme inactivation begins through dephosphorylation of the β-subunit. Evidence for the _in vivo_ occurrence of such a control mechanism is the finding that intravenous injection of adrenaline into rabbits produces a rapid phosphorylation of both the α and β subunits at sites identical to those phosphorylated _in vitro_.

Glycogen synthetase: It has been established for a number of years that glycogen synthetase can be isolated as a dephosphorylated enzyme which is almost fully active in the absence of glucose 6P, or as phosphorylated forms which are largely dependent on glucose 6P for activity. However, the nature of the protein kinases which catalyze these phosphorylation reactions is still debated[6].

The first protein kinase which has been shown to phosphorylate glycogen synthetase _in vitro_ is a cAMP-dependent protein kinase identical to the one which phosphorylates phosphorylase kinase.

Since adrenaline promotes a decrease in the activity of glycogen
synthetase in skeletal muscle in vivo, it has been suggested that
this hormone, through cAMP, promotes the breakdown of glycogen in
two ways: first, by activating phosphorylase kinase, phosphorylase
and the glycogenolysis pathway, and second by inactivating glycogen
synthetase and the glycogen synthesis pathway.

   Glycogen synthetase is composed of 4 subunits of uniform size
having each a MW of about 88,000. cAMP-dependent protein kinase
catalyzes the phosphorylation of two serine residues per subunit
(site 1 and site 2) with a concomitant decrease in the synthetase
activity in the absence of glucose 6P. However, it is difficult to
distinguish whether the phosphorylation of site 1 or site 2 causes
the change in the glucose 6P dependency of the synthetase. The results published until now are controversial[6,8].

   Several cAMP-independent protein kinases have been reported to
catalyze the phosphorylation of glycogen synthetase at sites different
from the 2 sites phosphorylated by the cAMP-dependent enzyme[9,10]. One
of these cAMP-independent protein kinases has been shown to be identical to phosphorylase kinase[11-14].

   The existence of several glycogen synthetase kinases capable of
phosphorylating the enzyme in vitro raises the question as to which
enzyme is operating under any given metabolic conditions in vivo.
The results obtained until now strongly favor the hypothesis that
changes in the activity of cAMP-dependent protein kinase underlie the
regulation of glycogen synthetase by adrenaline in skeletal muscle[15].

   Protein phosphatase 1: It is now established that β phosphorylase kinase phosphatase, phosphorylase phosphatase and glycogen synthetase phosphatase are multiple functions of the same enzyme which
has been termed protein phosphatase 1. α phosphorylase kinase phosphatase which is a different enzyme has in turn been termed protein
phosphatase 2[6]. Thus, in skeletal muscle, a single major activity,
protein phosphatase 1, carries out each of the dephosphorylations
that inhibit glycogenolysis or activate glycogen synthetase. It not
only reverses the activation of phosphorylase kinase and the inactivation of glycogen synthetase catalyzed by cAMP-dependent protein
kinase, but it also reverses the activation of phosphorylase catalyzed
by phosphorylase kinase and the inactivation of glycogen synthetase
catalyzed by cAMP-independent protein kinases (Fig. 2). This multifunctional enzyme is regulated by two heat-stable protein inhibitors
termed inhibitor 1 (MW = 20,000) and inhibitor 2 (MW = 30,000).

   Inhibitor 1: Inhibitor 1 is inhibitory only after it has been
phosphorylated by cAMP-dependent protein kinase. In contrast to all
known substrates of cAMP-dependent protein kinase which are phosphorylated on serine residues, inhibitor 1 is phosphorylated on a
threonine residue. It is phosphorylated at a similar rate to

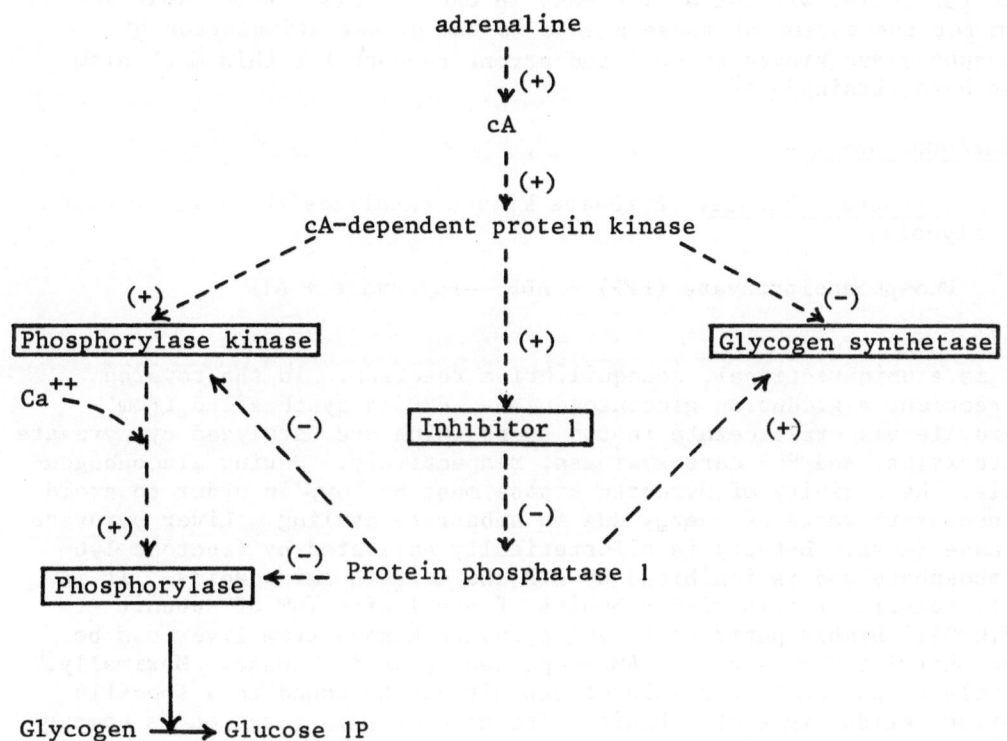

Fig. 2. Regulation of glycogenolysis and glycogen synthesis by
adrenaline in skeletal muscle through the simultaneous
phosphorylation of phosphorylase kinase, inhibitor 1 and
glycogen synthetase.

phosphorylase kinase. Moreover, it has been demonstrated recently
that this inhibitor is phosphorylated in skeletal muscle in vivo and
that its degree of phosphorylation is increased markedly by adrenaline[16]. It seems likely that the phosphorylation of inhibitor-1 may
be as important as the activation of phosphorylase kinase in the
elevation of phosphorylase a levels by adrenaline (Fig. 2).

Glycogen Metabolism in Liver

The studies about the effect of glucagon on liver are all in
keeping with a regulatory cascade scheme analogous to that of muscle.

On the other hand, it has been shown that α-adrenergic agents, vasopressin and angiotensin cause phosphorylase a formation in liver (or fat cells) without an increase in cAMP levels. A possible mechanism for the action of these agents is the direct stimulation of phosphorylase kinase by Ca$^+$ and strong support for this mechanism has been obtained[3,17].

Gluconeogenesis

L-Pyruvate kinase: Pyruvate kinase catalyzes the last reaction in glycolysis:

$$\text{Phosphoenolpyruvate (PEP)} + \text{ADP} \longrightarrow \text{pyruvate} + \text{ATP}$$

It is a unidirectional, nonequilibrium reaction. In the reverse direction, e.g. during gluconeogenesis, PEP is synthesized from pyruvate via oxaloacetate in two steps which are catalyzed by pyruvate carboxylase and PEP carboxykinase, respectively. During gluconeogenesis, the activity of pyruvate kinase must be low, in order to avoid unnecessary waste of energy due to substrate cycling. Liver pyruvate kinase (mostly L-type) is allosterically activated by fructose 1,6-diphosphate and is inhibited by ATP and certain amino acids. It is a tetrameric protein with subunits of equal size (MW of subunit = 57,000). Highly purified L-type pyruvate kinase from liver can be phosphorylated by ATP and cAMP-dependent protein kinase. Maximally, 1 mole of phosphate per mole of subunit can be bound to a specific serine residue in each subunit. The rate of this reaction is comparable to the rates of phosphorylation of phosphorylase kinase and glycogen synthetase. Maximal phosphorylation of L-pyruvate kinase results in a complete inhibition of the enzyme activity which can be reactivated by incubation with a phosphatase preparation from liver cell sap[5,18]. In the liver, glucagon rapidly increases the rate of gluconeogenesis and decreases the rate of glycolysis. It has been shown that pyruvate kinase is phosphorylated in vivo to some extent and that the incorporation of [^{32}P] phosphate into this enzyme is stimulated 2-3-fold by glucagon. At the same time, the hormone inhibited pyruvate kinase activity and elevated cAMP levels[19]. These observations offer strong support for the hypothesis that glucagon alters pyruvate kinase activity in the liver by a phosphorylation mechanism. Adrenaline has the same effects on carbohydrate metabolism in liver as glucagon. It increases the rate of gluconeogenesis and decreases the rate of glycolysis. These effects occur predominantly via an α-adrenergic stimulation and do not involve an increase in intracellular cAMP or activation of cAMP dependent protein kinase. Addition of epinephrine to hepatocytes results in decreased pyruvate kinase activity and also elicits phosphorylation of this enzyme. Preliminary studies indicate that regulation of liver pyruvate kinase activity by either glucagon or by catecholamines may occur through phosphorylation of the same amino acid residues[20].

Lipid metabolism

Hormone-sensitive lipase: cAMP mediates the lipolytic effects
of many hormones in adipose tissue. The rate-limiting step in the
hydrolysis of tri-cylglycerols is catalyzed by a hormone-sensitive
lipase. This lipase can be activated by cAMP-dependent protein kinase
in vitro and this mechanism is thought to be involved in the stimulation of lipolysis by catecholamines and other lipolytic hormones.
The problem is that hormone sensitive lipase has proved to be difficult to purify and it has therefore been a problem to show conclus-
ively that activation of the enzyme results from its phosphorylation[5].
Recently, Belfrage et al.[21,22] have purified hormone sensitive lipase
from adipocytes to an enzyme protein purity of about 35%. They have
shown that the lipase is phosphorylated in vivo and that the extent
of phosphorylation is increased when adipocytes are incubated with
noradrenaline (MW of the phosphopeptide = 84,000). This increase in
hormone sensitive lipase phosphorylation is concomitant to an increase
in the activity of the enzyme.

Acetyl-CoA carboxylase: Acetyl-CoA carboxylase (subunit MW =
220,000) catalyzes the carboxylation of acetyl-CoA to malonyl-CoA:

$$\text{Acetyl-CoA} + HCO_3^- + H^+ + ATP \rightleftharpoons \text{Malonyl-CoA} + ADP + Pi$$

This reaction is the first step toward fatty acid synthesis and is
one of the regulatory steps of this process. In vitro, acetyl-CoA
carboxylase is completely dependent on the allosteric activator
citrate for activity and is inhibited by very low concentrations of
palmityl-CoA. These metabolites promote polymerisation (active form)
and depolymerisation (inactive form) of the enzyme respectively[23].
The idea that the activity of acetyl-CoA carboxylase is also regulated by a phosphorylation-dephosphorylation mechanism has been debated for several years; evidence of the validity of this mechanism
has accumulated recently. In vitro, phosphorylation of the carboxylase is mediated by both cAMP-dependent as well as independent protein
kinases; cAMP-dependent phosphorylation takes place at a rate similar
to that of other well-established physiological substrates of this
enzyme[24]. Moreover, acetyl-CoA carboxylase subunit isolated from
adipocytes and from hepatocytes is phosphorylated and its degree of
phosphorylation is increased by hormones which elevate the intracellular level of cAMP (adrenaline in adipose tissue[25,26] and glucagon
in hepatocytes[27].) The increased phosphorylation is accompanied by
a decrease in the activity, measured in cell extracts in the presence
of citrate. The highly phosphorylated form with low specific activity
can be converted to a form of high specific activity by a dephosphorylation reaction[28,29]. However, the sites at which acetyl-CoA
carboxylase is phosphorylated in vitro as well as in vivo are still
to be identified.

ATP-citrate lyase: Fatty acids and sterols are synthesized in the cytoplasm from acetyl-CoA but acetyl-CoA is produced in the mitochondrion by the pyruvate dehydrogenase reaction. The carbon atoms of acetyl-CoA leave the mitochondrion as citrate which is then reconverted to acetyl-CoA in the cytoplasm by the enzyme ATP-citrate lyase which catalyzes the reaction:

$$\text{Citrate + ATP + CoA} \rightleftharpoons \text{acetyl-CoA + oxaloacetate + ADP + Pi}$$

ATP-citrate lyase isolated from rat liver contains 2 moles of serine-bound phosphate per tetramer (MW of subunit = 116,000) and a liver phosphatase has been isolated which removes this phosphate[30]. More-over, phosphorylation of ATP-citrate lyase is stimulated by glucagon and insulin in hepatocytes preparations and the effects of glucagon and insulin are additive[31,32]. Guy et al.[33] have purified to hom-ogeneity ATP-citrate lyase from lactating rat mammary gland and have shown that it can be phosphorylated by cAMP-dependent protein kinase at a rate sufficient to account for the effect of glucagon on the phosphorylation of the enzyme in hepatocytes. There is thus good evidence that glucagon-stimulated phosphorylation of ATP-citrate lyase is mediated directly by activation of the cAMP-dependent pro-tein kinase. The mechanism of insulin-stimulated phosphorylation remains to be established. It is still essential to demonstrate that glucagon stimulates the phosphorylation of ATP-citrate lyase at the same site phosphorylated in vitro by cAMP-dependent protein kinase and that phosphorylation of ATP-citrate lyase has an effect on its activity.

Concluding Remark

The reversible phosphorylation of enzymes is, without any doubt, and important mechanism by which cellular metabolism is regulated by hormones and neurotransmitters. A generality that seems to be emerging is that enzymes in biodegradative pathways are activated and enzymes in biosynthetic pathways inactivated by phosphorylation.

PROTEIN PHOSPHORYLATION IN MEMBRANES AND NEURAL FUNCTION

Nervous System

In the nervous system, cAMP-dependent, cGMP-dependent and Ca^{2+}-dependent protein kinases have been demonstrated and endogenous substrate proteins for each of these 3 types of protein kinases have been found[34].

Regulation of neurotransmitters release by phosphorylation of protein I: Proteins Ia and Ib, collectively referred to as protein I because their properties are similar, are endogenous substrates for cAMP-dependent and Ca^{2+}-dependent protein kinases. Protein Ia has a MW of 86,000 and an isoelectric point of 10.3 whereas protein Ib has a MW of 80,000 and an isoelectric point of 10.2. Ca^{2+} and cAMP stimulate the phosphorylation of different amino acid residues on these proteins. Protein I is only found in the nervous tissue (central and peripheral). It is located in neurons and appears to be associated primarily with synaptic vesicles. In brain slices and in whole animals its state of phosphorylation is affected by agents that modify the physiological state of nerve cells[35,36]. It is believed that protein I phosphorylation may be involved in the regulation of neurotransmitter release[37,38]. cAMP-dependent protein kinase seems to be implicated in two other processes in the nervous system namely the regulation of the biosynthesis of certain neurotransmitters and the functioning of microtubules.

Regulation of neurotransmitter synthesis by phosphorylation of tyrosine hydroxylase: Tyrosine hydroxylase catalyzes the conversion of tyrosine to 3,4-dihydroxyphenylalanine (DOPA) which is the rate-limiting step in the biosynthesis of the catecholamines dopamine and noradrenaline:

tyrosine   DOPA

dopamine   noradrenaline

Three different groups of workers[39,40,41] have demonstrated that tyrosine hydroxylase is a substrate for cAMP-dependent protein kinase in vitro and that there is a direct relationship between the amount of phosphate incorporated and the degree of activation of this enzyme. Because nerve stimulation results in elevation of intracellular cAMP, in stimulation of cAMP-dependent protein kinase and in activation of tyrosine hydroxylase, it has been proposed that this enzyme is a substrate of cAMP-dependent protein kinase in vivo.

Regulation of microtubule function by phosphorylation of MAP2: Microtubules prepared from brain homogenates by successive cycles of assembly-disassembly contain two high-molecular weight proteins designated as microtubules-associated protein 1 (MAP1) and microtubule-associated protein 2 (MAP2). MAP2 (MW = 300,000) is phosphorylated in vitro by an endogenous cAMP-dependent protein kinase which appears to be an integral component of the microtubules[42,43]; it is also phosphorylated in vivo[43]. It has been suggested that the phosphorylation of MAP2 may play a role in microtubule function.

Substrate for cGMP-dependent protein kinase: In contrast to cAMP-dependent protein kinases, for which numerous substrates have been found, it has been difficult to demonstrate the existence of endogenous substrates for cGMP-dependent protein kinase. A specific substrate for this enzyme has been found in the cytosol of mammalian cerebellum in experiments realized in an acellular system. This substrate which is a protein of 23,000 MW appears to be highly enriched in the Purkinje cells. Until now, no function is known for this cGMP-dependent phosphorylation[44].

Erythrocyte Plasma Membrane

Several polypeptides seem to be phosphorylated in the erythrocyte plasma membrane but until now it has been difficult to correlate any of these phosphorylations with a specific membrane function[45]. The phosphorylation of spectrin, a peripheral erythrocyte membrane protein, has been widely studied. Together with actin, spectrin lines the cytoplasmic surface of the red cell membrane. These proteins interact to form a tough but flexible cytoskeleton. Spectrin, as defined by SDS-polyacrylamide gel electrophoresis is composed of two major high MW polypeptide chains migrating as band I (MW = 240,000) and band II (MW = 220,000)[46]. This protein has only been found in erythrocytes and has been immunologically related to smooth muscle myosin[47]. Spectrin appears to be involved in determining cell shape and membrane deformability properties[48,49]. Spectrin and in particular band II is phosphorylated in vitro by cAMP-dependent and cAMP-independent endogenous protein kinases but its phosphorylation in vivo does not seem to require cAMP[50]. The casual relationship between spectrin phosphate levels and either red cell shape or spectrin binding to the membrane components is still debated[50-52].

LYSINE-RICH HISTONE PHOSPHORYLATION AND CONTROL OF TRANSCRIPTION AND MITOSIS

Histones are major structural components of the eukaryotic chromatin. The structural unit of the chromatin is the nucleosome which consists of DNA folded round a central core of protein made up of two copies each of histones H_2A, H_2B, H_3 and H_4 and of a length of linker DNA joining core particles. The lysine-rich histone H_1 is not involved in the structure of the core particles; it binds in part to the linker DNA region and seems to be involved in higher order packing of chromatin[53]. An attractive current hypothesis is that reversible phosphorylation of histones H_1 is a major mechanism for altering the conformation of the chromosome to allow such processes as DNA transcription, replication and mitosis.

Histone H_1 is the largest of the histones with a MW of 20,000. Its primary structure can be divided into essentially three regions:

the N-terminal region 1-41 which contains a very basic region
(segment 24-39), the central region 41-123 which contains 81% of the
apolar residues and the carboxyl half of the molecule 123-216 which
is nearly 90% lysine, alanine and proline. Cole[54] has suggested that
the basic regions are DNA binding sites of H_1 while the apolar central
region is required for other functions.

Phosphorylation of H_1 histones occurs <u>in vivo</u> in both dividing
and non-dividing cells. It has been shown that this phosphorylation
implicates more than one protein kinase, each of which catalyzes the
phosphorylation of a different site or set of sites in the molecule[55].
These sites of phosphorylation may be categorised as follows:

Serine-37 Phosphorylation in H_1 Histone

In non-dividing cells serine at position 37 can be phosphorylated by cAMP-dependent protein kinase in response to hormone action.
This has been illustrated by the work of Langan et al.[56] for the
system glucagon-liver and by the work of Lamy et al.[57] for the system
TSH-thyroid. Phosphorylation of serine-37 site, even in the presence
of cAMP stimulation, is not extensive (about 1% of the H_1 molecules
are phosphorylated) indicating that some selectivity may be involved.
Circular dichroism studies of H_1-DNA complexes and NMR studies have
shown that the interaction between H_1 histone and DNA is modified by
the presence of this single phosphate group[53,56]. Langan has proposed that this phosphorylation of H_1 could result in depression of
the template activity of the associated DNA, thereby allowing RNA
synthesis[56].

Growth-associated Phosphorylations in H_1 Histones

In addition to the limited amount of cAMP-dependent H_1 histone
phosphorylation which takes place in hormonally stimulated cells,
extensive phosphorylation of this histone, involving multiple sites
of phosphorylation of each H_1 molecule, takes place in dividing
cells. This phosphorylation occurs on both serine and threonine
residues localized in the amino-terminal and carboxy-terminal regions
of the molecule. Phosphorylation of these sites is catalyzed by a
cAMP-independent protein kinase which has been termed growth-associated
histone kinase and which is found only in proliferating cells. Phosphorylation of H_1 histone by this enzyme occurs throughout S-phase
and G_2, reaching high levels just prior to the onset of mitosis, and
falling abruptly as mitosis ends. Four major sites are located on
Thr 16 in the amino-terminal and on Thr 136, Thr 153 and Ser 180 in
the carboxy-terminal region of the molecule[55]. All the data obtained
until now are consistent with the proposal that growth-associated H_1
phosphorylation initiates chromosome condensation in prophase of the
cell cycle[58,59].

PROTEIN PHOSPHORYLATION AND THE REGULATION OF PROTEIN SYNTHESIS

In eukaryotes, gene expression is controlled not only at the level of transcription but also at the translational level. Regulation of translation takes place, in part, during the initiation of polypeptide chains. It is more and more accepted that control of translation involves protein kinases which, when activated, inhibit translation[60]. Since reticulocytes contain no nucleus they are widely used for the study of the regulation of protein synthesis at the level of translation. The main function of reticulocytes, the immediate precursors of the erythrocytes, is to synthesize globin, the protein component of hemoglobin. In reticulocytes, the synthesis of globin, together with that of other less abundant proteins, depends on the presence of heme, the prosthetic group of hemoglobin. Reticulocyte lysates, lacking mitochondria, do not synthesize heme and are markedly dependent on the addition of hemin for protein synthesis.

Regulation of Protein Synthesis by Hemin

In the absence of hemin, an inhibitor of polypeptide chain initiation is activated. This hemin-controlled inhibitor (HCI) is a cAMP-independent protein kinase (eIF-2α kinase) that specifically phosphorylates the small, α subunit (MW = 38,000) of the eukaryotic initiation factor eIF-2 interfering with its function[60]. The initiation factor eIF-2 is an ($\alpha\beta\gamma$) oligomer with a MW of about 122,000[61]. It is required for the binding of the initiator transfer RNA (Met-tRNA) to 40 S ribosomal subunit. Met-tRNA is not bound to 40 S ribosomal subunits directly but first forms a ternary complex with GTP and eIF-2. Farrell et al.[62] have presented evidence that phosphorylation of eIF-2α is associated with inhibition of protein synthesis. The phosphorylation of eIF-2α has been demonstrated in heme deficient lysates and is the primary event in the inhibition although other components could be involved. HCI is present in normal lysates as inactive precursor. Its activation seems to require its phosphorylation. It has been reported that phosphorylation of HCI is catalyzed by a cAMP-dependent protein kinase[60]. The model suggests that hemin maintains the cAMP-dependent protein kinase in a latent state by binding to the regulatory subunit of the holoenzyme in a manner which blocks its activation by cAMP. However, the role of cAMP in this activation is still debated[63].

Regulation of Protein Synthesis by Double-Stranded RNA

There are other mechanisms of translational regulation in reticulocytes. The best understood is the inhibition of protein synthesis by double stranded RNA. Addition of small amounts of ds-RNA to hemin-containing reticulocyte lysates provokes the activation of a cAMP-independent protein kinase that like HCI specifically phosphorylates the 35,000 subunit of eIF-2. This ds-RNA-dependent protein kinase and the HCI have been shown to be distinct proteins[64] although they seem to phosphorylate the same site(s) on eIF-2α. The ds-RNA-dependent protein kinase is present in normal lysates as inactive precursor. Its activation, like that of HCI, seems to require its phosphorylation. It probably involves self-phosphorylation upon interaction with ds-RNA. As will be discussed elsewhere in this volume, the chief enzyme induced by interferon appears to be a protein kinase that is activated by ds-RNA. This protein kinase is similar to that present, apparently in a constitutive manner, in rabbit reticulocytes[60].

There is evidence that phosphorylation of eIF-2 can control protein synthesis in cells other than reticulocytes[65] but although much work has accumulated which seems to indicate that the phosphorylation of the initiation factor eIF-2 is the direct cause of inhibition of its function, the precise mechanism of inhibition of translation is not yet perfectly understood.

Phosphorylation of Ribosomal Protein and of Initiation Factors Other than eIF-2

Incubation of ribosomes with protein kinases (both cAMP-dependent and independent), in the presence of ATP phosphorylates several proteins of the 60 S and 40 S subunits, whereas in vivo mainly one protein of the 40 S subunit (termed S6 in rat liver and S11 in reticulocytes[66]) is phosphorylated. In liver, this phosphorylation is markedly enhanced by cAMP and by glucagon which suggests the involvement of a cAMP-dependent protein kinase. No functional alteration of the ribosomes as a consequence of this phosphorylation has been detected[60].

The β subunit of eIF-2, several subunits of the initiation factor eIF-3 (composed of ten subunits) and the initiation factors eIF-5 and eIF-4B are phosphorylated by cAMP-independent protein kinases. There are thus far no indications that the phosphorylation of these factors play a part in translational regulation[60].

MYOSIN PHOSPHORYLATION IN SMOOTH MUSCLE CONTRACTION AND CELL MOTILITY

The two major proteins involved in muscle contraction as well as in the contractile activity of vertebrate non-muscle cells such as platelets and macrophages are actin and myosin. The energy required for this physical process is provided by ATP and is released by the interaction of actin with myosin, which activates the myosin ATPase activity. The concentration of Ca^{2+} controls the interaction of actin and myosin through different mechanisms. We will describe here one mechanism for regulating actin-myosin interaction which seems to be the dominant regulatory process in vertebrate smooth muscle cells as well as in non-muscle cells and which is based on the phosphorylation of myosin. This mechanism has been reviewed recently by Adelstein and Eisenberg[67]. The system is composed of two enzymes : myosin light chain kinase and phosphatase, and the substrate on which they act : myosin light chains (Fig. 3).

Phosphorylation of Myosin

The myosin molecule is composed of one pair of heavy chains (MW = 200,000) and two pairs of light chains (MW = 15,000-27,000). Both muscle and non-muscle myosins contain a pair of light chains (MW = 18,500-20,000) that are phosphorylated _in vivo_ and that are

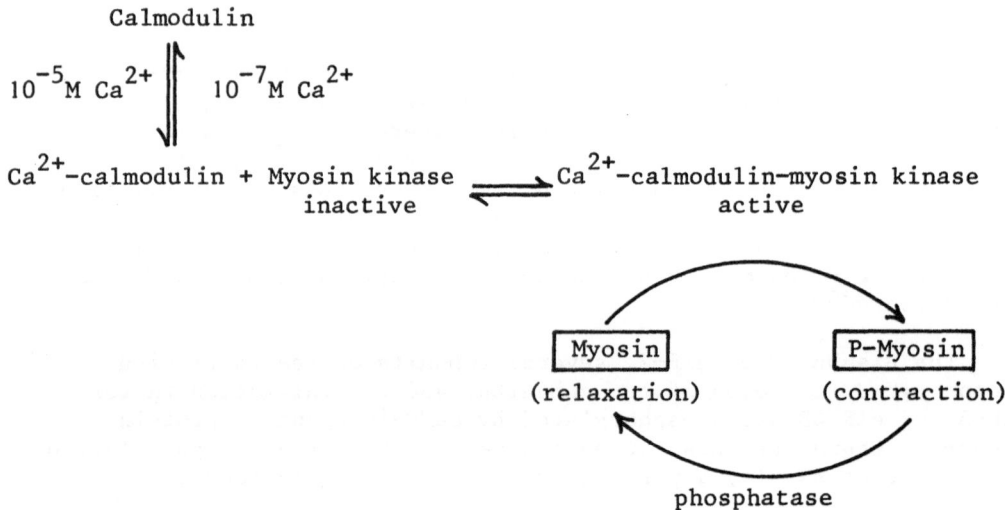

Fig. 3.   Schematic representation of the regulation of smooth muscle contraction by Ca^{2+}. Contraction is initiated by a rise in free Ca^{2+} from $10^{-7}M$ - $10^{-5}M$. The rise in Ca^{2+} activates the enzyme myosin kinase and results in the phosphorylation of myosin. The phosphorylated myosin can interact with actin and this interaction provokes contraction. A fall in Ca^{2+} concentration to $10^{-7}M$ inactivates myosin kinase. Myosin phosphatases which do not require Ca^{2+} for activity restore myosin to its dephosphorylated form and relaxation is achieved.

termed P-light chains. In vitro, these light chains can be phosphorylated up to one mole of phosphate per mole of myosin light chain.
This phosphorylation takes place on a particular serine residue
located near the amino-terminal end of the polypeptide chain (residue
13 for chicken gizzard and residue 15 for rabbit skeletal muscle).
Although phosphorylation of myosin light chains does not seem to be
essential in skeletal and cardiac muscle contraction[68], it is necessary for actin activation of the myosin ATPase activity in smooth
muscle, platelets and macrophages. Moreover, in vivo studies have
demonstrated a correlation between the phosphate content of these
myosin light chains and the contractile state of different smooth
muscles[69,70].

Regulation of Myosin Light Chain Kinase by Ca^{2+} and Calmodulin

Phosphorylation of the P-light chains is catalyzed by a highly
specific enzyme that has been isolated from muscle and non-muscle
cells. This kinase requires Ca^{2+} for activity but is cyclic nucleotide independent. The active kinase is composed of two proteins.
The first protein is a heavy chain with a MW of about 130,000 in
smooth muscle and non-muscle cells and of about 80,000 in skeletal
and cardiac muscle. The second protein is the smaller Ca^{2+}-binding
protein, calmodulin. The high molecular weight component of myosin
kinase is completely inactive in the absence of calmodulin. The
initial steps in activation of myosin kinase seem to be the binding
of Ca^{2+} to calmodulin and the binding of this complex to the kinase.
The only known substrate for myosin kinase is the myosin light chain.

Regulation of Myosin Light Chain Kinase by cAMP

A second mechanism of regulating myosin kinase activity involves
the phosphorylation of the enzyme by cAMP-dependent protein kinase.
Phosphorylation of myosin kinase results in decreased activity
apparently due to a decrease in the affinity of the phosphorylated
kinase for the Ca^{2+}-calmodulin complex. To date, this form of regulation has only been reported for myosin kinase isolated from smooth
muscle and platelets. It might explain how increased levels of cAMP
lead to relaxation in certain smooth muscles.

Myosin Light Chain Phosphatase

Much less is known about the phosphatase that dephosphorylates
myosin light chains. It has been purified from skeletal muscle where
it was found to be specific for myosin light chains. Two different
phosphatases have also been detected in smooth muscle and non-muscle
cells which show a marked preference for myosin light chains. Unlike
the kinase, the phosphatase is not dependent on Ca^{2+} for activity.

PROTEIN PHOSPHORYLATION IN VIRUS-TRANSFORMED CELLS

pp60^{src} Kinase

     Certain proteins responsible for transformation of virally in-
fected cells to the malignant state are protein kinases. These
kinases have the novel property of catalyzing the phosphorylation of
tyrosine residues in their substrate proteins rather than the usual
serine or threonine residues[71-73]. The system which has been the most
widely studied is that of the Rous or avian sarcoma virus, an RNA
virus in which a single viral gene, src, is responsible for the malignant transformation of cells in culture and the formation of tumors
in birds. A 60,000 MW protein kinase present in transformed cells
has been identified as the product of this gene and seems to be
essential for transformation of cells to the malignant state[72]. This
protein termed pp60^{scr} is a cAMP-independent protein kinase which is
itself phosphorylated in two distinct regions. One site is a serine
residue located in the amino-terminal 60% of the polypeptide. This
site is phosphorylated by a cAMP-dependent protein kinase in cell-
free extracts. The second site which seems to be a site of autophosphorylation, is a tyrosine residue located on the carboxy-terminal
40% of the polypeptide[71].

pp60^{sarc} Kinase

     It has recently been demonstrated that uninfected cells from a
variety of vertebrate species contain low levels of phospho-proteins
phosphorylated on tyrosine residues[72]. This modified amino acid
appears to have escaped earlier detection first because it is rare
(phosphoserine and phosphothreonine together are about 3000 times
more abondant) and secondly because both phosphotyrosine and phosphothreonine are difficult to separate by traditional electrophoretic
procedures. The protein kinase responsible for the phosphorylation
of tyrosine residues in normal cells is a highly conserved phosphoprotein that is closely related in both structure and sequence to
the pp60^{src} protein and which has been termed pp60 or pp60^{sarc}. This
pp60^{sarc} protein is the product of a gene, sarc, present in all vertebrate cells[74].

     The question that arises now is whether the protein kinase ac-
tivity of pp60^{src} has unique specificities for acceptor proteins
which could account for the altered phenotype of RSV infected cells
or whether the enzymatic properties of pp60^{src} and vertebrate pp60
are identical. In other words, transformation may either be due to
increased phosphorylation of tyrosine residues normally modified by
endogenous tyrosine protein kinases or else to aberrant phosphorylation of tyrosines. The answer to this question of functional equiv-
alence can come only from a knowledge of the in vivo substrates of
the two protein kinases pp60^{src} and pp60^{sarc}.

REFERENCES

1. J. Kuo and P. Greengard, 1969, Proc. Natl. Acad. Sci, USA,
   64:1349.
2. P. Greengard, (1978), Science, 199:146.
3. E. Krebs and J. Beavo, (1979), Ann. Rev. Biochem., 48:923.
4. E. Krebs, (1973), Endocrinology, Proceedings of the 4th International Congress, 17-29.
5. H. Nimmo and P. Cohen, (1977), Adv. Cycl. Nucl. Res., 8:145.
6. P. Cohen, (1978), Curr. Top. Cell Regul., 14:118.
7. S. Shenolikar, P.T. Cohen, P. Cohen, A. Nairn and S. Perry,
   (1979), Eur. J. Biochem., 100:329.
8. K. Huang and F. Huang, (1980), J.B.C., 255:3141.
9. D. Rylatt and P. Cohen, (1979), FEBS Letters, 98:71.
10. A. De Padi-Roach, P. Roach and J. Larner, (1979), J.B.C.,
    254:12062.
11. N. Embi, D. Rylatt and P. Cohen, (1979), Eur. J. Biochem.,
    100:339.
12. T. Soderling, A. Srivastava, M. Bass and B. Khatra, (1979),
    PNAS, 76:2536.
13. K. Walsh, D. Millikin, K. Schlender and E. Reimann, (1979),
    J.B.C., 254:6611.
14. P. Roach, A. De Paoli-Roach and J. Larner, (1978), J. Cycl. Nucl.
    Res., 4:245.
15. M. Dietz, J. Chiasson, T. Soderling and J. Exton, (1980), J.B.C.,
    255:2301.
16. J. Foulkes and P. Cohen, (1979), Eur. J. Biochem., 97:251.
17. H. De Wulf, S. Keppens, J. Vandenheede, F. Haustraete, C. Proost
    and H. Carton, (1980), in "Hormones and Cell Regulation",
    J. Dumont and J. Nunez, eds., Elsevier, Holland, 4:47.
18. L. Engström, (1978), Curr. Top. Cell Regul., 13:29.
19. J. Riou, T. Claus and S. Pilkis, (1978), J.B.C., 253:656.
20. M. Nagano, H. Ishibashi, V. McCully and G. Cottam, (1980),
    Arch. Biochem. and Biophys., 203:271.
21. P. Belfrage, G. Fredrikson, N. Nilsson and P. Stralfors, (1980),
    FEBS Letters, 111:120.
22. N. Nilsson, P. Stralfors, G. Fredrikson and B. Belgrage, (1980),
    FEBS Letters, 111:125.
23. K. Kim, (1979), Mol. Cell. Biochem., 28:27.
24. D. Hardie and P. Cohen, (1978), FEBS Letters, 91:1.
25. R. Brownsey, W. Hughes and R. Denton, (1979), Biochem. J.,
    184:23.
26. K. Lee and K. Kim, (1979), J.B.C., 254:1450.
27. L. Witters, E. Kowaloff and J. Avruch, (1979), J.B.C., 254:245.
28. D. Hardie and P. Cohen, (1979), FEBS Letters, 103:333.
29. G. Krakower and K. Kim, (1980), Biochem. Biophys. Res. Com.,
    92:389.
30. T. Linn and P. Srere, (1979), J. Biol. Chem., 254:1691.
31. M. Alexander, E. Kowaloff, L. Witters, D. Dennihy and J. Avruch,
    (1979), J. Biol. Chem., 254:8052.

32. A. Janski, P. Srere, N. Cornell and R. Veech, (1979), J. Biol. Chem., 254:9365.
33. P. Guy, P. Cohen and D. Hardie, (1980), FEBS Letters, 109:205.
34. P. Greengard, (1978), in "Cyclic Nucleotides, Phosphorylated Proteins and Neuronal Function", Raven Press, New York.
35. J. Forn and P. Greengard, (1978), PNAS, 75:5195.
36. V. Strömbom, J. Forn, A. Dolphin and P. Greengard, (1979), PNAS, 76:4687.
37. F.E. Bloom, T. Ueda, E. Battenberg and P. Greengard, (1979), PNAS, 76:5982.
38. P. De Camilli, T. Ueda, F.E. Bloom, E. Battenberg and P. Greengard (1979), Proc. Natl. Acad. Sci., USA, 76:5977.
39. T. Yamauchi and G. Fujisawa, (1979), J.B.C., 254:503.
40. T. Joh, D. Park and D. Reis, (1978), PNAS, 75:4744.
41. P. Vulliet, T. Langan and N. Weiner, (1980), PNAS, 77:92.
42. P. Sheterline, (1977), Biochem. J., 168:533.
43. R. Sloboda, S. Rudolph, J. Rosenbaum and P. Greengard, (1975), PNAS, 72:177.
44. D.J. Schlichter, J.E. Casnellie and P. Greengard, (1978), Nature, 273:61.
45. L. Waxman, (1979), Arch. Biochem. Biophys., 195:300.
46. G. Fairbanks, I. Steck and D. Wallack, (1971), Biochemistry, 10:2606.
47. M. Sheetz, R. Painter and S. Singer, (1976), Biochemistry, 15:4486.
48. S. Lux, (1979), Semin. Hematol., 16:21.
49. J. Palek and S. Lui, (1979), Semin. Hematol., 16:75.
50. C. Pinder, D. Bray and W. Gratzer, (1977), Nature, 270:752.
51. W. Birchmeier and S. Singer, (1977), J. Cell Biol., 73:647.
52. J. Anderson and J. Tyler, (1980), J.B.C., 255:1259.
53. H. Rattle, T. Langan, S. Danby and E. Bradbury, (1977), Eur. J. Biochem., 81:499.
54. R. Cole, (1977), in "Molecular Biology of the Mammalian Genetic Apparatus", P. T'So, ed., North-Holland, Amsterdam, pp 93.
55. T.A. Langan, (1978), in "Methods in Cell Biology", G. Stein, and J. Stein, eds., Academic Press, New York., 19:127.
56. T. Langan, (1973), in "Advances in Cyclic Nucleotide Research" P. Greengard and G. Robison, eds., Raven Press, New York, 3:99.
57. F. Lamy, R. Lecocq and J.E. Dumont, (1977), Eur. J. Biochem., 73:529.
58. H. Matthews and E. Bradbury, (1978), Exp. Cell Res., 111:343.
59. S. Corbett, E. Bradbury and H. Matthews, (1980), Exp. Cell Res. 128:127.
60. S. Ochoa and C. de Haro, (1979), Ann. Rev. Biochem., 48:549.
61. M. Lloyd, J. Osborne, B. Safer, G. Powell and W. Merrick, (1980), J.B.C., 255:1189.
62. P. Farrel, T. Hunt and R. Jackson, (1978), Eur. J. Biochem., 89:517.

63. D. Levin, V. Ernst and I. London, (1979), J.B.C., 254:7935.
64. D. Levin, R. Petryshyn and I. London, (1980), Proc. Natl. Acad. Sci., USA, 77:832.
65. R. Ranu, (1980), FEBS Letters, 112:211.
66. V. Du Vernay and J. Traugh, (1978), Biochemistry, 17:2045.
67. R.S. Adelstein and E. Eisenberg, (1980), Ann. Rev. Biochem. 49:921.
68. S.A. Jeacocke and P.J. England, (1980), Biochem. J., 188:763.
69. J.T. Barron, M. Barany, K. Barany and R.V. Storti, (1980), J.B.C., 255:6238.
70. R.A. Janis, B.M. Moats-Staats and R.T. Gualtieri, (1980), BBRC, 96:265.
71. M. Collett, A. Purchio and R. Erikson, (1980), Nature, 285:167.
72. T. Hunter and B. Sefton, (1980), PNAS, 77:1311.
73. B. Sefton, T. Hunter, K. Beeman and W. Eckhart, (1980), Cell, 20:807.
74. H. Oppermann, A. Levinson, H. Varmus, L. Levintow and J. Bishop, (1979), PNAS, 76:1804.

THE EFFECTS OF CYCLIC NUCLEOTIDE DERIVATIVES ON CELL METABOLISM

Bernd Jastorff

Fachbereich Biologie-Chemie – Universität Bremen
D-2800 Bremen 33
Federal Republic of Germany

INTRODUCTION

Analogs of cAMP and cGMP are commonly used to clarify the role of cyclic nucleotides in cell metabolism. Shortly after Sutherlands discovery of cAMP, chemists started to modify the molecule according the general rules of chemical synthesis[1]. Already at that time the best known cAMP analog – $^6N^{2'}$ O-dibutyryl cAMP – was synthesized by Posternak et al. and introduced into biochemical and biological studies. Many chemists from all over the world followed Posternak in the synthesis of analogs and thus up to today more than 600 derivatives of the natural cyclic nucleotides have been reported[2]. In principle modifications of these molecules can continue endlessly, since the multifunctional character of the nucleotide structure allows unlimited modifications[1]. Most of the derivatives were synthesized between 1968-1973 by chemists in the pharmaceutical industry, because at that time all the companies searched for a drug based on the structure of a cyclic nucleotide.

In basic research the overwhelming amount of analogs prohibited their systematic use. Only a very limited knowledge of the biological properties of all those analogs is available[2]. Therefore we asked the following questions: Are those few analogs commonly used in the field of cyclic nucleotide research meaningful and optimal for all the different types of experiments in which they are applied? Are they always used for a distinct purpose? Is there any explanation for the different behavior of analogs in different biological systems? Can analogs help us to elucidate the typical molecular interactions by which cAMP or cGMP are bound to their specific receptor proteins in all living systems they are involved in? Can these molecular interactions be correlated to a specific function?

Table 1. Use of Cyclic Nucleotide Analogs

Examples for application of analogs	Commonly used derivatives
Mimicking cyclic nucleotide action on enzymes, homogenates, cell cultures, organs, microorganisms, and animals	$6_N 2'$ O-dibutyryl cAMP, 8-bromo cAMP, 8-pchlorogphenyl thio cAMP, 8-bromo cGMP, cIMP; and unsystematically chosen derivatives available to the researcher.
Affinity chromatography for purification of protein kinases (It is not possible yet to purify PDE this way)	8- or 6-substituted derivatives of cAMP and cGMP coupled by an amino group to sepharose
Photoaffinity labeling of cyclic nucleotide depending proteins (protein kinases, cAMP receptor protein in cellular slime molds)	radioactive ^3H or ^{32}P 8-azido cAMP, 8-azido cIMP, 8-azido cGMP
Physical chemical (fluorescence studies) on purified proteins	1_N, 6_N-etheno cAMP
Screening for specific pharmacological potencies in several eukaryotic systems	unsystematic use of all analogs available to the different research groups

We try to answer these questions by a systematic approach, studying the effects of cyclic nucleotide analogs cooperatively with biochemists and biologists. The results of our approach are as follows.

APPLICATION OF ANALOGS IN CYCLIC NUCLEOTIDE RESEARCH

Table 1 gives a short incomplete list of applications of cyclic nucleotide analogs in biochemistry, biology and pharmacology. The most commonly used analogs given in Table 1 are partly commercially available. The choice of a certain compound seems not to be always related to its significant effects but often depends on fashion or availability of the analogs.

PITFALLS CONNECTED WITH THE USE OF ANALOGS

Scientists who want to use analogs in their experiments face several problems. Seldom a rational choice based on published experience is possible. An example of such application is 8-azido cAMP for photolabeling of cAMP-dependent protein kinases.

Mostly availability of compounds determines the selection. That this selection might be optimal for one system and ineffective in another is shown in Table 2, in which the action of analogs in four different cAMP-receptor systems has been presented.

This list could be extended considerably. It demonstrates however clearly that the different cAMP dependent proteins react differently on cAMP analogs. This difference may be due to the type of molecular interaction between cAMP and its receptor proteins.

Table 2. System Dependent Biological Activity of Analogs

	Biological activity			
	Protein kinase type 1	type 2	Dictyostelium discoideum	CPR-protein in E.coli
8-bromo cAMP	very high	very high	low	low
[6]N-butyryl cAMP	very high	very high	low	low
5'amino-5'deoxy cAMP	low	low	high	high
[1]N-oxide cAMP	high	high	high	low

Fig. 1 schematically shows the metabolic pathways of cAMP and two analogs in a liver homogenate. The analogs are metabolized differently from cAMP and from each other.

This again is only an examplaric example because all analogs have their distinct and typical catabolite pattern mostly different from that of the natural compounds.

If analogs are used in complex long term experiments catabolite analogs may accumulate and these intermediates may interfere with specific metabolic pathways of cAMP and thus lead to side reactions not due to cAMP itself. If one is not aware of this possibility one might give the wrong interpretation to the observed biological effects.

Another pitfall is the chemical instability of some analogs. The most prominent example is dibutyryl cAMP from which the 2'butyryl group is easily removed under physiological conditions. The butyrate itself may have its own biological activity (butyrate effect), which is not related to that of cAMP. Control experiments with butyric acid are therefore always necessary, if dibutyryl cAMP is used in long term experiments.

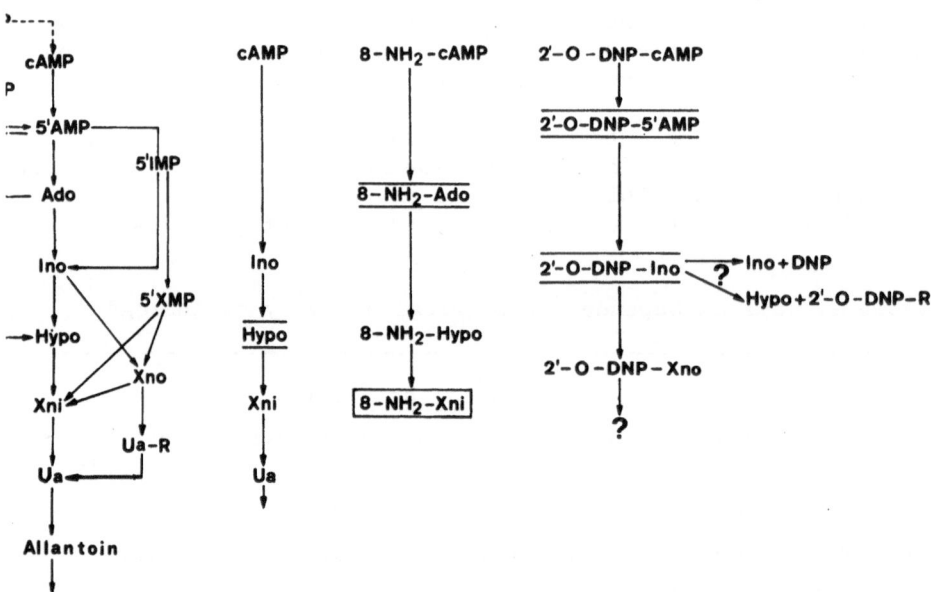

Fig. 1.   Scheme of the degradation of the cyclic nucleotides.
          Left: Part of the purine degradation pathway.
          Right: Products found in the liver homogenate by HPLC
          ═══   intermediate with a high concentration
          ▭   end product in the degradation.

The undesired effects of catabolites and instable groups may interfere particularly where high concentrations of analogs are used. The affinity of cAMP to its receptor proteins is rather high (from 10^{-9} - 10^{-6} M). Several analogs have the same affinity, which makes it not necessary to use them in concentrations of 10^{-3} - 10^{-2} M, even in experiments where the nucleotide has to penetrate through membranes. At these high concentrations several nucleotide analogs might unspecifically bind to other proteins. This again might lead to biological effects not due to cyclic nucleotide metabolism. Therefore it is necessary to measure dose response curves and apply the lowest effective concentration.

RATIONAL APPROACH FOR THE USE OF ANALOGS

Our approach to answer the introductory questions and to overcome pitfalls discussed is based on the assumption, that natural cyclic nucleotides - as all other small molecular ligands, - are recognized and bound to their receptor proteins by distinct chemical (molecular) interactions between certain atoms, atom groups or regions of the nucleotide and amino acid side chains or the backbone of the protein. X-ray analysis has proven this assumption to be true for several proteins and their specific ligands[3]. Protein-ligand-interaction is based on the following types of molecular forces:

- hydrogen bonding
- ion pairing
- van der Waals forces (dipole-dipole, dipole-induced dipole and dispersion forces)
- hydrophobic forces (entropic effects)

Mostly a combination of these forces is responsible for specific binding.

Intrinsic Properties of cAMP and cGMP for Specific Molecular Interactions

An analysis of the potential molecular interactions of the two natural cyclic nucleotides is shown in Fig. 2.

Both molecules allow the combination of several single types of interaction. They are able to donate (A) or accept (B) hydrogen bonds; an ion pair (C) can be formed towards a positively charged amino acid and the entire base moiety or a part of it can be bound via van der Walls or pure hydrophobic forces (D). From this analysis similarities and differences between cAMP and cGMP are obvious.

While the abilities for forming an ion pair and hydrogen bonds in the cyclic phosphate ribose region are identical, the hydrogen bonding pattern in the pyrimidine part of the base is quite different.

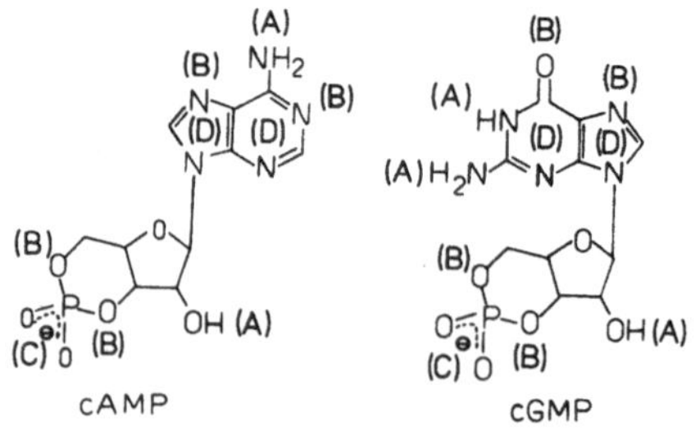

Fig. 2. Potential molecular interactions of cAMP and cGMP. (A: H-
         bond donor; B: H-bond aceptor; C: ion pair; D: hydro-
         phobic interactions).

Thus by forming specific hydrogen bonds towards this region protein
specificity can be obtained. Withought going into detail other differences will only be mentioned. The pyrimidine part of cAMP is much
more hydrophobic than the rather polar part of cGMP, allowing aspeci-
ficity by entropic effects. Guanine has a stronger polarizing
potency than adenine, thus forming more stable charge transfer complexes. The N-7 position of cGMP is more basic than that of cAMP.
cGMP prefers syn conformation, while cAMP has no special preference
for a syn or anti arrangement of the base.

These different and similar potentials for chemical interactions
with a protein allows either for specificity - typical for protein
kinases - or similar affinities, found for some phosphodiesterases.

To elucidate which specific interactions are formed with different cyclic nucleotide receptors, we developed the following systematic approach.

General Concept for Elucidation of Protein-Ligand-Interaction

As shown in Fig. 3 our approach depends on an interdisciplinary
cooperation of chemists and those groups working on the biological
functions of cAMP.

Systematically we want to elucidate which combinations of single
interactions are typical for the different cAMP receptor proteins
to answer the question whether there exist typical types of interaction selected during evolution or by chance for specific function.

Synthesis of definitely modified analogs of cAMP

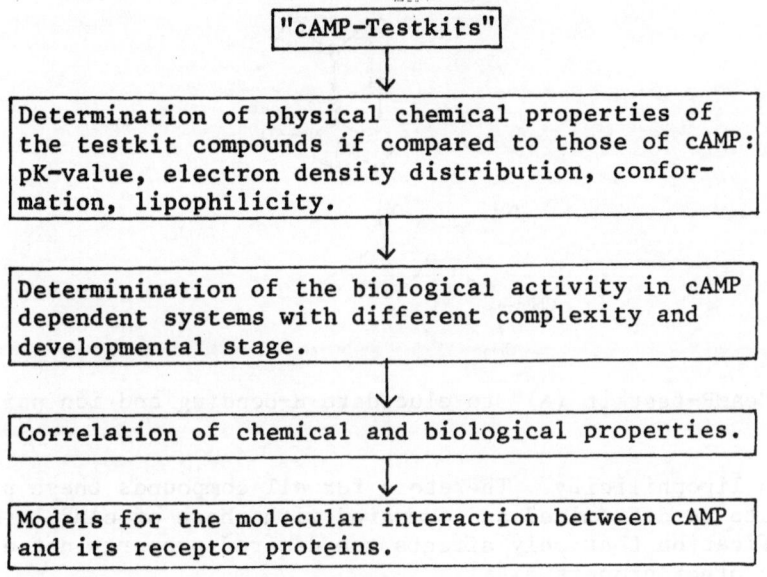

Fig. 3. Strategy for the elucidation of essential molecular interactions between cAMP and its receptors.

The Testkit Concept

Our selection of analogs to be tested in cAMP dependent systems
is based on the following assumptions: potentially interacting atoms
or atom groups on the nucleotide are modified in such a way, that
its typical mode of interaction (see Fig. 2) is prohibited. This
leads to a certain limited number of derivatives ("testkit"). Each
compound has a definitely reduced potency for molecular interactions.
Fig. 4 shows the "testkit A", which we developed to check whether a
hydrogen bond is involved and which one in binding to a protein and
whether an ion pair is formed.

To check the conformation, in which the base is bound we developed another "testkit". The derivatives are fixed in this kit either
in syn or anti conformation; the base of cAMP rotates easily. A
third "testkit" was developed to check the influence of polarizing
potencies on the molecular interaction and a fourth one, in which
the size of the molecule is systematically changed in all regions to
elucidate the general topography of the binding site. It is typical
for all testkits, that a chemical modification is not only influencing the specific position, where it is introduced, but also the
physical chemical properties of the whole molecule as for example:
pK-values, charge density, conformation, electron density distri-

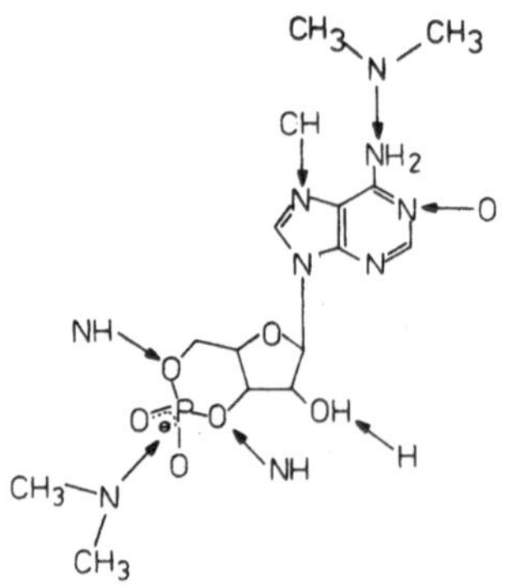

Fig. 4. "cAMP-testkit (A)" to elucidate H-bonding and ion pairing.

bution and lipophilicity. Therefore for all compounds these properties have to be determined and compared with those of cAMP, aiming
at a modification that only affects the desired property not essentially the other properties.

The Biological Model Systems "Dictyostelium discoideum" and Protein Kinase Type "1"

In Dictyostelium discoideum cAMP is an extracellular signal
inducing chemotaxis of the cellular slime mold[5]. The nucleotide is
bound by a highly specific receptor situated at the cell surface.
The signal is transferred into the cell and the following processes
are induced[6]: orientated movement (chemotaxis), protein synthesis
(i.e. phosphodiesterase) and cAMP excretion (relay system). In cooperation with the group of T.M. Konijn we investigated the influence
of the modifications of the "testkit A"-compounds on the chemotactic
activity in Konijn's small population assay (for details see cit.7).

In cooperation with J. Hoppe, the biological activity of testkit
A analogs was determined in the isolated protein kinase type 1 system
from rabbit muscle. The inhibition of the binding of ^{3}H-cAMP towards
the regulatory subunit of protein kinase by these analogs was
measured. The inhibitory constants represent the effect of a modification on affinity towards the kinase (for details see cit.8).

Both receptors exhibit a high affinity of around 10^{-8}M. To compare the biological activities of the analogs, in both systems, the
threshhold concentrations for chemotaxis[7] and the inhibition constants are rated in arbitrary values. Thus cAMP and derivatives
with similar biological activities receive a value of 6. When the
derivative is 10 times less active than cAMP it gets the value 5,

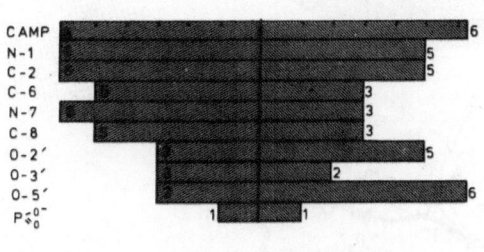

PROTEIN KINASE   DICTYOSTELIUM
DISCOIDEUM

Fig. 5. Schematic representation of the biological acrivity - rated
        in arbirary values - of the selected cAMP analogs shown in
        Fig. 3.

when 100 times less active the value 4. Fig. 5 schematically shows
the results. It clearly demonstrates that the two receptors have
a different sensitivity towards a distinct modification and thus
their specific molecular interactions towards cAMP are obviously different.

     In order to determine the chemical interactions that exist
between distinct atoms or atom groups of cAMP and distinct amino acid
side chains or the backbone of the receptor, the following assumption
has been made: If the biological activity is reduced at least 10^3-
fold, which means a drop in relative activity from 6 to 3 or less by
substitution of a distinct atom or atom group, this part of the molecule is considered to bind directly to the protein receptor and is
called an essential one. Drops in activity from 6 to 5 are considered
to be marginal effects depending on general stereochemical or electronic features, which have been changed by the modification of the
substrate.

     In Figs. 6 and 7 models for the two receptor binding sites are
shown, which explain the observed differences and similarities in
the biological activity of the analogs.

     While the regulatory subunit of protein kinase does not form any
hydrogen bond with the adenine moiety, the chemoreceptor of Dictyostelium binds the base by two hydrogen bonds towards NH-6 and N-7.
Whereas the 5'oxygen and the 2'hydroxyl group are involved in essential hydrogen bonding to the kinase. These atoms do not interact
essentially with the chemoreceptor.

     Only an ion pair towards the exocyclic oxygens on phosphorus
and a hydrogen bond towards O-3' correspond in both receptors.

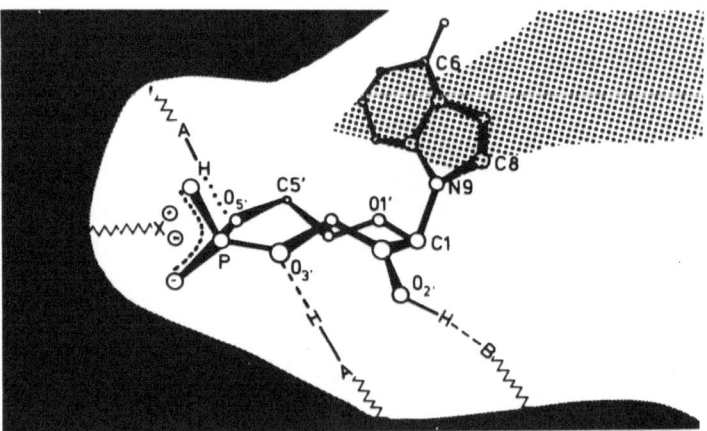

Fig. 6.   Model for cAMP binding to the regulatory subunit of Protein
          Kinase type 1

     The results obtained with the stereochemical testkit allowed to
determine the conformation of the base in the binding site; thus cAMP
binds in the anti conformation to Dictyostelium receptor and in the
syn conformation towards the kinase.

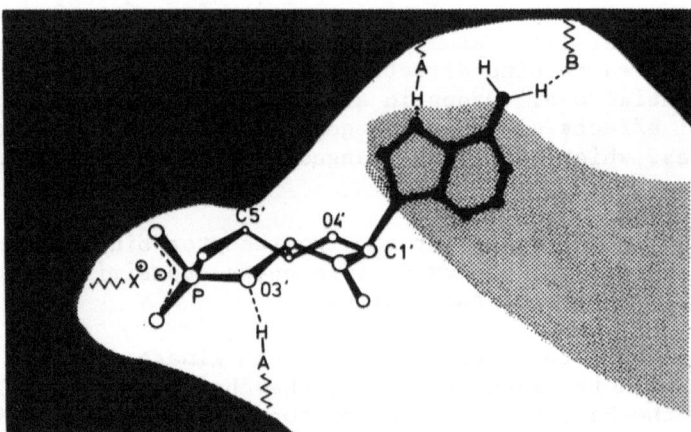

Fig. 7.   Model of cAMP binding to the chemotactic receptor of Dictyo-
          stelium discoideum.   A: part of an amino acid functioning
          as a hydrogen bond donor; B: part of an amino acid func-
          tioning as a hydrogen bond acceptor; X: positively charged
          amino acid side chains.

For the R-subunit of protein kinase Fig. 8 shows a refined picture of the binding obtained with additional testkits. The adenine moiety is bound by pure hydrophobic forces resulting in entropic effects in a cleft open to all sides, thus even very bulky substituents do not disturb the binding by steric hindrance. The ion pair is formed stereospecifically towards the equatorial situated oxygen atom.

The validity that the analogs used in our testkits described the topography of binding to both receptors was confirmed by the fact that all other analogs tested in both systems fitted the model[7,8]. This shows that a limited set of analogs can elucidate the specific molecular interactions between cAMP and its different protein receptors.

CONCLUSIONS

Our investigations demonstrate, that the overall binding effect observed between a given ligand and different proteins - although of the same order of magnitude - may nevertheless be caused by rather different binding increments attributable to individual atoms or atom groups.

This observation would seem to open the possibility of influencing discrete effector-enzyme interactions to ensure regulation of enzymes in a specific way.

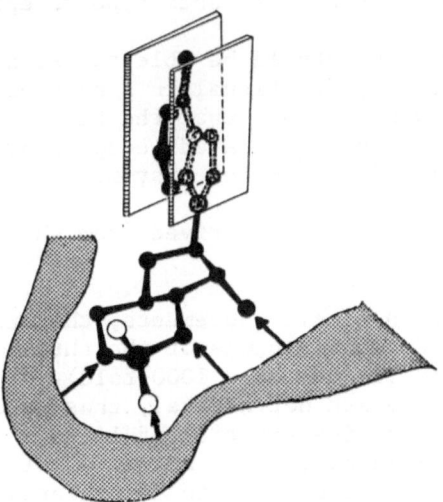

Fig. 8.   Schematic model for the arrangement of the base moiety in a hydrophobic pocket of protein kinase type 1.

Combining the knowledge of protein-ligand interactions as revealed by X-ray analysis and the knowledge of the chemical features of the ligand and its synthetic analogs will lead to a general approach in mapping binding sites of small effectors in proteins. In the case of "cAMP-receptors" we were able to demonstrate that a set of ten selected derivatives is sufficient to elucidate the molecular interaction with the protein. This would greatly simplify further attempts to characterize other binding sites. Synthesis should follow the general rule, that only the atom groups which could interact with the protein should be modified. This would minimize drastically the number of analogs which have to be synthesized to find e.g., an effective pharmaceutical drug.

Besides this possibility - namely the design of molecules with predictable biological or perhaps even pharmacological properties - such a testkit is useful for the following purposes:

- Comparison of binding sites of the same functional protein in different cell types of the same animal (in our system: protein kinase from bovine brain, bovine heart muscle, bovine kidney, etc.)
- Comparison of binding sites of the same functional protein in the same organs of different animals (in our system: protein kinase from brain of bovine, rat, rabbit, etc.)
- Comparison of binding sites with different evolutionary status (in our system: bacteria, amoebae, eukariotes)

These systematic comparisons may lead to recognition of general types of chemical interactions between substrates and proteins; it may lead to families of enzymes based on these types of interaction (in our system e.g., "syn type receptor" or "anti type receptor").

With this knowledge one should be able to use analogs of natural ligand which are essentially functional in a complex biological process. It should be proven however, that the biological activity of the different derivatives in an isolated receptor protein system is identical to the activity in a complex system.

Also other information can be derived from these systematic studies:

- Since binding depends on single increments, the exclusion of one element results in a dramatic change of the threshold concentration of biological activity (in our case 1000-fold). Those dramatic changes in concentration are not natural, thus nature can make use of this effect during the inactivation of ligands (e.g., deamination of adenosine yielding inosine means in molecular terms a change of a H-bond donating group into an accepting one: $-NH_2 \rightarrow O$. This results in inactivation of adenosine).
- The concentration dependency of binding can be used in regulating biological processes. A concentration just below the threshold

would not start the process; this would take place after an increase of the ligand concentration, and would be reduced again by
the above mentioned deactivating process.
The activity of a protein can be increased or decreased just by
adding or taking off one binding increment. This is either possible by a conformational change induced by an activator or inhibitor molecule or by the blockade of a certain interaction by phosphorylating or methylating processes.

This work can only be done by a very close cooperation of
chemists and biologists with continuous exchange of results.

Such a cooperation existed during the last years with
Drs. Konijn, Mato and van Haastert, Leiden, in the slime mold field,
with Dr. Wagner and Dr. Hoppe, Braunschweig-Stöckheim, and Dr. Bär,
Edmonton in the protein kinase system and with the chemists
Drs. Morr, Braunschweig-Stockheim and Dr. Freist, Gottingen and
especially with my coworkers Mr. T. Krebs, Dr. Schattka, Dr. Murayama
and Dr. Roesler.

REFERENCES

1. B. Jastorff, Nucleotide Analogues, in: "Eucaryotic Cell Function
    and Growth", pp 379-390, J. E. Dumont, B. L. Brown and
    N. J. Marshall, eds., Plenum Publishing Corporation, New York
    (1976).
2. J. P. Miller, Cyclic Nucleotide Analogues, in: "Cyclic 3', 5'-
    nucleotides Mechanisms of Action", pp 70-105, H. Cramer and
    J. Schultz, eds., Wiley Cie, London (1978).
3. G. E. Schulz and R. H. Schirmer, in: "Principles of Protein
    Structure", Springer Heidelberg (1979).
4. W. Kauzmann, Some Facotrs in the Interpretation of Protein
    Denaturation, Protein Chem., 14:1-63 (1959).
5. T. M. Konijn, Chemotaxis in the Cellular Slime Moulds, in Primi-
    tive Sensory and Communication Systems: the Taxes and Topism
    of Micro-organisms and Cells, pp 101-153, M. J. Carlisle, ed.,
    Academic Press, London (1975).
6. G. Gerisch and D. Malchow, Cyclic AMP Receptors and the Control
    of Cell Aggregation in Dictyostelium, Adv. Cyclic Nucleotide
    Res., Vol. 1, pp 17-31, P. Greengard and G. A. Robinson, eds.,
    Raven Press, New York (1976).
7. J. M. Mato, B. Jastorff, M. Morr and T. M. Konijn, A Model for
    Cyclic AMP-Chemoreceptor Interaction in Dictyostelium
    discoideum, Biochem. Biophys. Acta.544:309-314 (1978).
8. B. Jastorff, J. Hoppe and M. Morr, A Model for the Chemical
    Interactions of Adenosine 3',5'-Monophosphate with the R-sub-
    unit of Protein Kinase Type 1, Eur. J. Biochem. 101:555-561
    (1979).

THE CENTRAL ROLE OF CALCIUM IN STIMULUS-SECRETION COUPLING:
GENERAL CONCEPTS AND THE SPECIALIZED EXAMPLE OF THE POLYMORPHONUCLEAR
LEUKOCYTE

Elizabeth Schell-Frederick

Institut de Recherche Interdisciplinaire
Université Libre de Bruxelles

Secretion is a fundamental activity of complex organisms. Some cells are highly specialized in the synthesis and export of secretory products, e.g., the pancreatic exocrine cell, the β cell of the pancreatic islet, the mast cell, the chromaffin cell of the adrenal medulla, the acinar cell of the salivary gland. But the intracellular machinery for secretion appears to exist in all cell types except the erythrocyte.

The definition of the secretory pathway is the result in large part of the work of George E. Palade and his collaborators[1]. Attracted by the amazing structural organization of the pancreatic acinar cell (Fig. 1), they set out to define the role of subcellular organelles in the synthesis, processing and export of pancreatic proenzymes. Their analysis recognizes six successive steps:

1) synthesis on polysomes attached to the membrane of the rough endoplasmic reticulum.
2) segregation of the newly synthesized secretory proteins in the cisternal space of the rough endoplasmic reticulum.
3) intracellular transport to the Golgi complex, a step which at least in the pancreatic acinar cell requires energy, thus suggesting the existence of a "lock-gate" at this level of the secretory pathway.
4) concentration in the condensing vacuoles of the Golgi apparatus and subsequent conversion of these vacuoles into mature secretion vacuoles, termed zymogen granules in the pancreas.
5) temporary intracellular storage of secretory proteins in the secretion granules.
6) discharge by exocytosis, a process involving fusion of the granule membrane with the plasmalemma, also an energy requiring step. The minute by minute control of secretion is exercised at this level.

209

Release of prepackaged products follows a specific stimulus, the
attachement of a chemical agonist to the plasma membrane.

The basic features of secretion, thus defined for the pancreatic
acinar cell appear to hold true for all secretory cells, although in
specific cell types there are variations on this common theme. Thus,
for example, concentration and storage steps may be diminished or
absent, and discharge may occur either more diffusely along the
plasma membrane or preferentially into endocytic vacuoles. In addition, an analogue of secretion may be considered to be present even
in cells which do not produce proteins or other substances for export, i.e., the production of lysosomal enzymes. This process
utilizes the same cellular machinery and the same pathway as secretion, stopping short of discharge to the exterior in most instances.

The term "stimulus-secretion coupling", proposed in 1961 by
W.W. Douglas and R.P. Rubin[2], comprises the entire chain of events
linking the specific stimulus to the appearance of the characteristic
secretory product in the extracellular environment. As Palade and
his coworkers used the pancreatic acinar cell to formulate a model
of the secretory process per se, Douglas' group studied the system
of acetylcholine stimulation of the adrenal medulla as the first

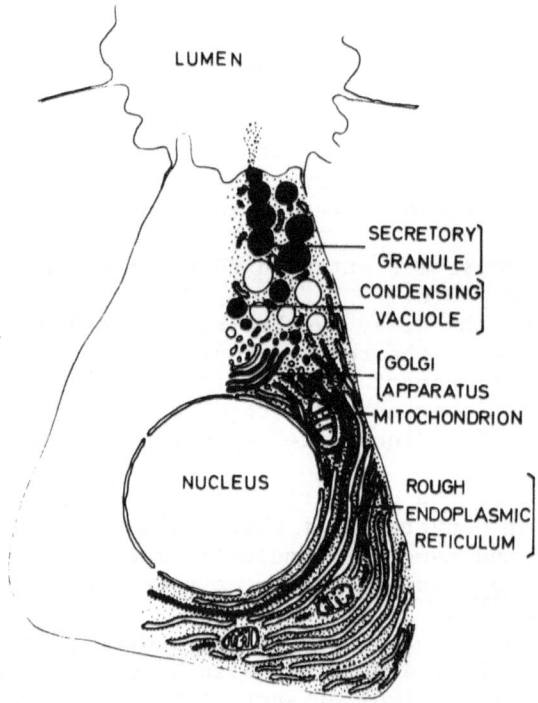

Fig. 1. Pancreatic acinar cell (adapted from reference 13).

example of stimulus-secretion coupling. It was, in fact, the demonstration of the central role of calcium in the secretory response to acetylcholine that led to the formulation of the new concept, in analogy to "excitation-contraction" coupling in muscle contraction[3]. The main findings implicating calcium in catecholamine secretion may be summarized as follows[4]:

1) There is an absolute requirement for calcium in the extracellular medium and its presence is sufficient for stimulus-evoked secretion; all other cations and anions can be omitted.
2) Acetylcholine stimulation induces ^{45}Ca influx into chromaffin cells. Manipulations which result in increased calcium entry into the cell, e.g., high potassium, or increased intracellular calcium, e.g., ouabain, also induce secretion. Under conditions where calcium can traverse the plasma membrane, it can act as the secretagogue. Barium, which has similar chemical properties to calcium and crosses the plasma membrane more easily is also a secretagogue.
3) Agents which block calcium entry, such as magnesium, suppress the secretory response. Local anesthetics inhibit secretion: they block inward calcium current in chromaffin cells without blocking sodium entry and depolarization.

The concept of stimulus-secretion coupling was intended to broaden the study of secretory systems based on the striking parallels between molecular events in the stimulated adrenal medulla and in muscle contraction.

Indeed, subsequent work in a variety of secretion systems, much of it by Douglas' own group, validated the concept of a general calcium-dependent mechanism by which widely varying stimuli evoke different cellular secretions.

In defining the central role of calcium in stimulus-secretion coupling, two basic questions must be addressed: 1) what are the mechanisms underlying the increase in intracellular calcium? 2) what are the molecular substrates for calcium's subsequent action?

The distribution of calcium in mammalian cells is complex and closely regulated, involving a number of distinct pools, binding proteins and membrane transport systems[5] (Fig. 2). The central and physiologically important pool is the free ionized calcium in the cytosol. In the resting state the concentration in this pool is between 5×10^{-8} and 10^{-7}M. As the calcium ion concentration in the extracellular medium is of the order of 10^{-3}M and the inside of the cell is negative with respect to the outside, calcium continually enters the cell along a steep electrochemical gradient. In order to maintain its low intracellular resting ionized calcium concentration, the cell must continually extrude calcium in addition to sequestering it in intracellular pools. Several specific points

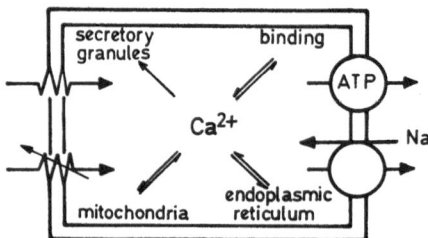

Fig 2.   Diagram adapted from reference 5 to illustrate the complexity
         of the mechanisms controlling intracellular ionized calcium
         concentration.   Indicated are: voltage dependent and inde-
         pendant Ca^{++} entry, "binding" to intracellular proteins,
         lipids, glycolipids and adenine nucleotides, sequestration
         by various intracellular organelles and extrusion by an
         energy dependent or Na-Ca exchange system.

about calcium distribution relative to secretory cells may be made[5],:
these cells are small in relation to the nerve and muscle fibres in
which so many elegant studies of calcium transport and distribution
have been performed.   This is of considerable importance because of
the rate at which intracellular changes in calcium concentration may
occur during stimulation and for the re-establishment of the basal
ionized calcium concentration.   Secretory cells contain numerous subcellular organelles which can take up and release calcium, i.e.,
mitochondria, endoplasmic reticulum and secretory granules.   In considering their respective functions in the control of intracellular
calcium concentration in the resting and stimulated state, one must
keep in mind not only uptake rate and capacity but also exchangeability.   It is net calcium exchangeability that determines fine,
rapid control of intracellular calcium.

     The possible sources of increased intracellular ionized calcium
as a consequence of stimulation are increased entry of calcium from
the exterior and/or release from intracellular stores.   The demonstration of the central role of calcium in stimulus-secretion coupling
was certainly facilitated by the fact that in the chromaffin cell
it is calcium entry which evokes secretion.   Except in such a case,
where stimulated secretion is entirely dependent on the presence of
calcium in the extracellular environment, it is often difficult or
impossible to identify the physiological source(s) of increased ionized intracellular calcium on the basis of in vitro experiments,
much less define the underlying mechanisms.   Stimulated secretion
can often be observed in the absence of extracellular calcium, demonstrating that external stimuli can mobilize internal calcium and
suggesting that such mobilization may be the primary event in the
control of secretion.   But the experimental finding is not necessarily
an accurate reflection of the in vivo situation in which calcium is

always present. An illustration of this problem is the mast cell.
Release of biogenic amines from the sentisitized mast cell, the
physiological situation, requires extracellular calcium, whereas
stimulation of the cell by the polyamine 48/80 does not[6]. Thus, it
is at least reasonable to suggest that in many systems calcium entry
is the primary event following specific stimulation. This does not
exclude a role for the intracellular binding and/or sequestering
systems. Clearly they must be involved in the termination of the
response. And in polar cells, where the stimulus acts at the baso-
lateral surface of the cell and exocytosis takes place at the luminal
surface (see Fig. 1), mechanisms must exist to control the activity
of the calcium sequestering systems so that the calcium signal is
transmitted across the cell[7]. A discussion of the role of cyclic AMP
in stimulus-secretion coupling is beyond the scope of this discussion. However, it should be noted that calcium is not always
the primary mediator of secretion; cyclic AMP plays this role in
enzyme secretion in the acinar cell of the parotid gland stimulated
by β-adrenergic agonists[8].

Little is known about the physiological mechanisms controlling
increased calcium entry into the cell. It is important to consider
the possible role of depolarization as voltage sensitive calcium
channels clearly exist in nerve axones and at the neuromuscular
junction and changes in membrane potential may often be detected
early in stimulus-secretion coupling. However, in many secretory
systems, depolarization does not appear to be an integral part of
the response. For example, in the adrenal chromaffin cell, the secretory response to acetylcholine is potentiated when sodium is removed from the extracellular medium. Under such conditions, membrane
potential falls very little. In addition, acetylcholine can stimulate secretion by previously depolarized cells[4]. In a given system,
the use of verapamil or its analogues may be useful in defining the
role of voltage-dependent calcium channels as they are specific
inhibitors of calcium entry by this route and apparently do not
affect what have come to be called "receptor-operated" channels[9].

Stimulation of secretion in exocrine and endocrine glands is
associated with increased turnover of phosphatidylinositol (PI)[10].
R.H. Michell has proposed that this PI response is important in the
generation of the calcium signal[11]. Evidence for this hypothesis
includes the finding that the PI response in general is insensitive
to calcium. In addition, M.J. Berridge and J.N. Fain have shown
that in the salivary gland of the blowfly, conditions which cause
a progressive depletion of membrane phosphatidylinositol also decrease calcium entry in response to 5-hydroxytryptamine[12].

It is tempting to extend the analogy betwen muscle and secretory
cells to include the mechanisms by which calcium may induce secretion.
As mentioned above, stimulation acts on the final step of the secretory process, i.e., exocytosis of products already contained in

secretory granules[13]. Thus, calcium may influence the arrival of
secretory granules at the external portion of the cell and/or the
fusion of the secretory granule membrane with an appropriate region
of the plasma membrane. The cortical cytoplasm of many non-muscle
cells contains actin filaments arranged in a three-dimensional network and myosin molecules[14].

There is growing evidence that these proteins regulate cell
movement and movement of at least some intracellular organelles, and
that actin and myosin contraction is regulated by calcium[15]. Calcium
appears to be necessary for fusion of secretory vesicles with the
plasma membrane[16]. In addition, calcium regulates the activity of
enzymes attacking polyphosphoinositides whose breakdown could lead
to increased membrane fusibility[17].

The polymorphonuclear leukocyte (PMN) is not a typical example
of a secretory cell. PMNs contain two types of granules, the azurophil
(or primary) granules and the specific (or secondary) granules which
differ in their time of appearance during PMN development, their content and their degranulation. During phagocytosis, PMNs recognize
bacteria and surround them with formation of a phagosome, an inside-
out vesicle of plasma membrane. Both types of granules fuse with
the phagosome and thus, granule contents are discharged selectively
into the phagocytic vesicle to kill and subsequently degrade the in-
gested bacteria. Nevertheless, granule fusion may occur with an as
yet incompletely closed phagosome leading to escape of granule contents into the medium (Fig. 3). This extracellular release of granule
contents has often been regarded as a non-physiological event.

However, although azurophil granules are certainly lysosome
"equivalents", containing hydrolases and serving intracellular diges-
tive functions, specific granules are not, and may function normally
as secretory granules. Human PMNs release specific granule contents
spontaneously when incubated in vitro in a physiologic medium. This
phenomenon is calcium dependent[18]. Spontaneous release is significantly increased when PMNs adhere to surfaces or when they migrate
across nitrocellulose filters in response to a chemotactic stimulus.
In addition, PMN isolated from sterile or non-sterile exudates have
selectively lost specific granule contents while retaining their
azurophil granule markers[19]. Thus, exocytosis of specific granules
may be a secretory response intrinsic to the normal function of PMNs
as inflammatory cells. In keeping with this, under conditions of
specific granule content release, a factor is liberated which activates the complement system generating the chemotactic factor C5a,
thus amplifying the inflammatory response[20].

Under particular experimental conditions PMNs may be transformed
entirely into secretory cells. In the presence of cytochalasin B
(CB), a fungal metabolite that reversibly inhibits cell movement and
phagocytosis, stimulated PMNs discharge a large percentage of their
granule contents into the external medium[21].

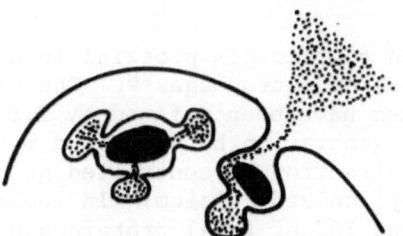

Fig. 3. Diagram showing selective discharge of granule contents into
a completely closed phagocytic vesicle and escape into the
medium via an incompletely closed phagosome.

Calcium is the mediator of stimulus-secretion coupling in the
leukocyte. Although stimulated PMN may release granular enzymes in
the absence of extracellular calcium, secretion is always enhanced
in its presence. The divalent cation ionophore A23187 acts as a
freely mobile carrier to equilibrate divalent cation concentrations
across membranes. It induces exocytosis of secretion granules in a
number of secretory cells, as expected according to the calcium hypothesis of stimulus-secretion coupling[6]. In the leukocyte, A23187
in the presence of calcium induces granule exocytosis[22] and,
interestingly, stimulates the oxidative metabolism of the cell in a
manner characteristic of phagocytosis[23]. The effect of the ionophore
is concentration dependent; at low concentrations, up to 10^{-6}M, it
selectively induces specific granule release, at higher concentrations
exocytosis from both granule types being observed[24]. These results
suggest that calcium entry is the primary event following stimulation
of the leukocyte.

The most extensive studies of calcium entry into PMN have been
carried out in cells stimulated with the synthetic chemotactic tri-
peptide formylmethionyl-leucyl-phenlyalanine (FMLP). In rabbit
leukocytes in the presence of cytochalasin B, FMLP induces ^{45}calcium
influx within 30 seconds and augments the cellular pool of exchangeable calcium. CB and FMLP each cause a small increase in calcium
entry by themselves[25].

During phagocytosis of zymosan or adherence of PMN to concanava-
lin A linked to sepharose beads, calcium appears to be displaced from
the plasma membrane in the regions of contact with the particle.
These experiments have been done by precipitating calcium ions <u>in situ</u>
using pyroantimonate[26].

Granules may be of particular importance in calcium homeostasis
in the PMN where mitochondria and the endoplasmic reticulum are

minimal in the mature cell. Indirect evidence suggests that calcium
may be actively taken up by granules[27]. Its degree of exchangeability
is not known.

Much of the work on contractile proteins in non-muscle cells
has been performed in PMN or macrophages[28]. The elements of movement
in the cortical cytoplasm have been defined by T.P. Stossel and
colleagues as, 1) force generation by actin and myosin; 2) orientation
of the force to confer direction by controlled gel-sol transform-
ations; 3) regulation by ionized calcium. In leukocytes, actin
compromises between 6 and 10% of total protein and a variable proportion is present as fibres oriented randomly in the cortex of the
cell. Numerous actin filaments can be seen to interact with the
plasma membrane by electron microscopy but the nature of the connec-
tion is unknown. Myosin represents 1% of total protein. In the
presence of $MgCl_2$ and ATP, leukocyte actin and myosin undergo a
"superprecipitation" reaction analogous to muscle contraction. In
contrast to muscle, orientation of contraction is not conferred by
the parallel arrays of actin and myosin filaments. In order to provide net movement one must postulate local alteration in the force
of contraction or on the mass on which it acts. Stossel has suggested
that such orientation results from changes in actin filament mass
controlled by crosslinking and gel-sol transformation. In leukocytes
actin is cross-linked by actin-binding protein, thus providing a
mechanism for controlled lattice formation. Gelsolin is a calcium
binding protein which reversibly dissolves gels of cross-linked actin
when the calcium concentration exceeds $2 \times 10^{-7}M$, apparently by
breaking actin filaments into shorter pieces. Localized changes in
calcium concentration thus permit contraction toward more highly
cross linked domains with movement of the cytoplasm and the membrane
to which actin filaments are attached toward these regions.

There is some indirect evidence that these elements are involved
in the secretory process. As mentioned above, cytochalasin B enhances
leukocyte exocytosis in the presence of secretagogues. CB has been
shown to dissolve actin-actin binding protein gels[29]. In the cell
this would be expected to lead to depletion of the cortex thus per-
mitting secretory granules to approach the plasma membrane.

The stimulus-secretion coupling concept has indeed been a fruit-
ful one. The idea that nature uses similar molecular mechanisms to
effect two seemingly very disparate responses, contraction and secretion, has been amply confirmed. A complete understanding of these
mechanisms and the interrelationships between calcium and other
mediators of the secretory process awaits further work.

BIBLIOGRAPHY

1. G. Palade, Intracellular aspects of the process of protein synthe-
     sis, Science 189:347 (1975).
2. W.W. Douglas and R.P. Rubin, The role of calcium in the secretory
     response of the adrenal medulla to acetylcholine, J. Physiol.
     159:40 (1961).
3. A. Sandow, Excitation-contraction coupling in muscular response,
     Yale J. Biol. Med. 25:176 (1952).
4. W.W. Douglas, Stimulus-secretion coupling : the concept and clues
     from chromaffin and other cells, Br. J. Pharmac. 34:451 (1968).
5. E.K. Matthews, Calcium translocation and control mechanisms for
     endocrine secretion, in : Secretory Mechanisms, Society for
     Experimental Biology Symposium XXXIII, C.R. Hopkins and
     C.J. Duncan, eds., Cambridge University Press, p. 225 (1979).
6. W.W. Douglas, The role of calcium in stimulus-secretion coupling,
     in : Stimulus-Secretion Coupling in the Gastrointestinal Tract,
     R.M. Case and H. Goebell, eds., MTP Press Ltd, Lancaster,
     England, p. 17 (1976).
7. H. Rasmussen, G. Clayberger and M.C. Gustin, The messenger function
     of calcium in cell activation, in : Secretory Mechanisms,
     Society for Experimental Biology Symposium XXXIII, C.R. Hopkins
     and C.J. Duncan, eds, Cambridge University Press, p. 161 (1979).
8. M. Schramm and Z. Selinger, Neurotransmitters, receptors, second
     messengers and responses in parotid gland and pancreas,
     in : Stimulus-Secretion Coupling in the Gastrointestinal Tract,
     R.M. Case and H. Goebell, eds, MTP Press Ltd, Lancaster,
     England, p. 49 (1976).
9. T.B. Bolton, Mechanisms of action of transmitters and other
     substances on smooth muscle, Phys. Revs. 59:606 (1979).
10. R.H. Michell, Inositol phospholipids and cell surface receptor
     function, Biochim. Biophys. Acta 415:81 (1975).
11. R.H. Michell, Inositol phospholipids in membrane function,
     Trends in Biochemical Science 4:128 (1979).
12. M.J. Berridge and J.N. Fain, Inhibition of phosphatidylinositol
     synthesis and the inactivation of calcium entry after prolonged
     exposure of the blowfly salivary gland to 5-hydroxytryptamine,
     Biochem. J. 178:59 (1979).
13. B. Satir, The final steps in secretion, Sci. Am. 233:28 (October
     1975).
14. E. Lazarides and J.P. Revel, The molecular basis of cell movement,
     Sci. Am. 240:88 (May 1979).
15. O.I. Stendhal and T.P. Stossel, Actin-binding protein amplifies
     actomyosin contraction and gelsolin confers calcium control
     on the direction of contraction, Biochem. Biophys. Res. Comm.
     92:675 (1980).
16. M. Gratzl and G. Dahl, Ca^{2+}-induced fusion of Golgi-derived
     Secretory vesicles isolated from rat liver, FEBS Letters
     62:142 (1976).

17. D. Allan and R.H. Michell, The possible role of lipids in control
    of membrane fusion during secretion, in : Secretory Mechanisms,
    Society for Experimental Biology Symposium XXXIII, C.R. Hopkins
    and C.J. Duncan, eds., Cambridge University Press, p. 323 (1979).
18. I.M. Goldstein, J.K. Horn, H.B. Kaplan and G. Weissmann,
    Calcium-induced lysozyme secretion from human polymorphonuclear
    leukocytes, Biochem. Biophys. Res. Comm. 60:807 (1974).
19. D.G. Wright and J.I. Gallin, Secretory responses of human neutro-
    phils : Exocytosis of specific (secondary) granules by human
    neutrophils during adherence in vitro and during exudation in
    vivo, J. Immunol. 123:285 (1979).
20. D.G. Wright and J.I. Gallin, A functional differentiation of
    human neutrophil granules : Generation of C5a by a specific
    (secondary) granule product and inactivation of C5a by azuro-
    phil (primary) granule products, J. Immunol. 119:1068 (1977).
21. R.B. Zurier, S. Hoffstein and G. Weissman, Cytochalasin B : effect
    on lysosomal enzyme release from human leucocytes, Proc. Natl.
    Acad. Sci. 70:844 (1973).
22. G. Zabucchi, R. Soranzo, F. Rossi and D. Romeo, Exocytosis in
    human polymorphonuclear leukocytes induced by A23187 and
    calcium, FEBS Letters, 54:44 (1975).
23. E. Schell-Frederick, Stimulation of the oxidative metabolism
    of polymorphonuclear leucocytes by the calcium ionophore
    A23187, FEBS Letters 48:37 (1974).
24. D.G. Wright, D.A. Bralove and H.I. Gallin, The differential
    mobilization of human neutrophil granules : Effects of phorbol
    myristate acetate and ionophore A23187, Am. J. Pathol. 87:273
    (1977).
25. P.H. Naccache, H.J. Showell, E.L. Becker and R.I. Sha'afi,
    Changes in ionic movements across rabbit pholymorphonuclear
    leukocyte membranes during lysosomal enzyme release : Possible
    ionic basis for lysosomal enzyme release, J. Cell Biol.
    75:635 (1977).
26. S. Hoffstein, Ultrastructural demonstration of calcium loss
    from local regions of the plasma membrane of surface-stimulated
    human granulocytes, J. Immunol. 123:1395 (1979).
27. A. Barthelemy, R. Paridaens and E. Schell-Frederick, Phagocytosis-
    Induced [45]calcium efflux in polymorphonuclear leucocytes,
    FEBS Letters 82:283 (1977).
28. T.P. Stossel, J.H. Hartwig, H.L. Yin and O. Stendahl,
    The motor of amoeboid leucocytes, Biochem. Soc. Symp. 45:51
    (1980).
29. J.H. Hartwig and T.P. Stossel, Interactions of actin, myosin,
    and an actin-binding protein of rabbit pulmonary macrophages
    III effects of cytochalasin B, J. Cell Biol. 71:295 (1976).

CALMODULIN

André Vandermeers, Marie-Claire Vandermeers-Piret and
Jean Christophe

Department of Biochemistry and Nutrition, Medical School
Université Libre de Bruxelles
Bld. de Waterloo 115, B-1000 Brussels, Belgium

Calmodulin is a small thermostable protein which has been involved in the regulation of a number of cellular activities such as
cyclic nucleotide and glycogen metabolism, smooth muscle contraction,
intracellular motility, and calcium transport. Calmodulin is apparently present in all nucleated cells. High concentrations from
1-50µM have been reported from a variety of mammalian tissues.

Calmodulin acts as an intracellular intermediary for Ca^{2+} ions
by modulating the activities of target proteins in a Ca^{2+}-dependent
manner.

DISCOVERY

In 1970, Cheung[1] discovered the presence of a heat-stable, dissociable activator in a crude cyclic nucleotide phosphodiesterase
preparation from bovine brain, that could be removed during purification by DEAE-cellulose chromatography.

At the same time, Kakiuchi and Yamazaki[2] documented a calcium-
dependent activation of brain cyclic nucleotide phosphodiesterase:
the sensitivity of the crude enzyme to Ca^{2+} was increased when a
heated brain supernatant was added to the assay mixture. The latter
authors concluded that the Ca^{2+}-dependent activity reflected the
presence of a heat-stable factor in the brain extract. This factor
turned out to be identical to the one discovered by Cheung.

The activation was subsequently purified from brain[3] and from
heart[4] and reported to be a Ca^{2+} binding protein[5] exhibiting many
similarities with skeletal muscle troponin C.

At that time, the biochemical interest so far focused upon troponin C was extended to the phosphodiesterase activator, especially as the latter was progressively recognized as a multifunctional Ca^{2+}-dependent regulator, abbreviated as CDR before being designated as calmodulin. Troponin C now appears as a specialized form of calmodulin involved in cardiac and skeletal muscle contraction.

PHYSICOCHEMICAL PROPERTIES

Calmodulin has been purified to apparent homogeneity from a variety of tissues including bovine brain[6], porcine brain[5], bovine heart[4], bovine adrenal medulla[7], bovine pancreas[8] and rat testis[9]. The amino acid composition of calmodulin derived from these various sources is nearly identical[8,10]. A striking feature is the large proportion of aspartic and glutamic residues which together account for 25% of the total composition, an observation that is in accord with a low isoelectric point around pH 4.0. Other notable features include the presence of a trimethyllysine residue in position 115, the absence of cysteine, the absence of tryptophan resulting in a low extinction coefficient ($E_{280}^{1\%}$ of 2.0), and a high phenylalanine to tyrosine ratio resulting in an unusual absorption pattern of the UV spectrum with five peaks at 251, 258, 264, 268 and 275nm that are characteristic of phenylalanine. Molecular weight determinations for calmodulin range from 15000–19000 daltons but the most reliable value, derived from the complete sequence of calmodulin from bovine brain[11] and rat testis[12], is 16700.

STRUCTURE

Comparison of the amino acid sequence of calmodulin from bovine brain[11] or bovine uterus[13] and from rat testis[12] indicates that the primary structure is highly conserved.

Calmodulin possesses 148 residues (Fig. 1) and displays a large homology with troponin C. Both proteins can be subdivided into four domains having a homologous sequence. Each domain contains about 30 amino acids and consists of a first α helix, a calcium binding loop, and a second α helix. When compared to bovine brain calmodulin, skeletal muscle troponin C is only seven amino acids longer at the amino terminal region and, in addition, it has three extra amino acids between the corresponding residues 78 and 79 of calmodulin, and a single extra residue at the C-terminus.

The internal homology in calmodulin is greatest when the first domain (residues 8–40) is aligned with the third domain (residues 81–113), and the second domain (residues 44–76) aligned with the fourth (residues 117–148). This level of internal homology is even greater than that observed in muscle troponin C. Conceivably, the calmodulin

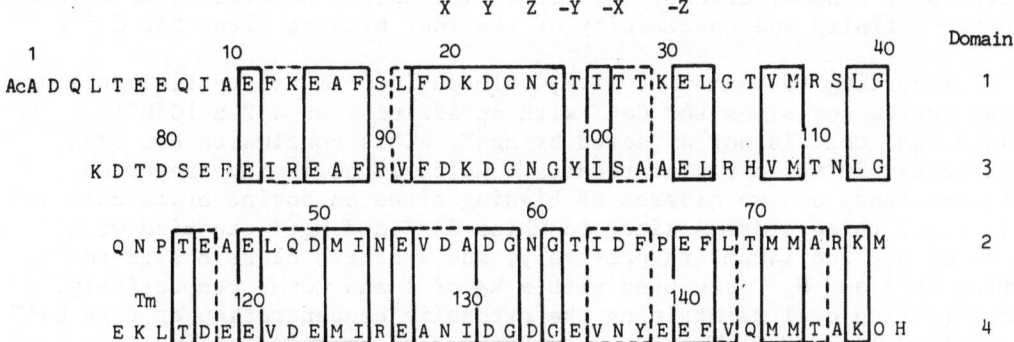

Fig. 1.   Amino acid sequence of bovine brain calmodulin according to
          Watterson et al.[11].   Solid line boxes indicate identical
          positions and dashed line boxes indicate conservative re-
          placements when domain 1 is aligned with domain 3 and domain
          2 with domain 4.   X, Y, Z, -Y, -X and -Z indicate putative
          calcium-binding residues (Ala A, Arg R, Asn N, Asp D, Gln Q,
          Glu E, Gly G, His H, Ile I, Leu L, Lys K, Met M, Phe F,
          Pro P, Ser S, Thr T, Tyr Y, Val V, Trimethyllysine TmK).

molecule plays a fundamental role in cellular function and cannot
suffer a great change in its amino acid sequence.

     The assignment of Ca^{2+} binding sites in calmodulin is based, as
for troponin C, upon the homology presented by each domain with corresponding regions of carp parvalbumin, in which the Ca^{2+} binding sites
have been located by X-ray crystallography[14].   According to
Kretsinger[15], each domain is characterized by a typical conformation
referred to as the "EF hand".   In this EF hand conformation, first
observed in the crystal structure of carp muscle parvalbumin, the
first helix, the calcium binding loop, and the second helix are related to one another like the forefinger, middle finger and thumb of
a hand, respectively.   Forefinger and thumb are extended at approximately a right angle: they point, respectively, toward the NH_2-
terminal direction of the first α helix (helix E) and the COOH-
terminal direction of the second helix (helix F).   The clenched
middle finger traces the course of the calcium binding loop (EF loop).
The calcium atom is coordinated by 6 oxygen atoms in an octahedral
arrangement (Fig. 1).

Ca^{2+} BINDING PROPERTIES

     The calcium binding properties of calmodulin have been studied
in several laboratories by equilibrium dialysis.   The results are
consistent with the prediction of four Ca^{2+} binding sites based on

structural considerations. Discrepancies exist, however, with regard to the affinity and specificity of the four binding sites for Ca^{2+}.

According to Dedman et al.[9], calmodulin from rat testis contains four equivalent sites for Ca^{2+} with an affinity of $4.2 \times 10^5 M^{-1}$ (K_d 2.5µM) that is not affected by Mg^{2+}. This conclusion has been challenged by three recent reports. Wolff et al.[16] have reported the existence of two classes of binding sites on bovine brain calmodulin i.e., three class A sites to which Ca^{2+} and Mg^{2+} can bind with a K_d of 0.2 and 140µM, respectively, and a single class B site to which Ca^{2+} and Mg^{2+} can bind with a K_d of 1 and 20µM, respectively. In a resting cell maintaining the cytosolic concentration of free Ca^{2+} between 0.1 and 0.01µM, the four cation binding sites of inactive calmodulin would be occupied, as a consequence, with Mg^{2+} since the concentration of the latter is about 1mM in vivo. Were the free Ca^{2+} concentration increased to the 1-10µM range, Ca^{2+} would then bind to class A sites with an app K_d of 3µM, leading to a $Ca_{3A} \cdot Mg_{1B}$-calmodulin active complex.

Klee[17] has also described two classes of Ca^{2+} binding sites but her evidence points towards each class binding 2 mol of Ca^{2+} (K_d of 4 and 12µM respectively). Seamon[18] has recently supported this inte̅ ̲cation by showing, by nuclear magnetic resonance evidence, a first conformational transition upon binding of two calcium ions at two high affinity sites and a second conformational transition upon calcium binding to the third and fourth lower affinity binding sites. Magnesium binding to calmodulin does not induce as large a conformational change as does calcium binding.

Whatever the precise specificity and affinity of each divalent cation binding site, the conformational transition associated with Ca^{2+} binding results in a more compact and more helical structure of calmodulin, that resists denaturation by boiling, 8M urea or 1% SDS[10]. This conformational change is required in order to regulate the calmodulin-dependent processes.

CALMODULIN-REGULATED PROCESSES

Cyclic Nucleotide Phosphodiesterase

The cyclic nucleotide phosphodiesterase from most tissues can be resolved into several forms one of them being Ca^{2+}-calmodulin-dependent. As an example, three peaks of phosphodiesterase activity are clearly separated when a rat pancreatic supernatant is submitted to gel filtration on Sephadex G 200 in the presence of 0.1mM EGTA (reference 8 and Fig. 2). Only phosphodiesterase P_3 is activated by calmodulin. The conditions required for observing a 3-fold activation include the presence of calcium, calmodulin, and 0.4µM cyclic

GMP as a substrate (Fig. 2). The apparent K_m is only slightly reduced (from 1.9 to 1.6µM) while V_{max} increases from 1.1 to 3.0 nmol cyclic GMP min^{-1} in agreement with other reports[10].

Half-maximal activation of the enzyme requires a 5nM calmodulin concentration in the presence of a large excess of Ca^{2+} ions.

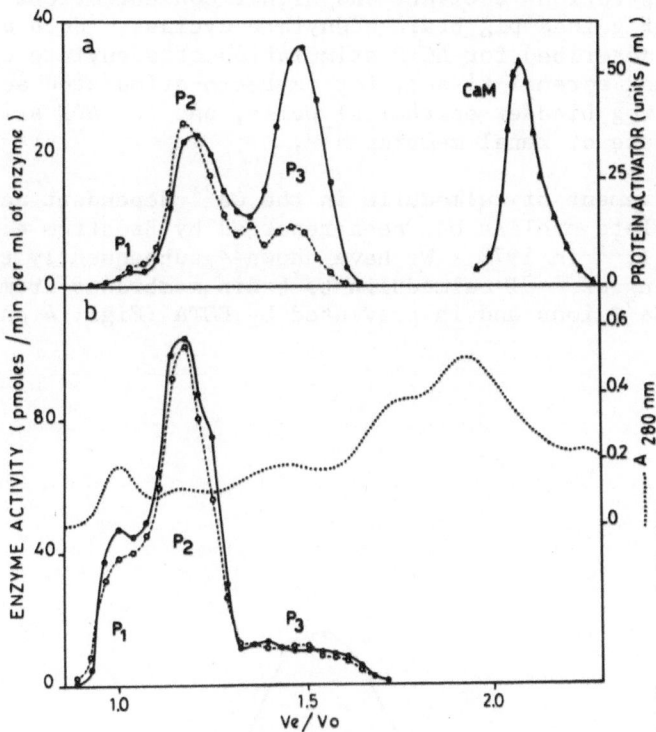

Fig. 2.   Partial purification of three cyclic nucleotide phosphodiesterases and of calmodulin from rat pancreas by gel filtration on Sephadex G-200. Rat pancreases were homogenised in 3 volumes of 10mM Tris-HCl buffer (pH 7.5) containing 1mM $MgCl_2$, 2mM 2-mercaptoethanol and 0.2mM EGTA. The homogenate was centrifuged at 105000 x g for 1h. 3ml of supernatant were applied to the column (150 x 2cm) equilibrated and eluted by 20mM Tris-HCl buffer (pH 7.5) containing 0.1M NaCl, 1mM $MgCl_2$ 0.1mM EGTA and 0.1mM dithiothreitol. Phosphodiesterase activity was assayed using either cyclic GMP (a) or cyclic AMP (b) in the presence of 0.2mM EGTA (o——o) or of 30µM Ca^{2+} and 30 units (i.e., a large excess) of purified bovine pancreas calmodulin (●——●). Rat pancreas calmodulin (∆—∆) was monitored in aliquots of fractions that had been boiled for 1 minute. Absorbance was recorded at 280nm (············)[8].

Adenylate Cyclase

      Adenylate cyclase activity from a wide variety of tissues is
inhibited by low concentrations of Ca^{2+}. However, Bradham et al.[19]
discovered in 1970 that a crude adenylate cyclase from brain exhibits
a biphasic response to Ca^{2+}. For example, Fig. 3 illustrates that
low Ca^{2+} concentrations activate and higher concentrations inhibit
the activity of guinea pig brain adenylate cyclase. Such a situation
has also been described for ACTH stimulation of adenylate cyclase
from adipose and adrenal tissue, for oxytocin-stimulated adenylate
cyclase from frog bladder epithelial cells, and for ADH stimulated
adenylate cyclase of renal membranes[10].

      The involvement of calmodulin in the Ca^{2+}-dependent activation
of brain adenylate cyclase has been reported by Brostrom et al.[20] and
by Cheung et al.[21] in 1975. We have shown[22] subsequently that the
specific binding of ^{125}I-calmodulin to brain membranes from guinea
pig requires Ca^{2+} ions and is prevented by EGTA (Figs. 4 and 5).

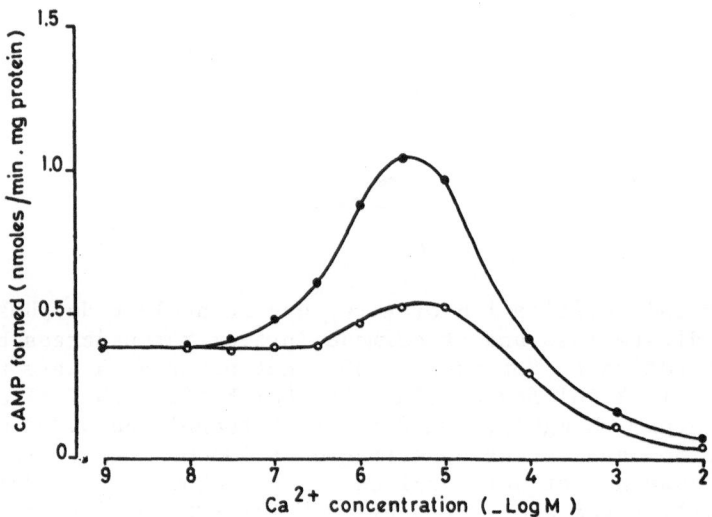

Fig. 3. The effect of free Ca^{2+} concentration on brain adenylate
         cyclase activity in the presence of 5mM $MgCl_2$. Brain mem-
         branes from guinea pig (15µg protein/assay) were incubated
         for 7 minutes at $37^{\circ}C$ in a medium (final volume 60µl) con-
         taining 30mM Tris-HCl (pH 7.4), 0.5mM 3-isobutyl- 1-methyl-
         xanthine, 1mM cyclic AMP, 0.5mM [α-^{32}P] ATP, 10mM phospho-
         (enol) pyruvate and 30µg/ml pyruvate kinase, without (o)or
         with 10^{-4}M exogenous calmodulin ($\bullet$)[22].

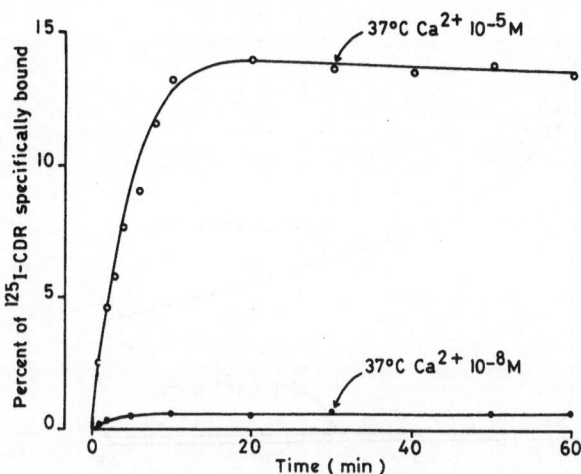

Fig. 4. Time course of specific binding of ^{125}I-calmodulin to brain
         membranes from guinea pig as a function of Ca^{2+} concen-
         tration. The membranes (60µg protein/assay) were incubated
         at 37°C in 120µl of a standard medium consisting of 1.5 x
         10^{-9}M ^{125}I-calmodulin, 20mM Tris-HCl buffer (pH 7.4), 5mM
         $MgCl_2$ and 0.2% bovine serum albumin, with 10^{-5}M free Ca^{2+}
         (o) or 10^{-8}M free Ca^{2+} (●)[22].

Myosin Light Chain Kinase

     Myosin light chain kinase is present in skeletal[23] and
smooth[24,25] muscle as well as in non-muscle cells[26]. The enzyme is
activated by calcium and calmodulin and catalyzes the phosphorylation
of the P light chains of myosin in both skeletal and smooth muscle.

     In the skeletal and cardiac muscle, the contraction cycle is
independent of phosphorylation and is controlled allosterically by
Ca^{2+} and troponin C. Thus, in this type of muscle, phosphorylation
of the P light chains does not switch on the system but may provide
a mechanism for modulating the contraction cycle.

     In smooth muscle and non-muscle cells, troponin C is probably
absent so that the mechanism of Ca^{2+} regulation via troponin C would
not apply. There is evidence that Ca^{2+} regulation is then fully
mediated by calmodulin and by the myosin light chain kinase. Indeed,
it is generally accepted[27] (but see ref. 28) that the phosphorylation
of myosin P light chain triggers smooth muscle contraction. The need
for such a covalent modification might explain why tension develops
slowly in smooth muscle. A similar mechanism would apply to non-
muscle contractile systems. The myosin light chain kinase present

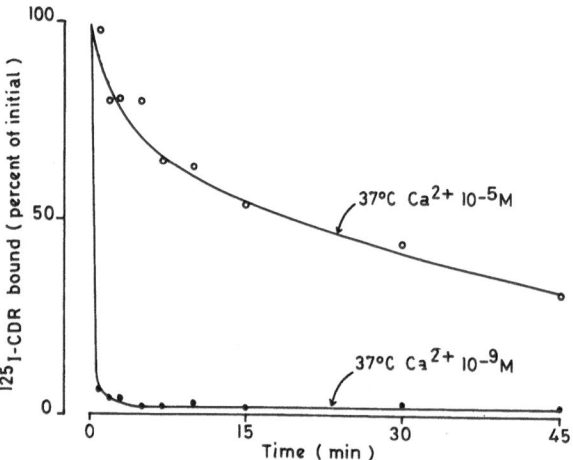

Fig. 5. Time course of dissociation of specifically bound ^{125}I-
calmodulin from brain membranes. Prelabeled membranes were
sedimented by centrifugation at $2^{\circ}C$ for 5 minutes at 15000 x
g, and washed with the original volume of standard buffer
medium containing $10^{-5}M$ calcium. After centrifugation, the
washed pellet was rehomogenized in the same buffer and 50μl
aliquots were distributed in tubes containing 5ml of initial
buffer at $37^{\circ}C$ (o), or of a buffer where the free Ca^{2+} concentration was reduced from 10^{-5} to $10^{-9}M$ (•)[22].

in blood platelets and in baby hamster kidney cells in culture is
also a calmodulin-dependent enzyme[29]. In addition, calmodulin has
been shown to be associated with actin filaments in a variety of
animal cells by an immunofluorescence procedure using a calmodulin
specific antibody[30]. Taken together, these observations suggest that
calmodulin might also activate non-muscle actomyosin ATPase through
a Ca^{2+}-calmodulin dependent myosin light chain kinase.

Glycogen Phosphorylase Kinase and Glycogen Synthase Kinase

        Glycogen phosphorylase kinase from skeletal muscle contains four
types of subunit and possesses the molecular structure $(\alpha\beta\gamma\delta)4$.
Calmodulin consitutes the δ subunit and is tightly bound to the
complex[31]. Phosphorylase kinase, when activated by calcium, in turn
converts inactive glycogen phosphorylase b into active phosphorylase
a. Although calmodulin is present in stoechiometric proportions
with the α-, β-, and γ-subunits, the activity of phosphorylase kinase
is increased up to 7-fold by the addition of further calmodulin to
the assay. This additional activation is caused by the interaction
of a second free molecule of calmodulin with each αβγδ unit[31].

Calmodulin also activates glycogen synthase kinase[32]. Phosphorylation by glycogen synthase kinase decreases the activity of glycogen synthase. Thus, calmodulin synchronizes the rates of glycogenolysis and glycogen synthesis so that glycolysis is increased during muscle contraction.

Other Calmodulin-dependent Kinase

Increasing evidence suggests that calcium-dependent protein kinases are present in many tissues and at least three cell compartments. Indeed, Schulman and Greengard[33] have reported a Ca^{2+}-calmodulin-dependent autophosphorylation in membranes derived from brain, adrenal gland, heart, lung, spleen and skeletal muscle. De Lorenzo et al.[34] have demonstrated that calmodulin permits calcium-dependent phosphorylation of proteins in synaptic vesicles associated with norepinephrine release.

Calmodulin also regulates the Ca^{2+}-dependent phosphorylation of cytosolic proteins from rat brain[35], of microsomal proteins from rat adipocytes[36], and of a specific protein from cardiac sarcoplasmic reticulum (vide infra).

In addition, calmodulin regulates NAD kinase activity in plants[37] and recently, calmodulin was found to stimulate cyclic GMP-dependent protein kinase from pig lung but this stimulation might not require Ca^{2+} ions[38].

Ca^{2+}-dependent ATPases

The Ca^{2+} pump of plasma membranes of red blood cells associated with the activity of a $(Ca^{2+} + Mg^{2+})$-ATPase is now recognized as the means whereby those cells maintain their low cytosolic Ca^{2+} concentration. Calmodulin activates dose-dependently, in the 1-10nM range, the $(Ca^{2+} + Mg^{2+})$-ATPase activity of red blood cell membranes[39] as well as Ca^{2+} uptake into inside-out red blood cell membrane vesicles[40].

Experimental data are still lacking to conclude that calmodulin stimulates the Ca^{2+} pump of plasma membranes in all cells. If so, calmodulin might then mediate various intracellular Ca^{2+} effects as well as intervene in the conclusion of the same effects.

Calmodulin also stimulates Ca^{2+} transport in cardiac sarcoplasmic reticulum[41], and might therefore facilitate the removal of calcium from the cytoplasm of muscle cells by stimulating the Ca^{2+} pump of the sarcoplasmic reticulum. A membrane bound Ca^{2+}-calmodulin dependent protein kinase catalyzes the phosphorylation of phospholamban, a small protein of the sarcoplasmic reticulum[42], and the

resulting conformational change might conceivably modify the hydrophobic microenvironment of the ATPase molecule, thereby enhancing
calcium uptake activity[42].

Microtubule Assembly/Disassembly

Calmodulin together with calcium is able to abolish microtubule
assembly _in vitro_ and also to promote their complete disassembly[30].
In vivo, in the intact cell, calmodulin is associated with the
mitotic apparatus and the pattern of calmodulin-specific immunofluorescence suggests its involvement in a similar assembly/disassembly
of microtubules during the movement of chromosomes from the metaphase
plate to the spindle poles[30].

STRUCTURE-FUNCTION RELATIONSHIP

One approach to relate the structure of calmodulin to its
function(s) is to explore whether isolated regions of the primary
sequence retain some of the properties of intact protein. The following conclusions can be drawn from the results so far available.

1. The sequence 78-106 is critical for the interaction of calmodulin
   with troponin I. This sequence is indeed common to peptides
   1-106 and 78-148 obtained by controlled tryptic cleavage 43, and
   to peptide 77-124 obtained by cleavage with CNBr 27, three
   peptides that are all capable to interact with troponin I 27,43.
   This sequence 78-106 includes the first helix and the calcium loop
   from the third domain.

2. The ability of calmodulin to stimulate cyclic nucleotide phosphodiesterase depends essentially on the integrity of the molecule. Indeed, among the various peptides obtained by controlled
   tryptic digestion only peptide 1-106 is active but 200-fold less
   than the native protein 43.

   The sequence 91-106, that corresponds to the third Ca^{2+} loop, is
   critical for the residual activity of peptide 1-106 since peptide
   1-90 is no longer able to stimulate the enzyme 43.

3. The sequence 71-77 (-Met-Met-Ala-Arg-Lys-Met-Lys-) which is located between the second and third domains is shifted toward the
   outside of the molecule upon Ca^{2+} binding. Indeed the bond of
   Lys 77 is not easily accessible to tryptic digestion in the presence of EGTA but becomes preferentially splitted when Ca^{2+} is
   present in the digestion mixture 43.

   Moreover, treatment of calmodulin with N-chlorosuccinimide results
   in selective oxidation of methionine residues at position 71, 72

and 76 (and possibly 109) provided that Ca^{2+} is present[44]. The oxidized protein no longer interacts with phosphodiesterase and troponin I, and has apparently lost its high affinity Ca^{2+} binding sites.

A second approach to relate the structure and function of calmodulin is to examine the consequences of a selective chemical modification of amino acid residues.

Walsh and Stevens[45] and (46) have compared the ability of several calmodulin derivatives to activate brain cyclic nucleotide phosphodiesterase. In addition, the displacement of ^{125}I-labeled calmodulin and the activation of adenylate cyclase by the same calmodulin derivatives were compared in a brain particulate fraction[46].

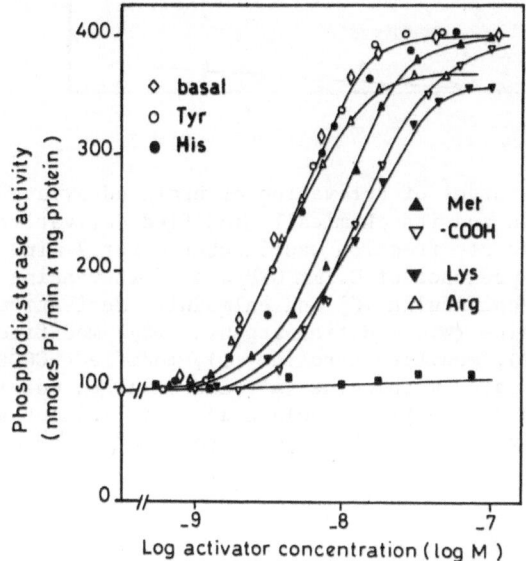

Fig. 6. Dose-effect curves of activation of brain soluble cyclic nucleotide phosphodiesterase by calmodulin and its chemically modified derivatives. The calmodulin-deficient phosphodiesterase was incubated for 30 minutes at 30°C in the presence of increasing concentrations of calmodulin (◇), and of its derivatives with modified lysine (▼), modified arginine (△), modified histidine (●), modified tyrosine (○), modified -COOH (▽), carboxymethylated methionine (▲), and oxidized methionine (■)[46].

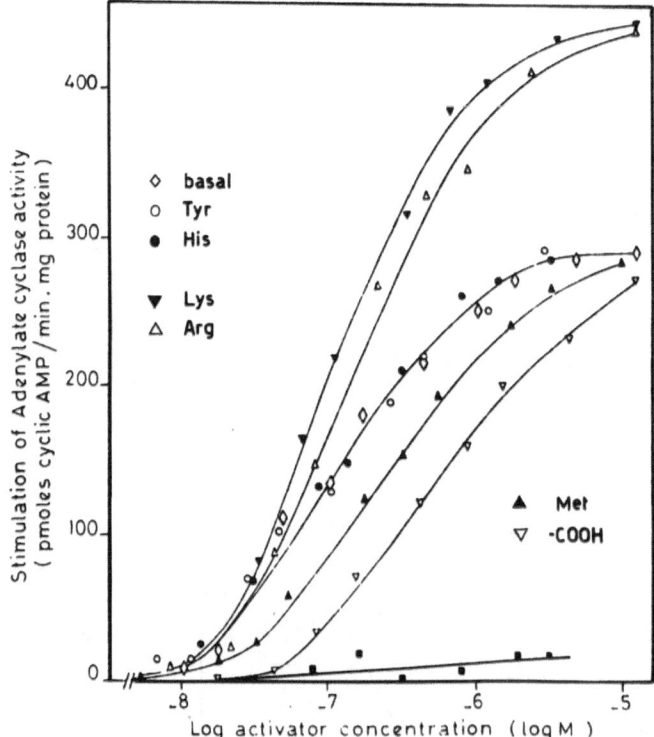

Fig. 7. Dose-effect curves of activation of brain adenylate cyclase
         by calmodulin and its chemically modified derivatives. The
         brain particulate fraction was incubated for 7 minutes at
         37°C in the presence of 0.1mM GTP and of increasing concen-
         trations of calmodulin (◊) and calmodulin derivatives with
         modified lysine (▼), modified arginine (△), modified
         histidine (●), modified tyrosine (○), modified –COOH (▽),
         carboxymethylated methionine (▲), and oxidized methionine
         (■). The basal adenylate cyclase activity without added
         protein activator was 399±27 pmol cyclic AMP formed x min^{-1}
         x mg protein^{-1},[46].

     Six chemical modification reactions do not affect the calcium
binding properties of calmodulin. Two of them (carbethoxylation of
the single histidine residue and nitration of both tyrosine residues)
exert no influence on the biological properties of calmodulin (Figs.
6 and 7). The amidation of carboxyl groups reduces the activating
properties of calmodulin i.e., the dose-effect curves are shifted
to the right (Figs. 6 and 7) but about twenty of the 40 free carboxyl
groups are estimated to be modified. Carboxymethylation with

iodoacetic acid in the presence of Ca^{2+} ions also reduces the efficiency of the calmodulin molecule (Figs. 6 and 7) and only affects 2 methionine residues at least one of them being most probably part of the 71-77 peptide sequence.

The modification of four arginine residues is responsible for a loss in affinity for the particulate adenylate cyclase system (Fig. 7) but not for phosphodiesterase (Fig. 6). On the contrary, the modification of seven lysine residues decreases the affinity for phosphodiesterase but not for adenylate cyclase. In addition, the lysine and arginine modifications result in a marked increase in the reaction rate of fully activated adenylate cyclase that contrasts with a moderate decrease in the reaction rate of maximally activated phosphodiesterase.

Thus, the role exerted by arginine and lysine side chains in calmodulin interaction with adenylate cyclase is not strictly comparable with their role in calmodulin interaction with cyclic nucleotide phosphodiesterase.

In conclusion, the interactions of the calmodulin derivatives with adenylate cyclase and cyclic nucleotide phosphodiesterase suggest a role for methionine, arginine, and lysine residues in the putative binding and activating area of calmodulin. This domain may conceivably include the 71-77 sequence (-Met-Met-Ala-Arg-Lys-Met-Lys-) lying on the surface of the molecule, between the second and third Ca^{2+} binding loops.

REFERENCES

1. W. Y. Cheung, Biochem. Biophys. Res. Commun. 38:533 (1970).
2. S. Kakiuchi and R. Yamazaki, Biochem. Biophys. Res. Commun. 41: 1104 (1970).
3. D. J. Wolff and F. L. Siegel, J. Biol. Chem. 247:4180 (1972).
4. T. S. Teo, T. H. Wang and J. H. Wang, J. Biol. Chem. 248:588 (1973).
5. T. S. Teo and J. H. Wang, J. Biol. Chem. 248:5950 (1973).
6. Y. M. Lin, Y. P. Lin and W. Y. Cheung, J. Biol. Chem. 249:4943 (1974).
7. J. C. Brooks and F. L. Siegel, J. Biol. Chem. 248:4189 (1973).
8. A. Vandermeers, M. C. Vandermeers-Piret, J. Rathé, R. Kutzner, A. Delforge and J. Christophe, Eur. J. Biochem. 81:379 (1978).
9. J. R. Dedman, J. D. Potter, R. L. Jackson, J. D. Johnson and A. R. Means, J. Biol. Chem. 255:8415 (1977).
10. D. J. Wolff and C. O. Brostrom, Adv. Cyclic Nucleotide Res. 11:27 (1979).
11. D. M. Watterson, F. Sharief and T. C. Vanaman, J. Biol. Chem. 255:962(1980).

12. J. R. Dedman, R. L. Jackson, W. E. Schrieber and A. R. Means, J. Biol. Chem. 253:343 (1978).
13. R. J. A. Grand and S. V. Perry, FEBS Lett. 92:137 (1978).
14. R. H. Kretsinger and C. E. Nockolds, J. Biol. Chem. 248:3313 (1973).
15. R. H. Kretsinger, in: "Calcium Binding Proteins and Calcium Function", R. H. Wasserman, R. A. Corraridono, E. Carafoli, R. H. Kretsinger, D. A. MacClenman and F. L. Spiegel, eds., pp. 63-72, Elsevier North-Holland Inc., New York (1977).
16. D. J. Wolff, P. G. Poirier, C. O. Brostrom and M. A. Brostrom, J. Biol. Chem. 252:4108 (1977).
17. C. B. Klee, Biochemistry 16:1017 (1977).
18. K. B. Seamon, Biochemistry 19:207 (1980).
19. L. S. Bradham, D. A. Hall and M. Sims, Biochim. Biophys. Acta 201:250 (1970).
20. C. O. Brostrom, Y. C. Huang, B. McL. Breckenridge and D. J. Wolff, Proc. Natl. Acad. Sci. USA 72:64 (1975).
21: W. Y. Cheung, L. S. Bradham, T. S. Lynch, Y. M. Lin and E. A. Tallant, Biochem. Biophys. Res. Commun. 66:1055 (1975).
22. A. Vandermeers, P. Robberect, M. C. Vandermeers-Piret, J. Rathé and J. Christophe, Biochem. Biophys. Res. Commun. 84:1076 (1978).
23. E. Pires, S. V. Perry and M. A. W. Thomas, FEBS Lett. 41:292 (1974).
24. A. Sobieszek and J. V. Small, J. Mol. Biol. 112:559 (1977).
25. R. Dabrowska, D. Aromatorio, J. M. F. Sherry and D. J. Hartshorne, Biochem. Biophys. Res. Commun. 78:1263 (1977).
26. J. L. Daniel and R. S. Adelstein, Biochemistry 15:2370 (1976).
27. S. V. Perry, R. J. A. Grand, A. C. Nairn, T. C. Vanaman and C. M. Wall, Biochem. Soc. Trans. 7:619 (1979).
28. S. Ebashi, T. Mikawa, M. Hirata and Y. Nomura, Ann. N.Y. Acad Sci. 307:451 (1978).
29. M. J. Yerna, D. J. Hartshorne and R. D. Goldman. Biochemistry 18: 673 (1979).
30. A. R. Means and J. R. Dedman, Nature 285:73 (1980).
31. S. Shenolikar, P. T. W. Cohen, P. Cohen, A. C. Nairn and S. B. Perry, Eur. J. Biochem. 100:329 (1979).
32. D. B. Rylatt, N. Embi and P. Cohen, FEBS Lett. 98:76 (1979).
33. M. Schulman and P. Greengard, Proc. Natl. Acad. Sci. USA 75:5432 (1978).
34. R. J. DeLorenzo, S. D. Freedman, W. B. Yohe and S. C. Maurer, Proc. Natl. Acad. Sci. USA 76:1838 (1979).
35. T. Yamauchi and H. Fujisawa, Biochem. Biophys. Res. Commun. 90: 1172 (1979).
36. M. Landt and J. M. McDonald, Biochem. Biophys. Res. Commun. 93: 881 (1980).
37. J. M. Anderson and M. J. Cormier, Biochem. Biophys. Res. Commun. 84:595 (1978).
38. T. Yamaki and H. Hidaka, Biochem. Biophys. Res. Commun. 94:727 (1980).

39. A. Mualhem and S. J. D. Karlish, Biochim. Biophys. Acta 597:631 (1980).
40. F. L. Larsen and F. F. Vincenzi, Science 204:306 (1979).
41. S. Katz and M. A. Remtulla, Biochem. Biophys. Res. Commun. 83: 1373 (1978).
42. C. J. Le Peuch, J. Haiech and J. G. Demaille, Biochemistry 18: 5150 (1979).
43. M. Walsh, F. C. Stevens, J. Kuznincki and W. Drabikowski, J. Biol. Chem. 252:7440 (1977).
44. M. Walsh and F. C. Stevens, Biochemistry 17:3924 (1978).
45. M. Walsh and F. C. Stevens, Biochemistry 16:2742 (1977).
46. P. Thiry, A. Vandermeers, M. C. Vandermeers-Piret, J. Rathé and J. Christophe, Eur. J. Biochem. 103:409 (1980).

CYCLIC NUCLEOTIDES AND CELL GROWTH

Jaques Otten

Institut de Recherche Interdisciplinaire
Université Libre de Bruxelles
1000 - Bruxelles, Belgium

INTRODUCTION

The discovery of cyclic AMP as an intracellular messenger for
many hormonally-controlled functions and more generally of its involvement in the regulation of cell response to exogenous effectors
led to the hypothesis that the adenylate cyclase-cyclic AMP system
might also have a role in the regulation of cell division. Indeed,
starting or stopping growth, increasing or decreasing growth velocity
are among the most important adaptive responses to a changing environment. The ubiquitous character of cyclic AMP in all animal cells made
of this substance a good candidate for a universal growth regulator.

The possible role of cyclic AMP and other cyclic nucleotides as
positive or negative growth signals has been mostly studied in cell
culture systems in which the many variables which may influence cell
proliferation are more easily controlled than in whole animals. Much
work has been carried out in normal and transformed fibroblasts and in
lymphocytes. The first evidence that cyclic AMP might be involved in
the regulation of cell growth came from studies by Bürck who showed
that 2 phosphodiesterase inhibitors, caffeine and theophylline, which
raise intracellular cyclic AMP levels, inhibit the proliferation of
normal and transformed baby hamster kidney fibroblasts. Soon after-
wards, other authors observed that cyclic AMP or its dibutyryl derivative had similar effects in several established fibroblast lines.
Another evidence for a role of cyclic AMP as a physiologic inhibitor
of proliferation was provided by the measurement of the cyclic AMP
content of randomly grown cell cultures. It was found that when fibroblasts are growing exponentially, there is an inverse correlation
between growth velocity and cyclic AMP concentration. When normal
cell growing in monolayer cultures approach confluency, their pro-

liferation slows down and then stops almost completely. This phenomenon, called density-dependent inhibition of growth, DDIG, was found
by some authors, but not by others, to be associated with a raised
cyclic AMP content. The discordance between the results from different
laboratories could be partially explained by the mixed composition of
some established cell lines. It was recently shown that the NRK cells,
a line of rat kidney fibroblasts, are a mixture of two different populations one of which displays DDIG and elevated cyclic AMP levels at
confluency, the other none of these two properties. In transformed
cells whose cyclic AMP content is lower than that of their normal
counterpart and which have lost DDIG, this property cannot be restored
by the addition of cyclic AMP analogs or substances raising cyclic AMP
levels in the cells. Thus, although in some experimental systems elevated cellular cyclic AMP levels are associated with DDIG, and although it is tempting to relate the loss of DDIG by transformed fibroblasts to their inability to maintain normal cyclic AMP concentration,
there is hitherto no clear cut evidence that cyclic AMP mediates DDIG.

If cyclic AMP functions as a negative growth signal in certain
conditions, in which part of the cell cycle does it act? In most
studies on fibroblasts, the addition of cyclic AMP or its analogs block
the cells in either early G_1 or G_2. In the systems in which growth is
arrested in early G_1, as in density inhibited cells, the stimulation
of growth by exogenous agents such as fresh serum is accompanied by a
brisk fall of cellular cyclic AMP. The adddition of cyclic AMP to the
medium before or at the time of the stimulation prevents the commit-
ment of the cells to DNA synthesis and proliferation.

While the bulk of the experimental work suggests that cyclic AMP
acts as an inhibitor of cell division in normal and transformed fibroblasts, and that a disordered adenyl-cyclase-cyclic AMP system may explain some of the abnormal properties of transformed fibroblasts other
less numerous observations indicate that in some systems cyclic AMP
has no effect on growth or may have a stimulatory effect on cell proliferation. Thus, even within the limited scope of fibroblasts in culture, the question whether cyclic AMP is a general physiologic negative
signal for cell growth is not definitely settled.

As far as other cyclic nucleotides are concerned, the data are
still less convincing. Fibroblasts which are stimulated to synthesize
DNA and divide raise their cyclic GMP content, while resting cells
generally have low levels of cyclic GMP. This observation has suggested that cyclic GMP could be a positive regulator of cell growth.
Alternatively, the hypothesis has been put forward that the most important regulating factor of proliferation was the ratio of cyclic GMP
to cyclic AMP. However, this hypothesis could not be supported by
attempts to stimulate growth with exogenous cyclic GMP or some of its
analogs.

In peripheral blood lymphocytes, cells which normally remain

definitely in a quiescent state but may be committed to DNA synthesis
and proliferation by specific antigens or lectins, the role of cyclic
nucleotides has been mostly studied in relation to calcium ion metabolism. In this system, the presence of Ca^{++} in the medium is necessary for the mitogenetic activity of lectin to take place. EGTA inhibits the stimulation by lectins. An interesting observation is that
PHA raises cyclic GMP levels of peripheral lymphocytes at low concen -
trations of $Ca^{++}(10^{-4}M)$ without inducing mitosis, suggesting that
cyclic GMP is not the signal for growth and its raising concentration
merely accompanies the growth-promoting processes. Conversely, cyclic
GMP could act directly on cell growth by facilitating the entry of Ca^{++}
into the cell at high (1,5 mM) exogenous concentration of Ca^{++}. Elevated concentrations of cyclic AMP and agents which increase cellular
cyclic AMP inhibit DNA synthesis in PHA-stimulated lymphocytes from
normal humans and from patients with chronic lymphocytic leukemia, in
the presence of Ca^{++} at 1 to 1,5 mM. This Ca^{++} concentration is of the
same order as that which is usually employed in experiments with fibroblasts. In thymocytes, which are a mixture of quiescent lymphocytes
and of activated lymphoblasts, in the presence of low Ca^{++} concentrations (0,5 mM), very low cyclic AMP concentrations (10^{-8} to $10^{-9}M$)
become mitogenic. At this level there is no perceptible change of
intracellular cyclic AMP content and the nucleotide is supposed to act
at the cell membrane. The mechanism of this stimulation effect is not
clearly elucidated.

Cyclic AMP has been shown to inhibit the growth of epidermal skin
cells and melanoma cells. It does not affect growth in some normal
liver cell lines (RLC) and in HTC hepatoma cells, it inhibits DNA synthesis in two other hepatoma cell lines (HTC). In several Morris hepatoma cell lines with very different growth velocities, no correlation
could be found between the population doubling time and either the
cyclic AMP or the cyclic GMP levels.

An interesting system is provided by endocrine glands under the
control of trophic hormone. In organs such as the adrenal or the
thyroid gland, trophic hormones, i.e. ACTH and TSH respectively, induce
immediate functional responses and long-term trophic effects. Most
immediate functional changes are mediated by cyclic AMP. One may
wonder whether this substance also mediates the trophic action of ACTH
or TSH. In recent experiments with normal dog thyroid cells maintained
in culture, it was observed that TSH stimulated the growth of the fol-
licular cells and that this effect could be reproduced by choleratoxin
and, partially at least, by cyclic AMP. In this system at least, the
effect of cyclic AMP appears to be quite the opposite of that observed
in most experiments with fibroblasts.

SUMMARY AND CONCLUSIONS

In most established fibroblast cell lines, exogenous cyclic AMP
inhibits cell growth and blocks cells in early G_1 or G_2. The reverse

correlation of intracellular cyclic AMP levels with growth velocity
in different experimental conditions suggests that the inhibitory
effect of cyclic AMP may have physiological significance. More data
are needed to attribute a physiologic role to cyclic GMP in the regulation of growth of fibroblasts. In lymphocytes committed to blastic
transformation and proliferation, the effect of cyclic AMP is concentration dependent. At Ca^{++} concentrations equal to those usually realized in fibroblast cultures, cyclic AMP inhibits DNA synthesis and
mitosis. At low Ca^{++} concentrations, very small amounts of cyclic AMP,
acting at the plasma membrane, stimulate DNA synthesis. Cyclic GMP
raises accompany PHA induced stimulation of growth, they do not directly mediate this stimulation. They could however indirectly induce
DNA synthesis by facilitating Ca^{++} entry into the cell. In other
systems, such as the thyroid, cyclic AMP may be a positive signal for
growth. At the present time, neither of the two cyclic nucleotides
here discussed can be considered as a universal regulator of growth
in all kinds of cells, although cyclic AMP intervenes either as a positive or as a negative growth signal in some particular systems.

REFERENCES

Chlapovski, F.S., Kelly, L.A. and Butcher, R.W., 1975, Cyclic nucleo-
        tides in cultured cells, Adv. Cyclic Nucleotide Res., 6 : 245.
Rebhun, L.I., 1977, Cyclic nucleotides, calcium and cell division,
        Int. Rev. Cytology, 49 : 1.
Rudland, P.S. and Jimenez de Asua, L., 1979, Action of growth factors
        in the cell cycle, Biochim. Biophys. Acta, 560 : 91.
Pastan, I., Johnson, G.S. and Anderson, W.B., 1975, Cyclic nucleotides
        and growth, Ann. Rev. of Biochem., 44 : 491.
Pastan, I. and Willingham, M., 1978, Cellular transformation and the
        "morphologic phenotype" of transformed cells. Nature, 274 :
        645.

MICROTUBULES AND MICROFILAMENTS : THE CELL CYTOSKELETON

Pierre Dustin

Department of Pathology, Faculty of Medicine
Université Libre de Bruxelles
B-1000 Brussels, Belgium

The cytoskeleton, the structural framework of the cell, was thought, about twenty years ago, to be represented by a few fibrillar proteins, the best known being those of muscle and keratinised cells (cf. De Robertis et al., 1960). New techniques - mainly electron microscopy, gel electrophoresis and immunohistochemistry - have demonstrated that most cells have an extensive number of fibrils in their cytoplasm, and many new cytoskeletal proteins are discovered each year. Structural proteins of the nucleus will not be considered here : our purpose is to give a summary of the main skeletal elements of the cytoplasm, as found in most cells. We will consider first those fibrillar structures which are found in nearly all Eukaryotic cells : microtubules, intermediary microfibrils, and contractile proteins of the actin type. A short list of other skeletal proteins will be given next, although the pace of recent discoveries - as evidenced for instance by the Second International Congress on Cell Biology held in September 1980 - is an indication of the growing complexities of the subject.

THE MICROTUBULAR FRAMEWORK

Microtubules (MT) are long (several microm.), slender structures with average diameter of about 24 nm (Dustin, 1978; Roberts and Hyams, 1979). They are present in all Eukaryotic cells, the single exception being the adult red blood cells of Mammals (with the exception however of the elliptic red blood cells of Camelidae). They are formed by the assembly of two molecules, tubulins α and β . These have closely related sequences of aminoacids, and result probably from the evolution of a primitive tubulin molecule. In some species other types of tubulins may be present, although one can consider that in all Vertebrates

239

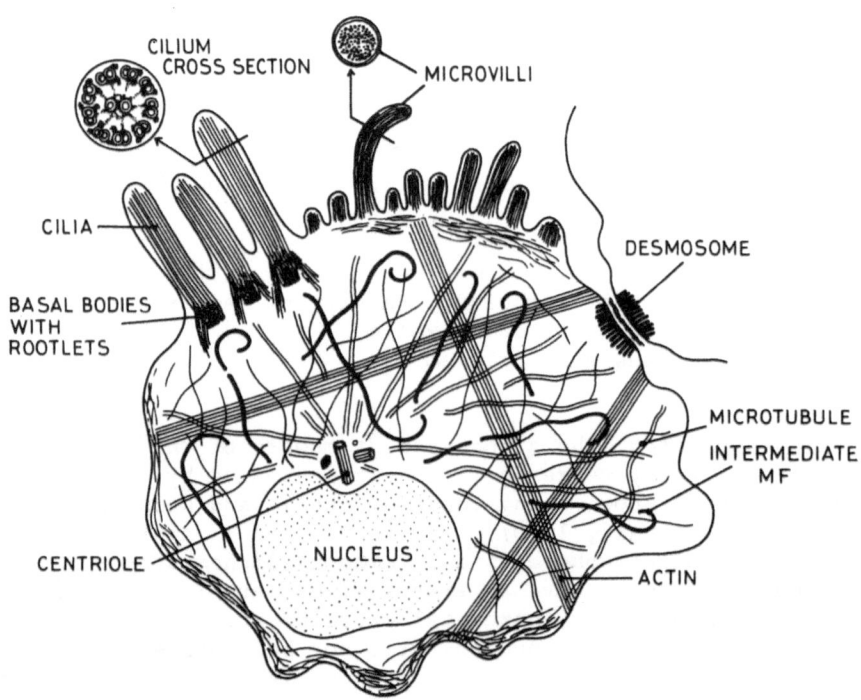

Fig. 1.   Principal constituents of the cytoskeleton. The microtubules
         radiate from the two centrioles located close to the nucleus.
         The intermediary filaments form also a dense network, the
         filaments being more curved than the microtubules. The actin
         microfibrils are often located close to the cell membrane,
         and form bundles attached to the cell membrane and which ex-
         tend from one side to the other of the cell. The cilia are
         complex microtubule organelles, attached to the basal bodies
         (which have the same structure as the centrioles), which are
         linked with the cytoplasm through rootlets (often striated).
         The axis of the microvilli contains many actin fibres. The
         desmosomes are complex fibrillar structures linking solidly
         neighbouring cells.

tubulins are identical. They assemble into dimers which are the
building-blocks of MT, the dimers forming, by longitudinal and side-by-
side assembly, first rings, then short filaments. These eventually

become grouped in a tubular structure made, most often, of 13 proto-
filaments. This assembly can be studied in vitro. It takes place in
the cell from a pool of unassembled tubulin (Kirschner, 1978).

The formation of microtubules in vivo involves always other proteins, the MAPs or "microtubule associated proteins". Two of these,
MAP1 and MAP2 have a molecular weight around 200,000 daltons, while
that of tubulins is about 53,000. Another MAP is the tau factor, with
a smaller molecular weight than the tubulins. MAPs appear to play several roles : they help in the assembly of MT, they stabilize their
structure, and build cross-bridges between MT and probably between MT
and other cytoplasmic structures - organelles such as mitochondria or
secretory granules, and other fibrillar proteins, in particular the
"intermediary" ones (vide infra).

The stability of MT in the living cell varies considerably. When
assembled in some complex structures such as centrioles and cilia,
they appear stable, but remarkable exceptions are known : in some Unicellulars the complex arrays of MT which build the axonemes of axopods
may disassemble in less than a second, to be later slower rebuilt from
the pool of tubulins. There exists thus an equilibrium between assembled and disassembled tubulins : the control of this remains poorly
understood, although several lines of research point to the role of
Ca^{2+}, which could disassemble MT even in micromolar concentrations,
and which is regulated by another protein, calmodulin (Klee et al.,
1980; Kumagai and Nishida, 1980).

The changes of the MT network are most evident in mitosis. In
the resting cell, a large number of MT extends from the pericentriolar
structures throughout the cytoplasm, curving gently, close to the cell
envelope. When mitosis starts, a dramatic change takes place : this
network disappears, and MT extend mainly from two structures so as to
form the polarized mitotic spindle : the pericentriolar structures
(or equivalent regions in plant cells which have no centrioles) localized at the poles of the spindle, and the kinetochores, specialized
zones of the chromosomes. The interrelations of these two types of MT
are most complex, and several theories have been recently proposed
(cf. Dustin, 1979; McIntosh, 1979). It is probable that the chromosome movements which are very slow result from the progressive disassembly of the kinetochore MT, while the cell elongates through the
growth of the polar (centriolar) MT. There are many reasons to think
that oppositely polarized MT - as those coming from each pole - may be
interrelated (through contractile proteins of the actin-myosin type?)
and that sliding movements between MT of opposite polarity may play a
role in mitotic movements.

Such MT-associated movements are better known in one type of
complex microtubular structures, the cilia. Here, 9 doublet MT (made
of one MT to which is attached an incomplete C-shaped one) and two
central MT, plus many other structures, are closely involved in motion :

ciliary beating. The MT act as skeletal elements, the energy being provided by the side-arms attached to the doublets and made of dynein a protein with ATPase activity (Warner and Mitchell, 1980). Observations on pathological abscence of the dynein arms show that the cilia have lost all motility.

Many other types of intercellular movements require the integrity of the MT and become impossible if the MT are disassembled either through the action of cold or high hydrostatic pressures, or that of drugs which become linked to the tubulin dimer and prevent its assembly into MT (colchicine, the Vinca alkaloids, benzimidazole compounds, griseofulvin, etc.). Such movements are for instance those of secretory granules towards the cell envelope, of leukocyte granules towards phagocytosis vacuoles, of pigment granules in melanophores, and the rapid movement of many proteins and sometimes secretory granules along the axones and dendrites of neurons (neuroplasmic flow) (Chan et al., 1980). The exact relation between the MT and these movements remain as poorly understood as those of mitosis. MT appear mainly as a framework or guides, along which other particles move. The source of energy and its relation to the MT and their MAPs is not as well known as that of ciliary motion. Proteins attached to the side of the MT and, temporarily, to the moving structures, act probably by some type of ratchet action, somewhat similar to the sliding of muscle filaments during contraction. The presence of proteins of the dynein and actin types in most cells suggest that these may provide the energy required. The MAPs mentioned above are however not contractile proteins.

One can summarize in the following way the skeletal function of MT. They form in resting cells a network which many observations indicate to be closely linked to the shape of the cell (Porter, 1966). They can however undergo in many conditions rapid disassembly, as evidenced in the first stages of cell division. They are indispensable for mitosis - hence their discovery in relation to the poisoning of mitosis by colchicine, through its specific fixation on tubulins (A.P. Dustin, 1939). Mitosis is only one type of cell motion in which MT are implicated, and they appear to serve as guides for the action of other (contractile?) proteins. Last, MT are involved in the assembly of much more complex structures, such as the centrioles, the basal bodies and the cilia and flagella, not to mention still more complex assemblies of hundreds of MT, linked by other proteins, as known in many Unicellulars and some spermatozoa. Through their relation with ciliary motion they play a considerable role in human physiology : abnormalities of cilia structure lead to complex disturbances and sterility, through the lack of spermatozoan movements (Afzelius, 1979).

THE INTERMEDIARY (10 NM) FILAMENTS

In various cells, long filaments with an average diameter of 10 nm, that is to say "intermediary" between the MT (25 nm) and the

finer (5 nm) actin fibrils, have been observed. The most well-known
are the neurofilaments of axones and dendrites, and the tonofilaments
of glial cells. The keratin fibrils of differentiated squamous epi-
thelia have also about the same dimension.

     A strange fact is that under some experimental conditions, the
amount of these fibrils may increase considerably in various cells,
and tightly-packed whorls of MF encircle the nucleus. These are birefringent (like MT) and some further stain by Congo red like amyloid
(which is also a group of fibrillar proteins, albeit extracellular
(Glenner, 1980). This takes place apparently without any protein
synthesis, as it is not affected by cycloheximide. It is found in
cells treated for some hours by poisons of MT such as colchicine.
However, the idea that these filaments are modified tubulin must be
rejected, for they do not stain by antitubulin antibodies. Their protein has a molecular weight of 55,000, and specific antibodies against
these filaments have been prepared and do not stain the MT. No explanation has yet been given for the increase of 10 nm MF after the action
of MT poisons. Another condition in which such filaments increase in
number in the nerve cells, is that of animals chronically intoxicated
by aluminium salts, which do not destroy MT.

     Since antibodies against these MF can be used for immunohisto-
chemical techniques, it has been demonstrated that the cytoplasm of
various cells contains a network of these filaments, which may exist
side by side with the network of MT. Lateral connections, similar to
those observed in the nerves between neurotubules (MT) and neurofilaments have been described. This filamentous network appears thus
as a second type of cytoskeleton; its purpose remains mysterious.

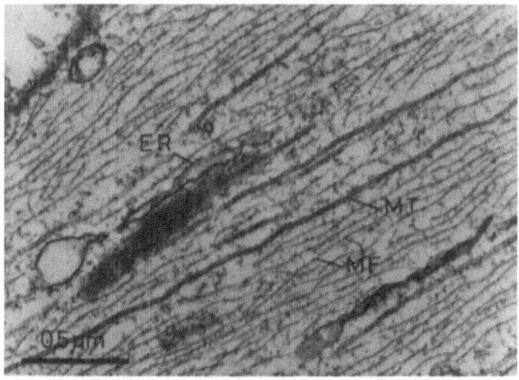

Fig. 2. Microtubules and neurofilaments. Longitudinal section of a
         nerve. MT : microtubules (neurotubules). MF : neurofilaments.
         ER : endoplasmic reticulum. Numerous proteins appear, to
         link the neurofilaments together, and the neurotubules with
         the neurofilaments.

The biochemistry of 10 nm MF has made great advances, and in epithelial cells these have been identified as prekeratin (or cytokeratin).
Keratin is present in far more cells than was believed, the squamous
cells of the skin being only a case of massive keratinization. In
pathology, the role of 10 nm MF in the formation of intracytoplasmic
bodies in liver cells known as Mallory's bodies, and found in al-
coholic cirrhosis, and also in animals chronically injected by griseofulvin (a mild MT poison), has been demonstrated (Franke et al., 1978).
These bodies are made of cytokeratin.

On the other hand, another protein, named vimentin (Franke et al.,
1979; Osborn et al., 1980) has been identified in connective tissue
cells, where it forms a similar network of 10 nm MF. Vimentin is anti-
genetically different from keratin. A surprising fact is that in some
cells, both vimentin and cytokeratin may be found side by side. This
means that such cells contain three different fibrillar cytoskeletal
components (tubulin, keratin, vimentin).

As mentioned above, the intermediary filaments were first studied
in the nervous tissue. Further analysis has shown that neuro- and
gliofilaments have a chemical structure different from vimentin or
keratin and that fibrils of glial cells are quite different from those
of neurons.

The function of the intermediary filaments remains poorly understood, perhaps because no specific poison, capable of disassembling
them is known. They do not undergo, during mitosis, the considerable
changes described for tubulins and MT. It is not known from what type
of "organizing center", comparable to the centrioles for MT, they grow.

THE ACTIN FIBRILS

These measure about 5 nm in diameter, and have now been found in
most cells, besides other proteins known from the study of muscle
(actinin, myosin, troponin). In fact, it appears now that muscle
fibers are a specialized type of cell which contains quantitatively
much more of these proteins found also in other cells, similarly to
keratinized cells as compared to other epithelial cells.

The study of these fibrils has been made easier by the use of a
more or less specific poison, cytochalasin, which destroys the 5 nm
MF without affecting either the MT or the 10 nm MF.

The actin fibrils appear in the cells as structures which can be
observed by electron microscopy, often as a mat of fine fibrils located
close to the cell membrane (and sometimes connected to the MT). The
use of immunofluorescence with actin antibodies has shown that bundles
of actin MF extend throughout the cytoplasm, often from one side to
another of the cell, and are linked to the cell membrane (through other

proteins such as vinculin). These bundles disappear after cytochalasin treatment.

The role of actin is certainly associated with cell motility, but while myosin has been detected in various cells, it is present in much smaller quantities than in the muscle, except in some locations, like the core of microvilli. This is considered to favor cell movements differing from those of muscle by their slowness and their absence of polarity. The role of actin in the spindle in relation to chromosome movements has been suggested (Forer, 1976), as some authors claim that antibodies to actin stain electively the spindle. This problem is still under discussion.

The location of actin close to the cell membrane is important not only in relation to changes of cell size and motility, but also in relation to movements of constituents of the cell membrane itself, such as proteins ("capping" in leukocytes). It is probable that the actin MF helps to transmit to the cell membrane changes related to the MT.

OTHER FIBRILLAR CYTOPLASMIC CONSTITUENTS

The complexity of cytoplasmic structures is increasing from day to day, and the techniques of electrophoresis combined with immuno-histochemistry have led to many new discoveries. Table 1, which does not claim to be complete, mentions the fibrillar proteins located in the cytoplasm or closely linked with the cell membrane, without listing the various muscle contractile proteins which are probably present in small quantities in most cells. Several other proteins, which play a role in disassembling various fibrous structures, could also be mentioned, such as calmodulin, or fragmin and plasticine, which decrease the size of actin filaments. It is evident that the study of the cytoplasm is becoming more and more complex - and no mention has been made here of the microtrabeculae of Wolosewick and Porter (1979), which structure all the cytoplasm except liquid spaces. All these findings demonstrate how far one is today from the old concept of cytoplasm as a more or less homogenous gel, containing only some organites such as mitochondria and Golgi vacuoles. The modern cell appears highly structured, with many fibrous proteins. These play a role in cell shape, motion, division, and in relations between cells. Their importance lies in their absence in some pathological conditions, the presence in some others (ageing) of abnormal fibrils such as the twisted filaments of neurons in Alzheimer's disease, the Mallory's bodies in cirrhosis, and various disturbances of mitosis and other forms of cell movement, such as ciliary beating.

Table 1.   Some protoplasmic fibrillar components

Name	Function
actin	contraction
ankyrin	attachment of structures to spectrin; receptor for lgG antibodies in red blood cells
clathrin	wall of the coated vesicles ("acanthosomes")
desmin	Z-discs of muscle
chondronectin	intercellular fibers in cartilage
fibronectin	intercellular relations (exoskeleton)
glial protein	filaments (10 nm) of glial cells
cytokeratin	filaments (10 nm) in epithelial cells (= prokeratin)
laminin	attachment of epithelial cells to type IV collagen
neurofilament protein	neurofibrils
MAPs	assembly and links between MT
tubulins	microtubules and microtubular structures (cilia, flagellae, axonemes, mitotic spindle)
spectrin	cytoskeleton of erythrocytes
vimentin	cytoskeletal fibers of 10 nm diameter
vinculin (focin)	links between actin bundles and fibronectin

REFERENCES

Afzelius, B., 1979, The immotile-cilia syndrome and other ciliary diseases. Intern. Rev. exp. Path., 19:1.
Chan, S.W., Worth, R., and Ochs, S., 1980, Block of axoplasmic transport in vitro by Vinca alkaloids. J. Neurobiol., 11:251.
De Robertis, E.D.P., Nowinski, W.W., and Saez, F.A., 1960, "General Cytology" (3rd edition), W.B. Saunders Co., Philadelphia, London.

Dustin, A.P., 1939, A propos des applications des poisons caryoclas-
        iques à l'étude des problèmes de pathologie expérimentale, de
        cancérologie et d'endocrinologie. Arch. exp. Zellforsch.,
        22:395.
Dustin, P., 1978, "Microtubules", Springer-Verlag, Berlin, Heidelberg,
        New York.
Dustin, P., 1979, Microtubules et mitose. Bull. Ass. Anat., 63:109.
Forer, A., 1976, Actin filaments and birefringent spindle fibers
        during chromosome movements, in: "Cell Motility", R. Goldman,
        T. Pollard, and J. Rosenbaum, ed., Cold Spring Harbor Laborat-
        ory.
Franke, W.W., Grund, C., Osborn, M., and Weber, K., 1978, Intermediate-
        sized filaments in rat kangaroo PtK$_2$ cells. 1. Morphology in
        situ. Cytobiologie, 17:365.
Franke, W.W., Schmid, E., Osborn, M., and Weber, K., 1978, Inter-
        mediate-sized filaments in rat kangaroo PtK$_2$ cells. 2. Struc-
        ture and composition of isolated filaments. Cytobiologie, 17:
        392.
Franke, W.W., Schmid, E., Winter, S., Osborn, M., and Weber, K., 1979,
        Widespread occurrence of intermediate-sized filaments of the
        vimentin-type in cultured cells from diverse vertebrates. Exp.
        Cell. Res., 123:25.
Glenner, G.G., 1980, Amyloid deposits and amyloidosis : the β-fibril-
        loses. New Engl. J. Med., 302:1283 and 1333.
Kirschner, M.W., 1978, Microtubule assembly and nucleation. Intern.
        Rev. Cytol., 54:1.
Klee, C. B., Crouch, T.H., and Richman, P.G., 1980, Calmodulin. Annu.
        Rev. Biochem., 49:489.
Kumagai, H., and Nishida, E., 1980, The interactions between calcium-
        dependent regulator protein (calmodulin) and microtubule pro-
        teins. Further studies on the mechanism of microtubule as-
        sembly inhibition by calmodulin. Biomed. Res., 1:223.
McIntosh, J.R., 1979, Studies on the mechanisms of chromosome movement
        in: "Motility in Cell Function", F.A. Pepe, J.W.Sanger, and
        V.T. Nachmias, ed., Academic Press, New York.
Osborn, M., Franke, W.W., and Weber, K., 1980, Direct demonstration
        of the presence of two immunologically distinct intermediate-
        sized filament systems in the same cell by double immunofluor-
        escence microscopy. Vimentin and cytokeratin fibers in cul-
        tured epithelial cells. Exp. Cell Res., 125:37.
Porter, K.R., 1966, Cytoplasmic microtubules and their functions, in:
        "Ciba Foundation Symposium on Principles of Biomolecular
        Organization", Churchill, London.
Roberts, K., and Hyams, J.S., ed., 1979, "Microtubules", Academic
        Press, London.
Second International Congress on Cell Biology, 1980, Europ. J. Cell
        Biol., 22:1.
Warner, F.D., and Mitchell, D.R., 1980, Dynein : the mechanochemical
        coupling adenosine triphosphatase of microtubule-based sliding
        filammet mechanisms. Intern. Rev. Cytol., 66:1.

Wolosewick, J.J., and Porter, K.R., 1979, Microtrabecular lattice of
    the cytoplasmic ground substance. J. Cell Biol., 82:114.

MECHANISM OF ACTION OF ACTH IN THE ADRENAL CORTEX

D Michael Salmon and Dennis Schulster

Division of Hormones
National Institute for Biological Standards and Control
Hampstead, London NW3 6RB, UK

INTRODUCTION

Cells in the anterior lobe of the pituitary are the origin of the 39-residue peptide hormone corticotropin ($ACTH_{1-39}$). The most sensitive known biological activity of this hormone is the stimulation of corticosteroidogenesis in zona fasciculata cells in the adrenal cortex. $ACTH_{1-39}$, however, has many other biological actions most notably in the liver, brain and adipose tissue[1]. Furthermore it is thought that different peptide fragments of $ACTH_{1-39}$ may elicit tissue dependent actions which involve both cyclic AMP-dependent and independent mechanisms. In this paper the possibility that different populations of receptors for ACTH may exist in adrenocortical cells, both capable of eliciting steroidogenesis to the same extent, but through different mechanisms, is considered.

Involvement of cyclic AMP in adrenal steroidogenesis

Cyclic AMP has been classically involved in the steroidogenic action of ACTH. Haynes[2] first demonstrated that ACTH added to incubation media caused the accumulation of cyclic AMP by bovine adrenal slices and Grahame-Smith et al[3] showed that increases in cyclic AMP induced by ACTH occurred before increases in the rate of steroidogenesis. Subsequent work considering the casual relationship between cyclic AMP and steroidogenesis has been extensively reviewed[4]. However the precise role of cyclic AMP in promoting steroidogenesis and whether cyclic AMP is obligatorily involved in ACTH action still remain unknown.

A similar problem exists with other neurohumoral agents capable of stimulating cyclic AMP with respect to their physiological re-

sponses. The evidence questioning the obligatory role of cyclic AMP
in steroidogenesis has been reviewed and include the following:

a)   studies in isolated, perfused cat adrenals showed that increases
     in tissue cyclic AMP were insufficient to elicit steroid release[5]

b)   kinetics of cyclic AMP production and steroid production using
     isolated adrenal cell column perfusion did not support an obliga-
     tory role for cyclic AMP as a mediator of ACTH[6].

c)   in isolated adrenocortical cells stimulated with cholera toxin,
     increased cyclic AMP concentrations were detected 20 minutes
     before increased steroidogenesis was observed[7].

d)   binding studies utilizing ^{125}I-labelled ACTH have provided evi-
     dence for two populations of ACTH receptors in adrenocortical
     cells[8]. This finding has been considerably expanded through the
     use of analogues of ACTH, some of which are agonistic towards
     cyclic AMP accumulation and steroidogenesis, and others antagon-
     istic.

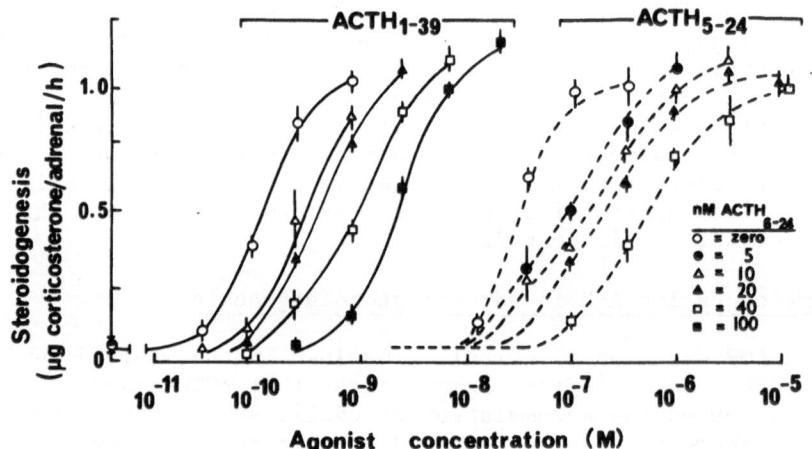

Fig. 1.   Inhibition of ACTH$_{1-39}$ and ACTH$_{5-24}$ stimulated steroido-
           genesis by ACTH$_{6-24}$. Dose-response characteristics for
           ACTH$_{1-39}$ (-) and ACTH$_{5-24}$ (--) were determined in the presenc
           of different concentrations of ACTH$_{6-24}$. Concentrations of
           ACTH$_{6-24}$ used were 0 (O), 5nM (●), 10nM (Δ), 20nM (▲), 40nM
           (□) and 100nM (■). Steroid was determined fluorimetrically
           after 1h incubations and error bars are the standard errors
           of the means of triplicate determinations. Reproduced with
           permission from the Biochemical Journal from Bristow et al.
           Biochem. J. 186: 599-603, 1980[9].

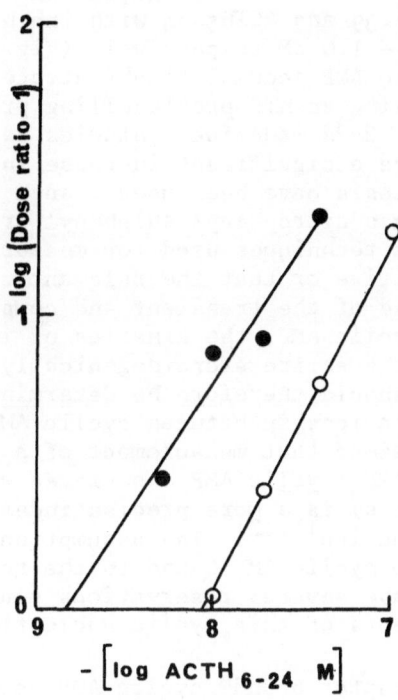

Fig. 2. Schild plot of the data in Fig. 1. The dose ratio (measured
as the ED_{50} for inhibited curve divided by the ED_{50} for the
uninhibited curve) was determined for the $ACTH_{6-24}$-inhibited
dose-response curve presented in Fig. 1. and a Schild plot for
$ACTH_{1-39}$ (O) and $ACTH_{5-24}$ (●) is shown. The intercept with
the x-axis gives the inhibitor constant, k_d (ie. the inhibitor concentration producing a 2-fold shift in dose-response
curve). Reproduced with permission from the Biochemical
Journal from Bristow et al. Biochem J. 186: 599-603, 1980[9].

Multiple receptor types for ACTH in adrenocortical cells

We have previously used $ACTH_{1-39}$ and $ACTH_{5-24}$ to show that although both analogues were equipotent with respect to steroidogenesis, this was not accompanied by a stimulation of total cyclic AMP[9]. It was also demonstrated that $ACTH_{6-24}$ inhibited steroidogenesis elicited by $ACTH_{1-39}$ and $ACTH_{5-24}$ with inhibition constants of 13.4 ± 3.1 nM and 3.4 ± 1.0 nM respectively (Fig. 1 and 2). In these studies total cyclic AMP accumulation (intracellular plus extracellular) was assessed using an ATP pre-labelling procedure using the cell permeable precursor $2-^{3}H$ -adenine. Studies using analogues which are claimed not to produce a significant increase in cyclic AMP concomitant with steroidogenesis have been used – such as those using $ACTH_{5-24}$[9] and those using o-nitrophenyl sulphenyl tryptophanyl ACTH[10] may be criticised in that techniques used for measuring cyclic AMP were insufficiently sensitive or that the relevant cyclic AMP pool had not been studied. Because of the transient and compartmentalized nature of increases in cyclic AMP, the kinetics of intracellular cyclic AMP concentrations over the entire steroidogenically-stimulatory dose range of ACTH analogues should therefore be determined in order to critically assess the relationship between cyclic AMP and steroidogenesis. It has been claimed that measurement of a fraction termed "receptor bound" cyclic AMP (cyclic AMP associated with material adsorbed to cellulose filters) is a more precise index of physiologically relevant cyclic AMP production[11,12]. The assumption has been made that this approximates to cyclic AMP bound to the regulatory subunit of protein kinase. We have several reservations concerning the interpretation of measurements of this cyclic nucleotide fraction.

The kinetics of "receptor bound" cyclic AMP accumulation in one report[11] shows a continuous rise over the initial 30 minute period following ACTH stimulation, whilst in another[12] it appears to be a close reflection of intracellular cyclic AMP concentration. In no case, however, can increases in "receptor bound" cyclic AMP be detected without increases in intracellular cyclic AMP. If "receptor bound" cyclic AMP represented only that fraction of the cellular nucleotide bound to protein kinase, then it would be expected that the rate of its accumulation should show a closer similarity to the rate of steroidogenesis.

With these considerations in mind, we examined the dose dependence of the time course of intracellular cyclic AMP production by $ACTH_{1-39}$ and $ACTH_{5-24}$ using a radioimmunoassay procedure for cyclic AMP[13], coupled with the acetylation modifications introduced by Harper and Brooker[14] to produce maximal sensitivity (Figs. 3 and 4). The results of these experiments substantiate our previous conclusions[9] that no cyclic AMP is formed in response to $ACTH_{5-24}$ over that dose range which stimulates steroidogenesis ($5 \times 10^{-8}M$ – $5 \times 10^{-7}M$). In contrast, using $ACTH_{1-39}$, close correlation was seen between intracellular cyclic AMP accumulation and steroidogenesis.

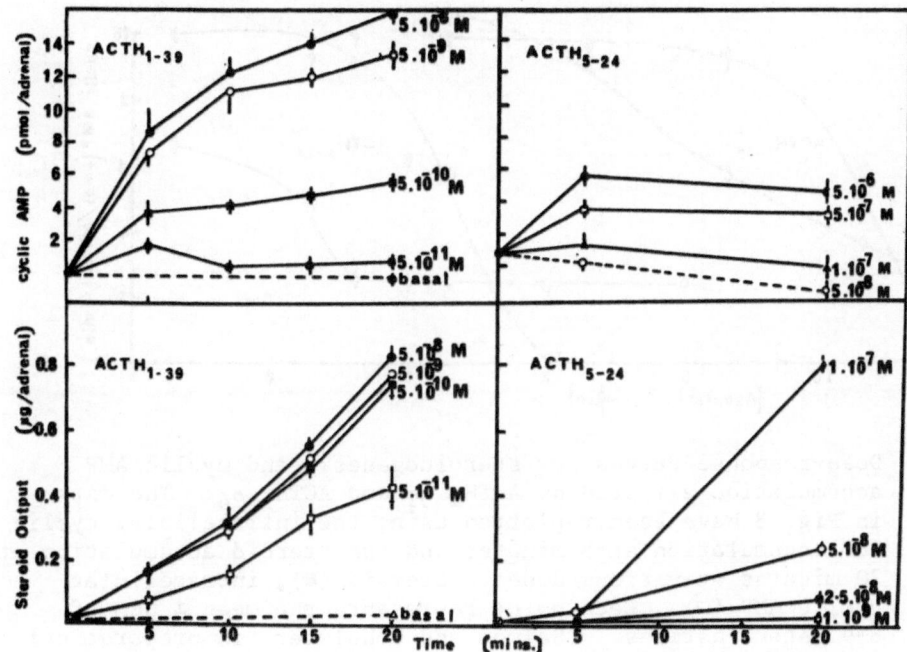

Fig. 3. Time courses and dose-dependence of steroidogenesis and
         cyclic AMP accumulation elicited by $ACTH_{1-39}$ and $ACTH_{5-24}$.
         Adrenocortical cells were incubated in the presence of ACTH
         analogues incubated for 60 minutes, after which time cyclic
         AMP was determined in the cell pellet and corticosterone
         measured fluorometrically in the supernatant after the
         indicated times. Each point represents the mean of 6-9
         determinations ± SEM (Salmon and Schulster, in preparation).

        More definitive information concerning the existence of multiple
receptor types for ACTH was obtained using the inhibitory analogue
$ACTH_{6-24}$. The dose dependence of the inhibition by $ACTH_{6-24}$ of intracellular cyclic AMP and steroidogenesis was determined for $ACTH_{1-39}$
and $ACTH_{5-24}$ (Fig. 5). These experiments provided an even more
striking dissociation between cyclic AMP and steroidogenesis. At
doses of $ACTH_{5-24}$ well in excess of those required to elicit maximal
steroidogenesis, in the presence of $ACTH_{6-24}$ (20 nM), there was no
detectable intracellular cyclic AMP formation. Two major conclusions
were reached from these experiments:

1)   cyclic AMP is not obligatory for the stimulation of steroido-
      genesis and

2)   ACTH may stimulate steroidogenesis through activation of differ-
      ent receptor types, both cyclic AMP-dependent and cyclic AMP-
      independent.

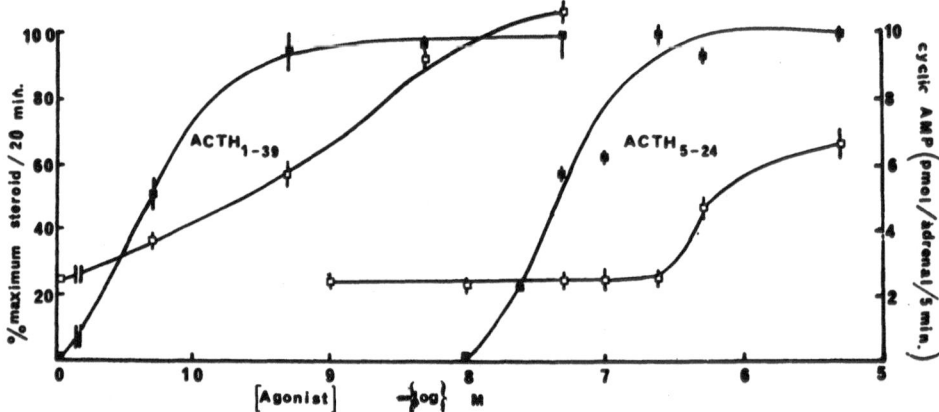

Fig. 4. Dose-response curves for steroidogenesis and cyclic AMP
         accumulation elicited by $ACTH_{1-39}$ and $ACTH_{5-24}$. The data
         in Fig. 3 have been replotted using the intracellular cyclic
         AMP accumulation at 5 minutes and the steroid accumulation at
         20 minutes at various doses. Steroid (■), intracellular
         cyclic AMP (□). Each point represents the mean ± SEM of
         6-9 determinations. (Salmon and Schulster, in preparation).

Alternative mechanism for ACTH action

        If it is accepted that steroidogenesis may be stimulated by ACTH
through mechanisms other than those involving cyclic AMP, then it is
reasonable to consider the nature of other secondary effector(s) for
ACTH. Over the past few years two major candidates have been under
consideration by different workers – cyclic GMP mediated phosphorylation mechanisms[15] and modulation of cellular Ca^{2+} distributions.
Most recently cyclic GMP has received little support and Hayashi et
al[16] have concluded that this is an unlikely mediator for ACTH action.
A requirement for Ca^{2+} in the actions of ACTH has been recognized
since The studies of Birmingham[17]. The situation is complicated by
the requirement of adrenocortical adenylate cyclase for Ca^{2+}, although
this effect is not specific for Ca^{2+} since lower concentrations of
other divalent cations (especially Mn^{2+} and Sr^{2+}) may substitute for
Ca^{2+} in this role[18]. The Ca^{2+} requirement of adenylate cyclase has
been reviewed by Halkerston[4]. However it is established that Ca^{2+}
is also required for steroid synthesis and release at steps subsequent
to cyclic AMP generation notably ACTH stimulated protein synthesis[19]
and pregnenolone synthesis by the P-450 system in isolated mitochondria[20].

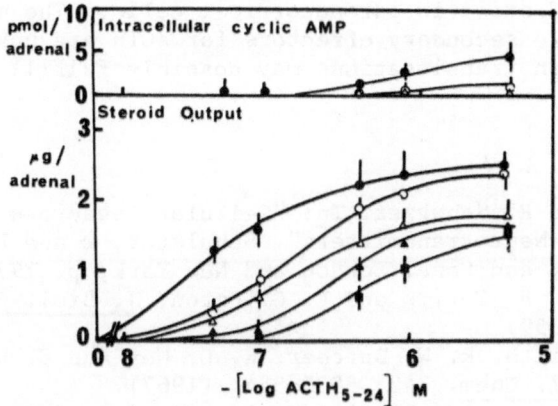

Fig. 5. Inhibition by $ACTH_{6-24}$ of $ACTH_{5-24}$ stimulated steroidogenesis
and intracellular cyclic AMP accumulation. Incubations were
carried out for 1 hour in the presence of $ACTH_{5-24}$ and
different concentrations of $ACTH_{6-24}$, O (●), 10nM (O), 20nM
(△), 40nm (▲) and 100nM (■). Error bars are the standard
errors of the means of triplicate determinations. (Salmon
and Schulster, in preparation).

Yanagibashi and collaborators[21] have presented evidence based on
changes in ^{45}Ca distribution following ACTH stimulation of adrenal
cells that ACTH stimulates Ca^{2+} influxes and similar conclusions were
reached by Leier and Jungman[22]. However studies on isolated, perfused
cat adrenals did not support this conclusion[23], but did suggest that
a redistribution of Ca^{2+} occurs within intracellular exchangeable
pools after ACTH stimulation of adrenocortical cells.

Experiments by a number of investigators[24,25] using verapamil,
an inhibitor of membrane Ca^{2+} translocation, have shown that this
agent will inhibit ACTH action at low concentrations. Further evidence
that this involvement of Ca^{2+} translocation must occur as a post-
receptor event is shown by recent studies demonstrating that cholera
toxin-evoked steroidogenesis is blocked by the addition of verapamil
at comparable concentrations to those which block ACTH.

CONCLUSIONS

    Although the concentration of cyclic AMP correlates well with
ACTH-induced steroidogenesis, it appears that cyclic AMP is not
obligatory for its action. Furthermore it is thought that multiple
receptor sites for ACTH exist in adrenocortical cells. The mechanism
of action of alternative secondary effectors for ACTH are unknown,
although cellular cation translocations may possibly fulfill this role.

REFERENCES

1.   D. Schulster and R. Schwyzer, In: "Cellular Receptors for
        Hormones and Neurotransmitters" Schulster, D and Levitski,
        A. eds. Wiley and Sons, London and New York, p. 197, 1980.
2.   R. C. Haynes, S. B. Koritz and F. G. Peron, J. Biol. Chem. 234:
        1421-1423 (1959).
3.   D. G. Grahame-Smith, R. W. Butcher, R. L. Ney and E. W. Suther-
        land, J. Biol. Chem. 242: 5535-5541 (1967).
4.   I. D. K. Halkerston, Adv. Cyclic Nucleotide Res. 6:99-136 (1975)
5.   R. A. Carcyman, S. D. Jaanus and R. P. Rubin, Mol. Pharmacol.
        7: 491-499 (1971).
6.   A. M. Hudson and C. McMartin, Biochem. J. 148: 539-544 (1975).
7.   J. W. Palfreyman and D. Schulster, Biochim. Biophys. Acta. 404
        221-230 (1975).
8.   R. A. J. McIlhinney and D. J. Schulster, Endocrinol. 64: 175-
        184 (1975).
9.   A. F. Bristow, C. Gleed, J. L. Fauchere, R. Schwyzer and D.
        Schulster, J. Biochem. 186: 599-603 (1980).
10.   W. R. Moyle, Y. C. Kong and J. Ramachandran, J. Biol. Chem.
        248: 2409-2417 (1973).
11.   E. J. Podesta, A. Milani, H. Steffan and R. Neher, Biochem. J.
        180: 355-363 (1979).
12.   G. B. Sala, K. Hayaski, K. J. Catt and M. L. Dufau, J. Biol.
        Chem. 254: 2816-2865 (1979).
13.   A. L. Steiner, C. W. Parker and D. M. Kipnis, J. Biol. Chem.
        247: 1106-1113 (1972).
14.   J. F. Harper and G. Brooker, J. Cyclic Nucleotide Res. 1: 207-
        218 (1975).
15.   J. P. Perchellet, G. Shanker and R. K. Sharma, Science 199:
        311-312 (1978).
16.   K. Hayashi, G. Sala, K. J. Catt and M. L. Dufau, J. Biol. Chem.
        254: 6678-6683 (1979).
17.   M. K. Birmingham, F. H. Elliot and P. H. L. Valere, Endocrin-
        ology 53: 687-689 (1953).
18.   D. D. Mahaffee and D. A. Ontjes, J. Biol. Chem. 255: 1565-1571
        (1980).
19.   R. V. Farese, Endocrinology 89: 1057-1063 (1971).
20.   E. R. Simpson, J. Waters and D. Williams-Smith, J. Steroid
        Biochem. 6: 395-400 (1975).

21.   K. Yanagibashi, N. Kamiya, G. Liu and M. Matsuba, Endocrinol.
          (Japan) 25:545-   (1978).
22.   D. J. Leier and R. A. Jungman, Biochem. Biophys. Acta. 329:
          196-210 (1973).
23.   S. D. Jaanus and R. P. Rubin, J. Physiol. 213: 581-598 (1971).
24.   E. J. Podesta, A. Milani, H. Steffen and R. Neher, Biochem. J.
          186: 391-397 (1980).
25.   K. Yanagibashi, Endocrinol. (Japan) 26: 227-232 (1979).

ACTIONS OF LUTROPIN (LH) AND cAMP IN THE REGULATION OF TESTICULAR STEROID PRODUCTION

H. J. van der Molen, B. A. Cooke[*], F. H. A. Janszen and
F. F. G. Rommerts.

Department of Biochemistry (Division of Chemical Endocrinology), Medical Faculty, Erasmus University, Rotterdam,
The Netherlands.

INTRODUCTION

The two main physiological functions of the mammalian testis are
localized in different tissue compartments; spermatogenesis occurs
within the seminiferous tubules and steroid hormone production within
the interstitial tissue. Within the interstitial tissue the Leydig
cells are the main, if not the only source of de novo steroid production. Trophic hormones from the pituitary are intimately involved
in the maintenance and regulation of these testis functions. Folli-
tropin (FSH) is known to have specific effects on the Sertoli cells
of the seminiferous tubules, but has no known effect on cells in the
interstitial tissue. Lutropin (LH) specifically interacts with Leydig
cells in the interstitial compartment, but has no direct effects on
the seminiferous tubules. The testis Leydig cells have become increasingly interesting as a model for studying the biochemical mechanisms
involved in the actions of trophic hormones on steroid production,
because recent technical developments have made it possible to study
specific hormone-biochemical parameters in specific isolated testis
cells.

STEROID PRODUCTION

The qualitative pattern of the pathways and intermediates in
steroid production in the testis of the human (as well as of most
other mammals) resembles those in other steroid-producing tissues,

---

[*]Present address: Department of Biochemistry, Royal Free Hospital
School of Medicine, University of London.

viz. the ovaries and adrenals. Cholesterol appears to be the obligatory precursor which can be converted intramitochondrially to pregnenolone, which is further metabolized extramitochondrially to several
other steroids via several Δ 5-intermediates (involving pregnenolone,
17α -hydroxy-pregnenolone, dehydroepiandrosterone and 5-androstene-
3 β,17 β-diol) as well as via several Δ 4-intermediates (progesterone,
17α -hydroxy-progesterone and 4-androstene-3,17-dione). The testicular
rate of conversion of pregnenolone through the Δ 4- and Δ 5-pathway is
different in different animal species. In the human testis the Δ 5-
pathway appears to be the most significant[25], whereas in the rat
testis the Δ 4-pathway via progesterone is quantitatively the most
important [21].

    In studying the cellular site of steroid production the ability
to dissect seminiferous tubules from the interstitial tissue as originally described by Christensen and Mason[3] has been very helpful. Enzym-
ic treatment of the testis followed by further cell purification also
makes it possible to isolate Leydig cells[14]. With such isolated
tissue fractions it has been possible to study the steroid production
in vitro and it has been shown[16] that testosterone is clearly produced by interstitial tissue in vitro, but not by seminiferous tubules.

    Subcellularly a crucial step in steroid production occurs in
Leydig cell mitochondria. Estimation of endogenous steroid production
during in vitro incubation of subcellular fractions shows that production of pregnenolone, an obligatory intermediate in the conversion
of cholesterol to testosterone, occurs only in mitochondria[23].

EFFECTS OF LH ON THE TESTIS

    Hypophysectomy of male animals results in a regression of the
testis including a decreased steroid production, resulting in decreased
testicular and plasma steroid levels. These effects of hypophysectomy
can be counteracted by administration of trophic hormones with lutro-
pin-like activity (LH and hCG) and it is known, that LH can stimulate
steroid production and secretion by testes in intact animals as well
as in isolated testicular tissue, isolated interstitial tissue and
purified Leydig cells[14]. All of the described steroidogenic effects
of trophic hormones on testis Leydig cells appear to be specific for
LH and hCG.

    During long-term administration of LH increased activities of
many of the enzymes involved in steroid metabolism have been observed.
Such long-term effects probably reflect a growth effect of LH on the
testis, resulting in an increased activity of enzymes.

    Short-term (seconds to a few hours) LH administration results
in concomitant changes of several different biochemical parameters.
The quantitatively most important point of attack of LH appears to be

the regulation of the mitochondrial cholesterol side-chain cleavage
activity involving the cytochrome P-450 system. This can be clearly
illustrated by the comparison of the endogenous steroid production in
mitochondrial preparations isolated from testes of normal and LH-
treated rats[24].

     For the biochemical parameters involved in the action of LH on the
testis Leydig cell, a model has been generally accepted which involves
several parameters known to be involved in the action of several other
non-steroid hormones. It is now established that lutropin (luteinizing
hormone, LH) specifically interacts with a plasma membrane receptor in
the testis Leydig cell, which results in a cascade of events which include increased cyclic AMP production, activation of cyclic AMP-dependent protein kinase and increased production of steroid hormones.
Stimulation of the testis in vivo with hCG or LH has also been shown
to lead to desensitization of the Leydig cells, resulting in loss of
LH receptors and adenylate cyclase activity and inhibition of certain
steroidogenic pathways[2]. Inhibition of Leydig cell protein synthesis
results in a rapid inhibition of LH-induced steroidogenesis[4] and has
also been reported to prevent down regulation of LH receptors[20,22],
suggesting that specific proteins are involved in both these processes.
Some of the observed actions of LH on testis Leydig cells are schematically summarized in Figure 1. For a proper interpretation of this
figure it is important to realize, that there is no certainty about
the sequence of effects observed after LH administration. Most changes
are correlated only in a time-related manner. Apart from the interrelated effects of adenylate cyclase, cAMP and activation of protein
kinase activity, there is, however, no absolute proof about the
possible interrelationships of the other parameters.

INVOLVEMENT OF CYCLIC AMP, PROTEIN KINASES AND PHOSPHOPROTEINS

     The available evidence indicates that cyclic AMP is the intra-
cellular messenger of LH action on Leydig cell steroidogenesis. No
other mechanism has been demonstrated to be involved. The major effect
of LH action via cyclic AMP appears to be the regulation of mitochondrial cholesterol side-chain cleavage involving the cytochrome
P-450 system[24]. In common with several other systems, it has been
demonstrated that stimulation of testosterone synthesis can occur in
rat Leydig cells without there being detectable changes in cyclic AMP
production[1,5,18,19]; LH was known to stimulate cyclic AMP-dependent
protein kinases[5] and therefore an investigation was undertaken to
determine if at low concentrations of LH, cyclic AMP-dependent protein
kinase activation in Leydig cells is a more sensitive parameter than
the cyclic AMP concentration itself[6]. During the course of that
investigation it was found that the Leydig cells contained both type
I and type II protein kinase. It was clearly demonstrated that all
levels of LH which stimulated testosterone production also stimulated
protein kinase activation; with the lowest amount of LH used (0-1ng/ml),

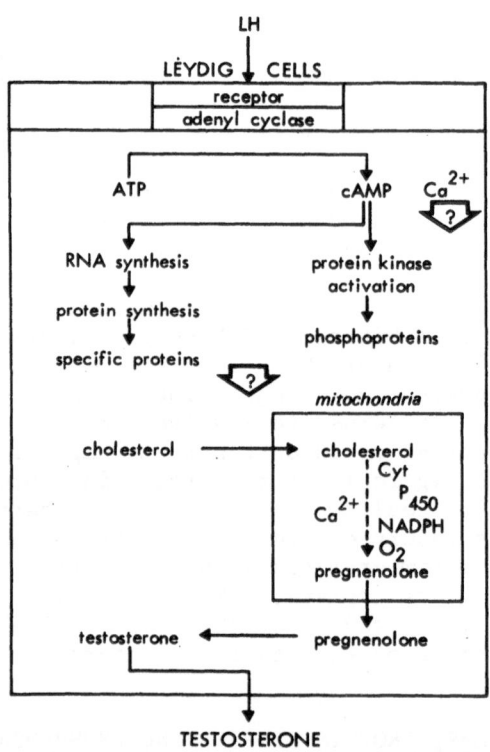

Fig. 1. Some LH actions on testis Leydig cells.

an 8.4 ± 0.9% (S.E.M., n = 6) stimulation of cyclic AMP-dependent
protein kinase activation occurred, increasing to 100% with 1,000
ng/ml, compared with 3.2 ± 1.0% (S.E.M., n = 7) for testosterone production with 0.1 ng/ml LH, which increased to 100% with 100 ng/ml LH.
With amounts of LH up to 1 ng/ml (which gave half maximal stimulation
of testosterone production), no detectable increases in net cyclic AMP
production were obtained. With higher amounts of LH, cyclic AMP levels
were increased, but maximum production was not reached with 1,000 ng/ml.

LH was also shown to stimulate the phosphorylation in testis Leydig cells of three endogenous proteins with apparent mol. wts. of 14,300, 57,000 and 77,600[7]. This effect was also found with dibutyryl cyclic AMP. A LH dose-dependent correlation was found between the phosphorylation of these proteins and stimulation of testosterone production.

Similarly in a Leydig cell tumour two proteins of mol. wts. of 14,000 and 57,000, whose phosphorylation was increased by LH, also showed a good dose-dependent correlation of phosphorylation and testosterone production[12].

The above studies on phosphorylation in intact cells were extended to cell-free systems of Leydig cells[10]. The homogenates and subcellular fractions were incubated with (^{32}P)-ATP and then subjected to polyacrylamide slab gel electrophoresis, followed by autoradiography and densitometry. The addition of cyclic AMP to incubations of Leydig cell homogenates increased the phosphorylation of proteins of mol. wts. of 58,000, 55,000, 47,000 and 37,000. The lowest amount of cyclic AMP required to give a detectable increase in phosphorylation of the proteins was 0.1 µmol/l and reached a maximum with 10 µmol/l. The rate of phosphorylation was very high; maximum phosphorylation was found already after 10-30 sec and thereafter a rapid decrease in phosphorylation of the proteins of mol. wts. 58,000, 55,000 and 47,000 took place. In contrast, the degree of phosphorylation of the 37,000 mol. wt. protein remained constant during incubation for required cyclic AMP and Mg^{++}, but was not increased in the presence of cyclic GMP, Zn^{++}, Ca^{++}, NaCl or testosterone. The 58,000, 55,000 and 47,000 mol. wt. proteins were found to be localized in the cell cytosol, whereas the 37,000 mol. wt. protein was associated with the 500 g and 1,500 g particulate fractions. The 58,000 mol. wt. protein was further purified by affinity chromatography and was found to be identical to the R subunit of the type II protein kinase.

The above data are consistent with an important role for cyclic AMP, cyclic AMP-dependent protein kinases and phosphoproteins in the control of testis Leydig cell functions.

EFFECTS OF LH ON SPECIFIC PROTEIN SYNTHESIS

It has been shown that inhibitors of protein synthesis will interfere also with the effect of LH on testosterone production in rat testis[4,13,17]. Addition of cycloheximide to Leydig cells was shown to inhibit rapidly testosterone synthesis and the decrease followed first order kinetics with a half-life of 13 min[4]. Further work showed that LH stimulates the synthesis of a protein of mol. wt. 21,000 in adult rat Leydig cells[15]. This protein was detected by labelling the Leydig cell proteins with (^{35}S)-methionine, followed by separation by polyacrylamide gel electrophoresis and radiography of the dried gel.

Incubation of Leydig cells with dibutyryl cyclic AMP or cholera toxin
also resulted in the stimulation of synthesis of the protein. Actinomycin D prevented the LH-induced protein synthesis when added immediately or one hour after the start of the incubation, but not when added
after five to six hours. This was interpreted as reflecting that after
induction of mRNA coding for LH-induced protein, LH had no influence
on the synthesis of the protein in the presence of actinomycin D. From
these results it was concluded that specific stimulation of protein
synthesis by LH is most probably mediated via cyclic AMP and involves
the action of the cyclic nucleotide at the level of the nucleus.
Similar studies were carried out on immature and tumour Leydig cells[15]
and it was found that LH induces the synthesis of proteins with mol.
wts. of 27,000 and 29,000 in tumour Leydig cells and proteins with
mol. wts. of 11,000, 21,000, 27,000 and 29,000 in Leydig cells from
immature rats. Thus in similarity with the rat adult Leydig cell, LH
induces a protein of similar mol. wt. (21,000) in the immature cell,
but not in the tumour cell. Cyclic AMP-induced phosphorylation of
these different proteins in the various Leydig cell preparations could
not be detected, nor could an effect of cycloheximide be detected on
the proteins which were phosphorylated during stimulation with cyclic
AMP. Addition of cycloheximide to the Leydig cells showed that the
LH-induced proteins had half-lives of more than 30 min. In addition,
the lag period before induction of the proteins was demonstrated to
be 2 hours compared with less than 5 minutes for the stimulation of
steroidogenesis. It is therefore unlikely that the demonstrated LH-
induced proteins play a role in the short-term stimulation of testosterone production by LH. In fact they could even be inhibitors of
steroidogenesis and possibly be involved in desensitization mechanisms.
It is also possible that the demonstrated stimulation of LH-induced
protein synthesis is unrelated to steroidogenesis and it is only one
of the first trophic effects of LH on Leydig cells.

THE ROLE OF SPECIFIC PROTEINS IN LH-STIMULATED TESTOSTERONE SYNTHESIS

     In a study in which the role of RNA synthesis in Leydig cell
steroidogenesis was investigated[9], it was demonstrated that the inhibitory effect of the RNA synthesis inhibitors actinomycin D and cor-
dyceptin on LH-stimulated testosterone was found to decrease if the
cells were pre-incubated after isolation and before the addition of
LH. Furthermore, inhibition of testosterone synthesis in the pre-
incubated cells did not occur until one hour after addition of the
inhibitor, whereas testosterone synthesis was increased within 5 min
by LH. These results indicated that during pre-incubation, synthesis
occurs of substances involved in steroidogenesis and this synthesis
is not necessarily dependent on LH.

     These results and the rapid stimulation of testosterone produc-
tion by LH led to the proposal that the RNA(s) and regulator protein(s)
required for stimulation of Leydig cell steroidogenesis are synthesized

continuously and independently of LH[8]. If this hypothesis is true,
then it should have been possible to inhibit the production of the
short half-life regulator protein(s) in the absence of LH and to
have demonstrated a subsequent lower LH stimulation of steroidogenesis.
However, it was found that pretreating Leydig cells with cycloheximide
had no effect on subsequent LH-stimulated testosterone production[11].
In order to demonstrate an effect of cycloheximide, it was found
necessary to add dibutyryl cyclic AMP during the pretreatment (this
compound was added in place of LH, because it can easily be washed out
of the cells). This resulted in a 60% inhibition of subsequent LH-
stimulated testosterone synthesis compared with the control during the
first 5 min incubation. It was concluded that the effect of cycloheximide does require the presence of LH (or cyclic AMP) and that LH does
influence the regulator protein(s) directly or indirectly. The hypothesis summarized in Figure 2 was therefore put forward[11]. In the
absence of LH the regulator protein(s) is present in a stable form
with a long half-life and is synthesized continuously. Addition of
cycloheximide will prevent its further synthesis, but because of its
long half-life, there will be a sufficient pool of this protein(s) for
subsequent stimulation of steroidogenesis when LH is added. In the
presence of LH the stable protein is converted to an unstable form
with a shorter half-life. Thus in the presence of cycloheximide
further synthesis of this short half-life protein will be inhibited,
the pool will be depleted and steroidogenesis will cease. An additional effect of this transformation may be that it is an activation
step. It remains to be investigated which mechanisms are involved,
but they could include direct effects of lutropin on the protein or
indirect effects by activation of proteolytic enzymes and/or phosphorylation.

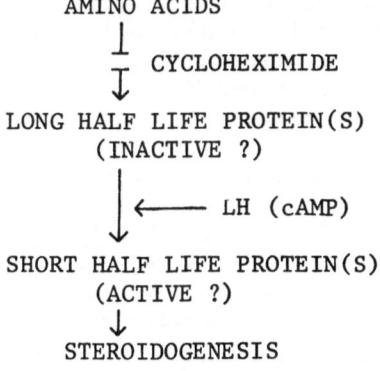

Fig. 2. Hypothesis for the role of protein synthesis in the
         mechanism of LH action.

REFERENCES

1. K. J. Catt and M. L. Dufau, Spare gonadotrophin receptors in
    rat testis, Nature (London) New Biol. 244: 219 (1973).
2. K. J. Catt, J. P. Harwood, G. Aguilera and M. L. Dufau, Hormonal
    regulation of peptide receptors and target cell responses,
    Nature 280: 109 (1979).
3. A. K. Christensen and M. M. Mason, Comparitive ability of semin-
    iferous tubules and interstitial tissue of rat testes to
    synthesize androgens from progesterone-4-^{14}C in vitro, Endo-
    crinology 76: 646 (1965).
4. B. A. Cooke, F. H. A. Janszen, W. F. Clotscher and H. J. van der
    Molen, Effect of protein synthesis inhibitors on testosterone
    production in rat testis interstitial tissue and Leydig cell
    preparations, Biochem. J. 150: 413 (1975).
5. B. A. Cooke and J. W. C. M. van der Kemp, Protein kinase activity
    in rat testis interstitial tissue. Effect of luteinizing
    hormone and other factors, Biochem. J. 154: 371 (1976).
6. B. A. Cooke, L. M. Lindh and F. H. A. Hanszen, Correlation of
    protein kinase activation and testosterone production after
    stimulation of Leydig cells with luteinizing hormone, Biochem.
    J. 160: 439 (1976).
7. B. A. Cooke, L. M. Lindh and F. H. A. Janszen, Effect of lutro-
    pin on phosphorylation of endogenous proteins in testis
    Leydig cells: Correlation with testosterone production,
    Biochem. J. 168: 43 (1977).
8. B. A. Cooke, F. H. A. Janszen and M. J. A. van Driel, Inhibition
    of Leydig cell steroidogenesis: Effect of actinomycin D before
    and after preincubation of Leydig cells in vitro, Int. J.
    Androl. suppl. 2: 240 (1978).
9. B. A. Cooke, F. H. A. Janszen, M. J. A. van Driel and H. J. van
    der Molen, Evidence for the involvement of lutropin-indepen-
    dent RNA synthesis in Leydig cell steroidogenesis. Molec.
    cell Endocrinol. 14: 181 (1979).
10. B. A. Cooke, L. M. Lindh and H. J. van der Molen, Cyclic AMP
    dependent phosphorylation of endogenous proteins in rat
    testis Leydig cells, J. Endocrin. 83: 32P (1979).
11. B. A. Cooke, L. M. Lindh and H. J. van der Molen, The mechanism
    of action of lutropin on regulator protein(s) involved in
    Leydig cell steroidogenesis, Biochem. J. 184: 33 (1979).
12. B. A. Cooke, L. M. Lindh, F. H. A. Janszen, M. J. A. van Driel,
    C. P. Bakker, M. P. I. van der Plank and H. J. van der Molen,
    A Leydig cell tumour – A model for the study of lutropin
    action, Biochim. Biophys. Acta 583: 320 (1979).
13. P. F. Hall and K. B. Eik-Nes, The action of gonadotropic hormones
    upon rabbit testes in vitro, Biochim. Biophys. Acta 63: 411
    (1962).
14. F. H. A. Janszen, B. A. Cooke, M. J. A. van Driel and H. J.
    van der Molen, Purification and characterization of Leydig
    cells from rat testes, J. Endocrin. 70: 345 (1976).

15. F. H. A. Janszen, B. A. Cooke, M. J. A. van Driel and H. J. van
     der Molen, The effect of lutropin on specific protein syn-
     thesis in tumour Leydig cells and in Leydig cells from im-
     mature rats, Biochem. J. 172: 147 (1978).
16. F. H. de Jong, A. H. Hey and H. J. van der Molen, Oestradiol-
     17β and testosterone in rat testis tissue: Effect of gonad-
     otrophins, localization and production in vitro. J. Endocr.
     60: 409 (1974).
17. C. Mendelson, M. L. Dufau and K. J. Catt, Dependence of gonado-
     tropin-induced steroidogenesis upon RNA and protein synthesis
     in the interstitial cells of the rat testis. Biochim. biophys.
     Acta 411: 222 (1975).
18. W. R. Moyle and J. Ramachandran, Effect of LH on steroidogenesis
     and cyclic AMP accumulation in rat Leydig cell preparations
     and mouse tumour Leydig cells. Endocrinol. 93: 127 (1973).
19. F. F. G. Rommerts, B. A. Cooke, J. W. C. M. van der Kemp and
     H. J. van der Molen, Effect of luteinizing hormone on 3',5'-
     cyclic AMP and testosterone production in isolated inter-
     stitial tissue of rat testis, FEBS Lett. 33: 114-118.
20. J. M. Saez, F. Haour and A. M. Cathiard, Early hCG-induced
     desensitization in Leydig cells, Biochem. Biophys. Res. Comm.
     81: 552 (1978).
21. L. T. Samuels, L. Bussman, K. Matsumoto and R. A. Huseby, Organ-
     ization of androgen biosynthesis in the testis, J. Steroid
     Biochem. 6: 291 (1975).
22. R. M. Sharpe, Relationship between testosterone fluid content
     and luteinizing hormone receptors in the rat testis, Biochem.
     Biophys. Res. Comm. 75: 711 (1977).
23. G. J. van der Vusse, M. L. Kalkman and H. J. van der Molen, Endo-
     genous production of steroids by subcellular fractions from
     total rat testis and from isolated interstitial tissue and
     seminiferous tubules, Biochim. Biophys. Acta. 297: 179 (1973).
24. G. J. van der Vusse, M. L. Kalkman and H. J. van der Molen,
     Endogenous steroid production in cellular and subcellular
     fractions of rat testis after prolonged treatment with
     gonadotrophins, Biochim. Biophys. Acta 380: 473 (1975).
25. T. Yanaihara and P. Troen, Studies of the human testis. I.
     Biosynthetic pathways for androgen formation in human test-
     icular tissue in vitro, J. Clin. Endocr. Metab. 34: 783 (1972).

REGULATION OF GASTRIC ACID SECRETION

W. Joseph Thompson and Gary C. Rosenfeld

Department of Pharmacology
The University of Texas Medical School at Houston
Houston, Texas   77030

INTRODUCTION

Despite the fact that increased gastric acid secretion is not diagnostic for peptic ulcer disease, it is generally believed that where there is no acid there is no ulcer. Thus, there remains a strong medical interest in the understanding of the cellular regulation of gastric acid secretion by the major secretagogues, histamine, gastrin and acetylcholine and therapy designed to interfere with this regulation. Released as a result of various physiological stimuli, including food, these paracrine, endocrine, and neurohumoral agents act by both direct and indirect pathways to effect the production and secretion of hydrogen ions from parietal cells located in the lower portion of the oxyntic glands in the fundic stomach. Recently, the study of gastric acid secretion has been advanced by the advent of methods to isolate gastric oxyntic glands and gastric parietal cells. These preparations have allowed the formulation of new concepts to explain in vivo and in vitro physiological and pharmacological data derived from studies on the interactions of hormones that regulate the acid secretory process[14,17,20].

Gastric parietal cells also produce intrinsic factor, a 55,000 molecular weight glycoprotein, which binds and transports vitamin B_{12} to the ileum. The lack of intrinsic factor production leads to pernicious anemia, a disease with an additional parietal cell autoimmune component, which is normally accompanied by achlorhydria indicating the close functional association of intrinsic factor and hydrogen ion secretory processes. Patients with Zollinger-Ellison syndrome show elevated acid secretion due to excessive gastrin production.

The physiological stimuli that influence the release of parietal cell regulatory hormones do so through a complex interaction with vagal

269

efferent and afferent fibers, local mucosal refex arcs, antral G
(gastrin) cells, duodenal endocrine cells, and gastric mast and
enterochromaffin cells. Stimulation occurs in three phases, cephalic,
gastric and intestinal, and results in the release of both the secretagogues, acetylcholine, gastrin and possibly histamine as well as
several acid secretory inhibitors, the physiological significance of
which are still not clear.

Acetylcholine, which is released by the vagus in response to
chewing, swallowing, hypoglycemia, smell etc. stimulates acid secretion directly, as well as indirectly by releasing gastrin into the
circulation. Food entering the stomach tends to neutralize the pH
of the gastric juice which aids in gastrin release and the stimulation
of acid production. Also, food causes stomach distention which stimulates both local vagal reflexes and afferent sensory fibers to the
central nervous system and the subsequent release of acetylcholine
from the efferent vagus nerve. Digested peptides also release gastrin
from intestinal G cells, as well as unidentified intestinal hormones
termed entero-oxyntins. When, due to acid secretion, the pH of the
juice contents falls below 3, antral gastrin release ceases and acid
secretion is inhibited. Secretion is also inhibited by secretin which
is released from the duodenum into the circulation and inhibits parietal cells and gastrin release by G cells. Other digestion monomers
(e.g. fatty acids) release polypeptides such as gastric inhibitory
peptide, cholecystokinin, and uncharacterized enterogastrones which
are possible physiological inhibitors of hydrogen ion production[10].

The role of histamine as a physiologically important regulator of
acid secretion is not clear. Acid secretion is considered an H_2-type
histamine response, being differentiated from H_1-type histamine responses, such as smooth muscle contraction by its insensitivity to
classical antihistamines such as tripelennamine, and its inhibition
by the newer histamine H_2-receptor antagonists such as cimetidine[7].
Historically, investigations of the actions of histamine have figured
significantly in concepts regarding the regulation of the acid secretory process. In 1938 MacIntosh proposed that histamine was an essential mediator of the actions of acetylcholine and gastrin on acid
secretion[12], a hypothesis later supported by Code[4] which became known
as the final common mediator hypothesis. In the absence of isolated
parietal cell preparations, however, it was difficult, if not impossible, to know which of the secretagogues act directly on the parietal
cell and which act by indirect mechanisms. In addition each secretagogue is known to potentiate the action(s) of the other; that is,
together they result in a secretory response greater than that produced
by the sum of each used singularly at maximal concentrations.

The final common mediator hypothesis, as well as the physiological significance of histamine, gradually lost favor, but not without
argument, until the recent synthesis of the pharmacologically specific
H_2-receptor antagonists, such as burimamide, metiamide, and the now

clinically useful antagonist, cimetidine[3]. In addition to being more
potent, gastrin and acetylcholine produce higher maximal responses
than histamine and data on its in vivo release was difficult to in-
terpret[9,13]. Also, the final common mediator hypothesis did not
explain secretagogue potentiation on the inhibition by atropine, a
specific anticholinergic antagonist, to reduce histamine and acetylcholine induced acid secretion. Nevertheless, cimetidine blocks acid
secretion effected by all three secretagogues and upon repeated stimulation, in vitro gastric mucosa was shown to become refractory to
gastrin and acetylcholine, but not to histamine[11].

        Grossman and Konturek proposed an alternative to the final common
mediator hypothesis in which parietal cells were endowed with distinct
receptors for histamine, gastrin and acetylcholine[8]. This model suggests that histamine is the major acid secretagogue and its actions
are modified by acetylcholine and gastrin through secretagogue-receptor
interactions. With the advent of isolated gastric glands and parietal
cells significant progress has been made toward an understanding of
receptor content, as well as pathways subsequent to secretagogue receptor interactions.

        Parietal cells have been successfully isolated from gastric mu-
cosa from a variety of species[20]. Physiological responses, pharmacological sensitivity, and biochemical and morphological criteria are
varied and not complete with the different cell preparations. Parietal cells comprise approximately 10% of total gastric mucosal cells
and are the largest ($\approx 20\mu$) in this tissue. They have distinctive
morphology consisting of numerous mitochondria and a prominent vesi-
culated smooth endoplasmic reticulum. Upon stimulation, the vesicles
fuse to form canaliculi with numerous microvilli that are the sites
of acid secretion. In the intact tissue the parietal cells show
polarity in that their serosal and mucosal membranes are differentiated
to achieve active transport of hydrogen, chloride, and sodium ions and
to maintain electrochemical, osmotic and pressure gradients. Isolated
cells are prepared by enzymatic digestion of mucosal surface cells,
mechanical dispersal usually with divalent cation chelators, and
purification by density gradient or counter-flow centrifugation[15,19].

        Which secretagogues act directly on parietal cells have been
defined by biochemical criteria of direct receptor binding and adenyl
cyclase activation and functional criteria of cyclic AMP accumulation,
oxygen consumption and aminopyrine uptake, the former two processes
being indirect measures of hydrogen ion secretion. Muscarinic cholinergic receptor sites have been directly demonstrated by binding studies
using the tritiated antagonist quinuclidinyl benzilate[5]. Studies of
aminopyrine accumulation showed close agreement of affinity constants
calculated for cholinergic antagonist inhibition with those calculated
from the direct binding studies[6]. Cholinergic binding and functional
activation were shown to be specific, as neither was influenced
significantly by histaminic antagonist.

Receptors of histamine were demonstrated by activation of adenylyl cyclase, as well as function. Histamine induced cellular responses and adenylyl cyclase activation were selectively inhibited by H_2-receptor agonists and antagonists effects on aminopyrine accumulation were remarkably similar to those determined for activation and inhibition of adenylyl cyclase[14,19,20]. Histamine activation of parietal cell adenylyl cyclase was dependent upon guanine nucleotide cofactor which showed hysteretic kinetic properties observed for the enzyme in other tissues. It has been proposed that GTP is an important factor in the regulation of acid secretion by histamine which acts to couple histamine occupied receptor sites with catalytic sites to convert ATP to cyclic AMP[1]. Histamine, but not acetylcholine or gastrin, elevated cyclic AMP content in cells isolated from all specific studies[18,20].

Studies on gastrin receptors are much more limited and not definitive. A small increase in oxygen consumption has been observed in isolated canine cells which was not blocked by H_2-receptor antagonists. However, full gastrin response required histamine[16]. Studies of secretagogue actions in isolated glands have been in good agreement with functional analyses in isolated cells[2]. However, gastrin responses were not observed in the glands except in the presence of theophyline. These results were interpreted as indicating the secondary release of histamine which acts singularly to enhance aminopyrine uptake in parietal cells.

Studies with isolated cells have led to the hypothesis by Soll that secretagogues act on separate receptors[17]. This proposal differed from that of Grossman and Konturek in that no interactions between secretagogue receptors were hypothesized. Cholinergic and histaminergic antagonists were proposed to act with specificity at the level of the parietal cell, the apparent conflicting in vivo non-specificity of the antagonists caused by the inhibition of a potentiated response due to the local release of histamine and acetylcholine. Our studies have led us to hypothesise that secretagogue interactions reflect cellular interactions of pathways regulated by receptor initiated second messenger production, cyclic AMP and calcium[20]. We proposed that secretagogues act through separate receptors. Acetylcholine and histamine cause mobilization of calcium ions, and histamine alone increases cyclic AMP. These second messengers promote multiple levels of phosphorylation of regulatory phosphate acceptor proteins via calmodulin sensitive and cyclic AMP-dependent kinases which are the basis of potentiating secretagogue effects.

To recapitulate, our understanding of the regulation of gastric acid secretion has progressed immeasurably with the use of isolated parietal cells. A basis for the explanation of seemingly contradictory data obtained in vivo is becoming apparent. With respect to the major acid secretagogues, parietal cells have been shown to contain receptors for muscarinic cholinergic agonists by direct binding and functional analyses and receptors for histamine H_2-agonists by adenylyl cyclase and functional analyses, but gastrin receptors remain elusive. The conflicting nature of in vivo findings stems from the potentiation of response obtained by multiple receptor pathway stimulation. In the future, studies on acid secretory inhibitors will yield information on the potentiation aspect of hormone regulation of acid secretion and studies on parietal cell biochemistry will yield information on the mechanism of potentiation, as well as ion transport and structural protein modification and subcellular compartmental integration. This approach toward understanding a complex hormone regulatory process should aid in better treatment and diagnosis of peptic disease.

REFERENCES

1.   J. Abramowitz, R. Iyengar and L. Birnbaumer, 1979, Guanyl nucleotide regulation of hormonally-responsive adenylyl cyclases. Mol. Cell. Endo. 16: 129-246.

2.   T. Berglindh, H. F. Helander and J. Obrink, 1976, Effects of secretagogues on oxygen consumption, aminopyrine accumulation and morphology in isolated gastric glands. Acta Physiol. Scand. 97: 401-414.

3.   J. W. Black, W. A. M. Duncan, C. J. Durant, C. R. Ganellin and M. E. Parsons, 1972, Definition and antagonism of histamine H_2-receptors. Nature 236: 385-390.

4.   C. F. Code, 1965, Histamine and gastric secretion: A later look. Fed. Proc. 24: 1311-1321.

5.   R. Ecknauer, W. J. Thompson, L. R. Johnson and G. C. Rosenfeld, 1980, Isolated rat gastric parietal cells: Cholinergic response and pharmacology. Life Sic. (submitted).

7.   M. Feldman and C. T. Richardson, 1978, Histamine H_2-receptor antagonists. Adv. in Int. Med. 23: 1-24.

8.   M. I. Grossman and S. J. Konturek, 1974, Inhibition of acid secretion in dog by metiamide a histamine antagonist acting on H_2-receptors. Gastroenterology 66: 517-521.

9.   L. R. Johnson, 1971, Control of gastric secretion: No room for histamine? Gastroenterology 61: 106-118.

10. L. R. Johnson, ed., 1977, "Gastrointestinal Physiology" C. V. Mosby, St. Louis, Mo.

11. K. D. Kasbekar, H. A. Ridley and J. G. Forte, 1969, Pentagastrin and acetylcholine relation to histamine in H^+ secretion by gastric mucosa. Am. J. Physiol. 216: 961-67.

12. F. C. MacIntosh, 1938, Histamine as a normal stimulant of gastric secretion. O. J. Exp. Physiol. 28: 87-98.

13. P. K. Rangachari, 1978, Histamine as the final common mediator: A view from the fence. Acta Physiol. Scand. Spl. Suppl. p. 209-218.

14. G. C. Rosenfeld, S. J. Strada, E. J. Dial, C. F. Bearer and W. J. Thompson, 1980, Histamine, cyclic nucleotides, and gastric parietal cell secretion. Adv. Cyclic Nucl. Res. 12: 225-265.

15.   A. H. Soll, 1978a, The actions of secretagogues on oxygen uptake
      by isolated mammalian parietal cells. J. Clin. Invest. 61:
      370-380.
16.   A. H. Soll, 1978b, The interaction of histamine with gastrin and
      carbamylcholine on oxygen uptake by isolated mammalian parie-
      tal cells. J. Clin. Invest. 61: 381-389.
17.   A. H. Soll and M. I. Grossman, 1978, Cellular mechanisms in acid
      secretion. Ann. Rev. Med. 29: 495-507.
18.   A. H. Soll and A. Wollin, 1979, Histamine and cyclic AMP in iso-
      lated canine parietal cells. Am. J. Physiol. 237(5): E444-450.
19.   W. J. Thompson, L. K. Chang and G. C. Rosenfeld, 1980, Regulation
      of adenylyl cyclase of purified rat gastric parietal cells.
      Am. J. Physiol. (In press).
20.   W. J. Thompson, E. D. Jacobson and G. C. Rosenfeld, 1980, Cyclic
      AMP in the gastrointestinal tract: receptor control of hydro-
      gen ion secretion by mammalian gastric mucosa. Handbook of
      Experimental Pharm. (In Press).

MOLECULAR BASIS OF INTERFERON ACTION

Jean Content and Martine Verhaegen-Lewalle

Institut Pasteur du Brabant
28, rue du Remorquer
B-1040 Brussels, Belgium

1. INTRODUCTION

Interferon was discovered 24 years ago by Isaacs and Lindenmann (1957). Allantoic fluid from chick embryonated egg previously inoculated with inactivated influenza virus contained an antiviral activity detectable when transferred to fresh cells. This potent activity was pH 2 resistant, inactivated by trypsin, and not sedimentable at 100,000 x g. This substance was rapidly found to have important other properties: it had a broad spectrum of antiviral activity i.e. it was active against most animal viruses; it was species specific and was relatively non toxic. All these fundamental properties have been amended recently (section 7 and 10) although the basic observations remain essentially true. Extension of these early observations to other animal systems have led to considerable development of interferon research in several directions. It is widely recognized now that the antiviral activity of interferon (IFN), which led to its discovery, is only the "visible part of an iceberg", since IFN has a large spectrum of other activities (Stewart II, 1979b; De Clercq, 1980): it can exert an anticellular activity, or antitumour activity when injected in the animal and interacts with the immune system, exerting an important surveillance control against tumour or infectious cells by stimulation of natural killer cell activity. Interferons have been reported to inhibit the differentiation process in Friend virus-infected erythroleukemia cells (Rossi et al., 1977) and to inhibit the conversion of 3T3-L1 mouse fibroblasts into adipocytes (Keay and Grossberg, 1980), suggesting a possible function in the regulation of eukaryotic cell differentiation. It should be noted that interferon has really multiple effects on the animal cell itself ranging from plasma membrane and cell motility to lysosomes and

275

cytoskeleton function and organization. On lymphocyte and other cell surfaces, it increases the expression of histocompatibility antigens and in some cases the expression of Fc receptors for immunoglobulins. It also enhances its own production in some circumstances (priming effect) and reduces cell growth, protein synthesis, DNA synthesis and thymidine incorporation. Of these multiple effects some have been recently ascribed to interferon itself since the experiments could be repeated with electrophoretically pure interferon. (Gresser et al., 1979).

Although we shall refer to "interferon" (IFN) very often in this text multiple varieties of interferons exist. They are produced by different cell types stimulated with different inducers (Table I), mainly in tissue culture systems. For all these interferons the amount of material produced has always been limiting and since interferons have extremely high specific activity their purification has been problematic for a long time. Only recently several murine and human interferons were purified to homogeneity. This was confirmed by partial amino acid sequence determination at the NH_2-site of the molecules. All sequences were further extended and confirmed by nucleic acid desequencing of the entire cloned gene for various α and β HuIFN's. In addition to this structural knowledge, cloning has indicated that α- and β-IFN genes are - besides histone genes - the second known example of eukaryotic genes devoided of intervening sequences. It has also opened possibilities for large scale production of these expensive molecules which can be further purified from bacterial extracts by affinity chromatography on the very recently developed monoclonal antibody columns.

Another important consequence of this work has been the finding that both at the genomic level and at the protein level considerable sequence heterogeneity exists among HuIFN's.

This is important when considering the mechanism of interferon action:

1. very few biochemical studies have been done with pure IFN (most studies were done with preparations containing 0.1 to 1 % IFN).
2. even pure HuIFN-α may contain multiple molecular species. Each of them may have a distinct spectrum of activity when the various actions of interferon are considered.
3. furthermore different subspecies of IFN might have complex interactions. It seems established that small amounts of HuIFN-β can potentiate certain activities of γ-IFN's (Fleischmann et al., 1979, de Ley et al., 1980).

In this brief review we will limit ourselves to a few mechanisms apparently triggered by IFN, which may play a role in the expression

Table I. The diversity and nomenclature of human interferons

Interferon type	sub-types	Cell source	Inducer	Specific activity	M W	Stability at pH2
HuIFN-α Leukocyte	$\alpha_1 \ldots \ldots \alpha_8$	Blood buffy coats	Sendai virus or other paramyxoviruses	$2-4.10^8$U/mg	15,000 & 20,000	+
HuIFN-β Fibroblast	not found	Foreskin or foetal muscle and skin fibroblasts (diploid)	$(I)_n \cdot (C)_n$ or other double-stranded RNA's	"	20,000	+
HuIFN-α (+β) Lymphoblastoid		Lymphoblastoid cell lines (e.g. Namalva)	Sendai or measles virus or other paramyxoviruses	"	15,000 & 20,000	
HuIFN-γ Immune or type II	not found	Blood buffy coats (T lymphocytes)	Lectins (PHA)* or SEA**	10^8 U/mg	44,000	

* PHA : Phytohemagglutinin

** SEA : Staphylococcal enterotoxine A

of the antiviral state. For other aspects of IFN research, the reader may consult some recent reviews on interferon (Stewart II, 1979a; De Clercq, 1980), on its multiple non antiviral actions (Stewart II, 1979b) and clinical applications (Cantell, 1979b); Scott and Tyrell, 1980), on antitumoral activity in humans (Sikora, 1980) and on the mechanisms of interferon induction (Torrence and De Clercq, 1977; De Clercq, 1980). Finally on the subject treated here, excellent reviews include those from Baglioni (1979), Revel (1979), Hovanessian (1979), Williams and Kerr (1980) and Sen (1981).

2. EARLY EVENTS IN INTERFERON ACTION

a) Even at low temperature (4°C) IFN binds to receptive cells and modifies the cell membrane. This binding can be inhibited competitively with TSH or cholera toxin and vice versa. Very recently a "domain" in the carboxy-terminal region of α and β HuIFN was found to be very similar to a region of the β subunit of cholera toxin (Derynck, 1981). This region of both molecules could be involved in receptor recognition.

IFN plasma membrane receptors have not yet been isolated. They probably contain gangliosides and glycoprotein structure(s). Their role in the species specificity of IFN is not yet clearly established. Aguet (1980) has recently described a binding assay (using I^{125}-labelled electrophoretically pure IFN) and found an IFN resistant murine cell line apparently missing such a receptor. It is not known whether IFN has to be internalized after binding to exert its effects.

b) Beyond this stage very little is known concerning the mechanism of interferon action. Plasma membrane changes including morphological and physiochemical changes have been described (Friedman, 1977). The earliest biochemical event detectable in IFN-treated cells consists in a rapid rise in cGMP concentration (Tovey et al., 1979), within a few minutes after IFN treatment.

Whether any of these modifications is responsible for the transmission of a cytoplasmic message to the nucleus where transcription of certain mRNA's species would be derepressed is not known. A few hours after interferon treatment new proteins appear in cytoplasm of treated cells (see section 8c) and two enzymes, a phosphoprotein kinase and a 2-5A synthetase are induced (see 4 and 5). However, as soon as 3 hr after IFN treatment of murine or human cells a mRNA coding for the 2-5A synthetase (as detected by translation in Xenopus oocyte) can be extracted from their cytoplasm (Shulman and Revel, 1980). Another interesting information is that continuous interaction between IFN and its putative receptor should proceed for at least 3-4 hours in order to obtain maximal 2-5A synthetase induction. Beyond this time IFN may be washed out or neutralized with specific antibodies without affecting the further development of an antiviral state (Shulman and Revel, 1980).

3. REQUIREMENTS FOR THE DEVELOPMENT OF THE ANTIVIRAL STATE

Any proposed biochemical mechanism should take the following basic observations into consideration:

a) Interferon has a very broad spectrum of action and prevents the replication of various classes of animal viruses whose reproduction implies totally different pathways (Friedman, 1977). Furthermore, it may be very naive to consider a single mechanism for all viruses. Rather recent observations indicate on the contrary that the metabolic pathways described in the following sections (the IFN-induced, double-stranded RNA (dsRNA)-dependent enzymes) may or may not play a role in preventing viral replications, depending on the kind of virus considered (Nilsen et al., 1980b) (see section 6). It should be stressed again that, although it has always been tempting to explain all antiviral actions of IFN by a translation inhibition (a "final common path" for various animal viruses), numerous situations exist where IFN exerts its effects differently. Some systems indicate an inhibition of primary transcription of viral RNA (Review in Friedman, 1977; Hovanessian, 1979). The SV40 system is of special interest since in this case the kind of effect observed depends on the timing of IFN treatment in the infectious cycle. If IFN is added late in the lytic SV40 cycle, viral protein translation is inhibited selectively, indicating the remarkable discriminative inhibition of viral protein synthesis by IFN in an in vivo situation (Review in Revel, 1979). Other recent works indicate that budding of retroviruses is affected in IFN-treated cells and that particles released have a lower infectivity. Interestingly this observation has been extended now to other RNA viruses including VSV, a typical cytopathic RNA virus.

b) It has been calculated that one or very few molecules of IFN per cell suffice to protect a tissue culture. Is this phenomenon due to cell spreading of a diffusable secondary signal (Blalock and Baron, 1977) or to the "recycling" of IFN molecules jumping from cell to cell ("hit and run")?

c) On the other hand, it is known that the expression of the antiviral state requires several hours to be established and that during this period protein and RNA synthesis are required (Taylor, 1964).

d) This fundamental observation has led several groups to search for a discriminative inhibition of protein synthesis by using cell free systems derived from IFN-treated cells. Very early Marcus and Salb (1966) described such a system. The selective inhibition of viral mRNA translation was dependent upon a postulated ribosome-bound antiviral protein synthesised after IFN treatment. When better cell-free translation systems became available, these experiments were not confirmed except for the fact that cell-free systems derived from IFN-treated cells were usually found to be less active than those obtained from control cells (Review in Friedman, 1977; Hovanessian,

1979). No discrimination between translation of viral and non-viral mRNA's was found. The concept of IFN-induced translation inhibitor had a strong impact on the research in this field. Such a molecule was looked for in different subcellular fractions. One of the most rewarding observations in this context was that from Kerr et al. (1974) indicating an increased sensitivity of IFN-treated cell lysate to the inhibitory effect of low concentration of dsRNA. The physiological relevance and origin of this observation were that it could explain some selectivity of the IFN effect at the cellular level since only virus infected cells may contain enough dsRNA to activate the inhibitor(s) of protein synthesis. But the possibility to reach an enzymatic approach came from the observation of a strong parallelism between the IFN-treated cell lysate and the rabbit reticulocyte lysate. This highly efficient translation system was known very well at the time for its remarkable sensitivity to dsRNA inhibition and at the same time it was known that dsRNA addition activated a cascade of phosphoprotein kinases in the reticulocyte resulting in the phosphorylation of the α subunit of eIF-2 (Farrel et al., 1977); it is not yet known whether this phenomenon is responsible in part or in totality for the inhibition of protein synthesis observed in the reticulocyte and, to further complicate the situation, the 2-5A system (section 5), is entirely present in the reticulocyte lysate and could thus also play a role in this case (Williams et al., 1979). But the discovery of a phosphoprotein kinase activity in IFN-treated cells (Review in Baglioni, 1979; Hovanessian, 1979; Torrence, 1981; Sen, 1981) was not the only benefit of this comparison of the two systems : ATP also was required in addition to dsRNA for synthesis during preincubation of interferon-treated cell (Roberts et al., 1976). Once produced this potent inhibitor exerted its action in the absence of dsRNA in a lysate of control cells. It was heat stable, of low molecular weight and active at subnanomolar concentrations. It was identified as an ATP polymerisation product (section 5). The enzyme responsible for its synthesis (the 2-5A synthetase) was isolated shortly after and initially characterized by its ability to function after binding to a poly(I). poly(C)-sepharose matrix (Roberts et al., 1976).

4. THE INTERFERON-INDUCED PHOSPHOPROTEIN KINASE

     Three different groups observed that IFN induced an ATP-and dsRNA-dependent phosphoprotein kinase able to transfer selectively the γ-phosphate group from ATP to one or two proteins in lysates from IFN-treated cells (Roberts et al., 1976); Zilberstein et al., 1976b; Lebleu et al., 1976). The first protein, of M_r 67,000 (often referred to as 67 K) in the murine cells or 69-73 K in human cells, is the predominant substrate for the endogenous reaction. In a few cases, a phosphoprotein of M_r 35,000 was also observed (Lebleu et al., 1976; Roberts et al., 1976). The reaction is best detected by addition to the system of an exogenous substrate, such as calf thymus histones or purified rabbit reticulocyte eIF-2.

a) Most of the recent results support the following mechanism of action of the IFN kinase. The enzyme strongly requires ATP and dsRNA but in very low concentration (10-100 ng/ml). GTP cannot replace ATP in this reaction, and the enzyme is not activated by cAMP or cGMP. Two steps are required for the phosphorylation : first the dsRNA and ATP-dependent activation of the protein kinase and second, the substrate phosphorylation itself. The first phase could be in fact an autophosphorylation of the enzyme in the presence of dsRNA or, alternatively, ATP could be responsible for a conformational change. However most studies argue in favour of the first concept since the most highly purified enzyme preparations mainly contain the 67 K protein itself. Once activated in the presence of ATP and dsRNA, the kinase does not seem to require dsRNA for the second phase. The known target for the activated kinase in IFN-treated cell is the α subunit of eIF-2. Phosphorylation of this protein occurs at the same sites with both IFN-induced and reticulocyte lysate protein kinase. After this phosphorylation eIF-2 seems unable to function normally in (eIF$_2$-GTP-met tRNA$_f$) ternary complex formation. The reason for this failure is not known but the initiation blocking is reversed in vitro when fresh eIF-2 is added to the system. In summary, the so-called 67 K protein could be (if confirmed after complete purification) an eIF-2 kinase which requires for its activation a dsRNA- and ATP-dependent autophosphorylation step.

b) The second controversial question is the reason for dsRNA requirement. Epstein et al. (1980) suggested that poly(I).poly(C) functions as inhibitor of an antagonistic phosphatase present in crude cell free extracts. IFN-induced kinase preparation apparently loses the dsRNA dependence after binding to poly(I).poly(C) sepharose column. This could reflect either the removal of a phosphatase, or the removal of a regulatory subunit (Hovanessian, 1979) or the permanent association with minimal amount of dsRNA leaking from the chromatographic poly(I).poly(C) matrix. This last explanation seems the most likely since other extensive purification procedures led to several thousand fold purified enzyme preparation, still totally dsRNA dependent (Sen et al., 1978; Kimchi et al., 1979b).

c) The third controversial question is whether the possible cascade of protein phosphorylations leading to the phosphorylation of eIF-2 is the explanation for the initiation block of translation observed in IFN-treated cell lysate (Content et al., 1975). This question is in fact also to be answered for the mechanism of regulation described in the reticulocyte lysate (Review in Hunt, 1980). Two kinds of experiments might answer this question : first, partially purified IFN-induced kinase strongly inhibited protein synthesis when added to control cell extracts (Farrell et al., 1978, Kimchi et al., 1979b). Second, specific inhibition of the nuclease pathway in IFN-treated cell lysates did not prevent the dsRNA-mediated inhibition of translation which can thus be explained at least in part by the kinase pathway (Miyamoto and Samuel. 1980).

d) The relevance to the in vivo situation is also controversial and no direct proof exists until now that kinase induction and activation are required for the expression of IFN antiviral action (Gupta, 1979; see section 6).

e) Finally, it is not definitely proved that interferon induces a dsRNA dependent kinase as detected in IFN-treated cell extracts. The increase of enzymatic activity detected rarely exceeds 2-4 fold and for this reason this is an effect difficult to quantitate accurately specially if one attempts to correlate with the development of amplitude of the antiviral state (West and Baglioni, 1979). This is mainly due to the fact that in many cell free lysates the 67K-kinase is already present at a non-negligible level before any interferon treatment. If the phenomenon described in a) is not an autophosphorylation and if interferon induces the "kinase pathway", does it affect the enzyme itself or the substrate or both? Earlier work from our group (Vandenbussche et al., 1978) based on mixing experiments of active 67K containing and non-containing crude lysates indicated that the enzyme either acts exclusively in cis or autophosphorylates.

5. THE 2',5'- OLIGOISOADENYLATE SYSTEM

Elements of the system

The second dsRNA-dependent enzymatic system implies a 2',5'-oligoisoadenylate (2-5A) synthetase and an endonuclease. When activated by ds-RNA, the 2-5A synthetase is capable of polymerizing ATP into an unusual series of oligonucleotides with the general structure $ppp(A2'p)_n5'A(n = 1-14)$ (Kerr and Brown, 1978; Ball and White, 1978; Zilberstein et al., 1978; Farrell et al., 1978; Baglioni, 1979; Samanta et al., 1980). These products of low molecular weight are thermostable and in turn activate, at subnanomolar concentrations, the latent endonuclease which cleaves single stranded-RNA (Clemens and Williams, 1978; Baglioni et al., 1978; Farrell et al., 1978). 2',5'-oligoisoadenylates are degraded mostly by a specific phosphodiesterase (Schmidt et al., 1979).

Purification of 2-5A Synthetase

2-5A synthetases have a wide distribution. They are found in extracts of various IFN-treated cells and in extracts of rabbit reticulocytes and normal mouse lymphocytes (see section 9). The enzyme from L cells has been partially purified by affinity chromatography on dsRNA bound to Sepharose columns (Hovanessian and Kerr, 1979). When immobilized on the column, the enzyme can be used for the synthesis of 2-5A in a two phases system. On the other hand, the enzyme from Ehrlich ascites tumor cells treated with IFN has been purified to homogeneity (Dougherty et al., 1980). This procedure involves differential precipitation of the ribosomal salt wash fraction with ammonium sulfate and chromatography on DEAE-cellulose and CM-cellulose. The apparent molecular weight of the enzyme is 105,000 as determined by gel electrophoresis in SDS and about 85,000 when determined by centrifugation through a glycerol gradient (Dougherty et al., 1980).

Mechanism of 2-5A Synthesis

Samanta et al. (1980) have determined that, in the presence of dsRNA, the purified 2-5A synthetase from Ehrlich ascites tumor cells can convert about 97% of the ATP into 2-5A and pyrophosphate; the stoichiometry of this reaction can thus be formulated as:

$$(n + 1) \text{ ATP} \longrightarrow \text{pppA}(2'p5'A)_n + n \text{ pyrophosphate}$$

(the extent of the reverse reaction being negligible)

The mechanism of chain elongation has been investigated in order to determine whether it is processive or not. Three groups have found that the 2-5A synthetase could elongate added dimer (Minks et al., 1980; Justesen et al., 1980., Johnston et al., 1980). One of them (Justesen et al., 1980), using 1000-fold purified enzyme from rabbit reticulocytes, has noted an accumulation of dimers during the first phase of the reaction, indicating that this enzyme catalyses the de novo synthesis of the ologonucleotide from ATP and that the mechanism of elongation of the 2', 5' oligonucleotides is not processive. This is at odds with the results of Minks et al., (1980) who did not note such an accumulation of dimers using crude Hela extracts, perhaps on account of the presence of significant levels of 2-5A degrading enzymes (Minks et al., 1979a).

Substrate Specificity of 2-5A Synthetase

ATP is not the only substrate that 2-5A synthetase is able to polymerize. In fact, the 2',5'-oligoadenylate synthetase is a 2',5'-nucleotidyltransferase that also incorporates a wide range of other nucleoside triphosphates into co-oligonucleotides. However, the purified enzyme from rabbit reticulocytes catalyzes only the addition of one unit of CMP, GMP, UMP, 2'-dAMP, 3'-dAMP, dCMP, dGMP or TMP to the 2'-OH end of a preformed oligoadenylate (Justesen et al., 1980). In this case, a further polymerization is not possible. The adenosine triphosphates 2'-dATP and 3'-dATP are not incorporated efficiently into 2-5A and inhibit its synthesis (Minks et al., 1980).

On the other hand, the acceptor site where the primer "binds" appears to require that the 2'-terminal nucleotide be adenosine with a free 2'-OH group. Furthermore, diribonucleoside monophosphates, linked either 2'-5' or 3'-5', can serve as primer for 2-5A synthetase, but only if they contain an adenosine residue at their 3'-terminus (NpA) (Ball and White, 1979).

Natural Occurrence of Various 2-5A Containing Oligonucleotides

The 2-5A synthetase can also elongate other nucleotides with a blocked p5'A terminus, such as A5'ppppp5'A and NADH (Minks et al., 1980), ADP ribose and also rRNA's and tRNA's (Ball and White, 1979; Knight et al., 1980; Thang et al., personal communication, 1980), where nucleotide transfer can occur at the 3' terminal-CCA end. All these observations have been made in cell-free lysates. Recent observations on the intracellular concentration of these derivatives (Knight et al., 1980) indicate that extracts from IFN-treated, EMC infected cells which contain substantial levels of trimer or tetramer (100 nM, instead of 2.5 nM in control cells) do not contain detectable levels of either NAD2'p5'A2'p5'A or A5'p4A2'p5'A2'p5'A. It is thus too early to decide if these 2-5A modified nucleotide derivatives represent only a biochemical curiosity or may have a functional impact. In this case, it would be rather easy to conceive the multiplicity of metabolic consequences for the cell of an alteration of essential cofactors of the respiratory metabolism or inactivation of some tRNA's, rRNA's or of a growth regulator such as A5'p45'A. However nobody has yet shown the consequence of 2'5' adenylation on the functional activity of these components.

Activation of 2-5 Synthetase

The optimal concentration of double-stranded RNA (poly(I).poly(C)) or reovirus dsRNA increases with the concentration of the enzyme and has been calculated to correspond to about 79 RNA base pairs per molecule of enzyme (Samanta et al., 1980). This estimation is in agreement with the finding that both the 2-5A synthetase and the protein kinase isolated from Hela cells could not be activated by dsRNA shorter than 30 nucleotides and required dsRNA longer than 65 to 80 base pairs to be maximally activated (Minks et al., 1979b). Thus a relatively long stretch of base pairs, uninterrupted by either a mismatch or a discontinuity in one of the complementary strands, is required for the activation of the two dsRNA-dependent enzymes. The fact that the concentration of 2-5A trimers (tetramers) increases by a factor of at least 40 fold in non-infected, IFN-treated mouse L cells (Knight et al. 1980) can be interpreted in two ways:

    i) either non-infected cells contain enough accessible dsRNA
structure
    ii) or the enzyme depends on other(s) yet unknown natural activator(s).

Inducers of 2-5A Synthetase. Does Interferon act like a Hormone?

Very recently, other molecules than interferon have been proposed as inducers of 2-5A synthetase activity, and of an antiviral state similar to that resulting from IFN treatment.

It has been observed (Krishnan and Baglioni, 1980b) that treatment
of Daudi or Raji cells with hydrocortisone results in an increase in
2-5A synthetase activity which is a function of the time of treatment
and the hormone concentration. This increase does not seem to be accounted for by the production of interferon and is probably due to de
novo synthesis of (2',5')A polymerase. These results are interesting
in view of the hypothesis of Blalock and Stanton (1980). These authors
have proposed that the action of IFN is not unique but shared with
other hormones. Results of Blalock and Harp (1981) concerning the IFN
and adrenocorticotropic hormone (ACTH) induction of steroidogenesis,
melanogenesis and antiviral activity seem to support this hypothesis.
Indeed, these authors have observed that IFN could have species-
specific hormonal activity and that ACTH could have cell-specific antiviral activity. On the other hand, previous studies (Blalock and
Smith, 1980) have shown strong antigenic relatedness among human leukocyte interferon, ACTH, and endorphins, implying that there are underlying structural similarities between these molecules.

Substrate Specificity of the 2-5A Endonuclease

In most cell free systems from IFN-treated cells the inhibition
of mRNA translation is not discriminative: it does not spare host cell
mRNA's (Review in Revel, 1979). The same observation applies to the
2-5A dependent endonuclease in in vitro systems. Very recently Wreschner et al. (1981) investigated the sequence specificity of the 2-5A
dependent ribonuclease (from different animal cells) by sequencing the
products of partial digestion of a terminally labelled RNA substrate.
They found that the enzyme cleaves on the 3' side of UpN sequence
(preferentially at UpA and UpU sequences). Similar observations by
Lengyel and his colleagues and the fact that the activated endonuclease
preferentially cleaves poly U (whereas poly C, poly G or poly A are
spared) (Lengyel, personal communication) suggest that:

i) the 2-5A activated endonuclease might recognize its cleavage
sites by base pairing of the enzyme-bound 2-5A oligomers to oligo U
stretches in the RNA substrate.
ii) although the 2-5A dependent nucleases recognize specific RNA
sequences these are too short to allow discriminative destruction of
viral mRNA's.

Nilsen and Baglioni (1979) have suggested that a discriminative
action of this nuclease may depend on topological factors. In their
model 2-5A synthetase would be activated by viral dsRNA (replicative
form or replicative intermediate). Only in the surrounding of this
viral replicative form would the 2-5A concentration be sufficient
to activate the nuclease. Experimental evidence for such a model has
been provided more recently by Nilsen et al. (1980a). On the other
hand, the observation that 2-5A synthetase could selectively accumulate
in VSV and Moloney murine leukemia virions also indicates that the
compartmentalization of this enzyme may have an important role for the
discriminative action of the endonuclease (Wallach and Revel, 1980).

<u>Structural Requirements for Nuclease Activation by 2-5A</u>

A minimum size of the oligomer is required. Dimers are non-active in inhibition of protein synthesis or activation of the nuclease. Trimers and higher molecular weight oligomers are active except in the reticulocyte lysates where trimers are inactive (Williams et al., 1979a).

The 5'-terminal triphosphate is required. Bacterial alkaline phosphatase treated 2-5A (cores) are biologically inactive in cell-free assays but 5'-diphosphorylated 2-5A tetramers are as active as 5'-tri-phosphorylated 2-5A tetramers in the nuclease binding assay (Knight et al., 1980). The 3' terminal region seems less essential for this activity since covalently added pCp at the 3'-end of 2-5A tetramers does not prevent its capacity to bind and activate the 2-5A dependent nuclease (Knight et al., 1980).

6. ARE THE TWO INDUCED ENZYMES 2.5 SYNTHETASE AND 67-73K PROTEIN KINASE SUFFICIENT FOR THE ESTABLISHMENT OF AN ANTIVIRAL ACTIVITY?

Several examples from the very recent literature have been assembled in Table II. They clearly show that in many known instances there is really no correlation between the presence or induction of the two dsRNA-dependent enzymes and the establishment of antiviral and/or anticellular effect of interferon. Other comparisons include those of chromosome 21-mono, di- and trisomic human fibroblasts. The sensitivity of these cells to the antiviral effect of IFN-β increases with the number of chromosome 21 copies; however this order of sensitivity is not reflected by their 2-5A synthetase inductibility when studied as a function of IFN dose. (Verhaegen-Lewalle,M., Content,J., Zhang, De Clercq,E.; unpublished results).

Another interesting situation is that occurring upon human cell treatment with HuIFN-γ. When compared to the HuIFN-β, at equivalent antiviral units, HuIFN-γ has a more pronounced anticellular effect although it induces significantly less 2-5A synthetase than HuIFN-β does (Verhaegen-Lewalle et al., in preparation) (see section 10).

7. COULD THE 2-5A OLIGONUCLEOTIDES HAVE MORE THAN A SINGLE TARGET?

As described earlier the only known target for 2-5A in cell free systems is the latent endoribonuclease which is activated by the various oligomers (see section 5). Numerous observations have indicated that 2-5A nucleotides could have an anticellular growth effect. If 2-5A or 2-5A cores are introduced in living cells, one usually observes a decrease in DNA synthesis which seems specific for this kind of nucleotides (Williams and Kerr, 1978). 2-5A cores can also inhibit DNA synthesis (mimicking the effect of IFN) if incubated in

Table II.   Induction of different activities in IFN-treated cells.

Cell type	Treatment	Antiviral effect	Anticellular effect	2-5A synthetase activity	63 - 73K phosphoprotein kinase A	Reference(s)
RSa	IFN α, β	+	+	+	+	Vandenbussche et al., 1981
IF^r	IFN α, β	+	-	+	+	
HEC-1	--	- (a)	-	+(b)	+	Verhaegen et al al., 1980
	IFN α,β,γ	-	-	+	+	
L1210 S	IFN α, β	+	+	+	+	Marti et al., 1981
L1210 R	IFN α, β	-	-	-	-	
EC undifferentiated	IFN α, β	-/+ (c)	-	+	-	Wood & Hovanessian, 1979 ; Nilsen et al., 1980b
EC differentiated	IFN α, β	+	+	+	+	
3T3/NIH	IFN α, β	+	+	+	-	Hovanessian et al., 1981
K/Balb	IFN α, β	+	+	(++) (d)	+	
K/Balb (+ anti IFN, 3 weeks)	IFN α, β	+	+	+	+	
MRC-5	IFN α or β	+		-	-	Meurs et al., personal communication
Hela	IFN α or β	+		++	+	
Daudi	IFN α or β	+	++	±	-	
Daudi or Raji	hydrocortisone		+	+	?	
Namalwa	--	-		++ (e)		Stark et al., 1979 ; Johnston et al,1980
Friend cells	--			+		Stark et al., 1979

(a) with VSV, Sindbis, vaccinia or EMC virus
(b) HEC-1 cells are not inducible but constitutive for both 2-5A synthetase and 73 K phosphoprotein kinase activity ; prolonged treatment with anti α or β IFN has no effect on this pattern.
(c) In this case IFN induces a partial antiviral state : the antiviral effect depends from the kind of virus used. Undetectable with standard VSV, it can be observed with some ts mutants of VSV or with EMC virus (Nilsen et al., 1980b)
(d) K/Balb cells have a very high level of 2-5A synthetase in the absence of IFN treatment. This activity is reduced by prolonged antibody treatment when the cells become normally inducible by IFN.
(e) These cells are constitutive for IFN synthesis but the level of 2-5A synthetase remains elevated even if the cells are grown in the presence of antibody to IFN.

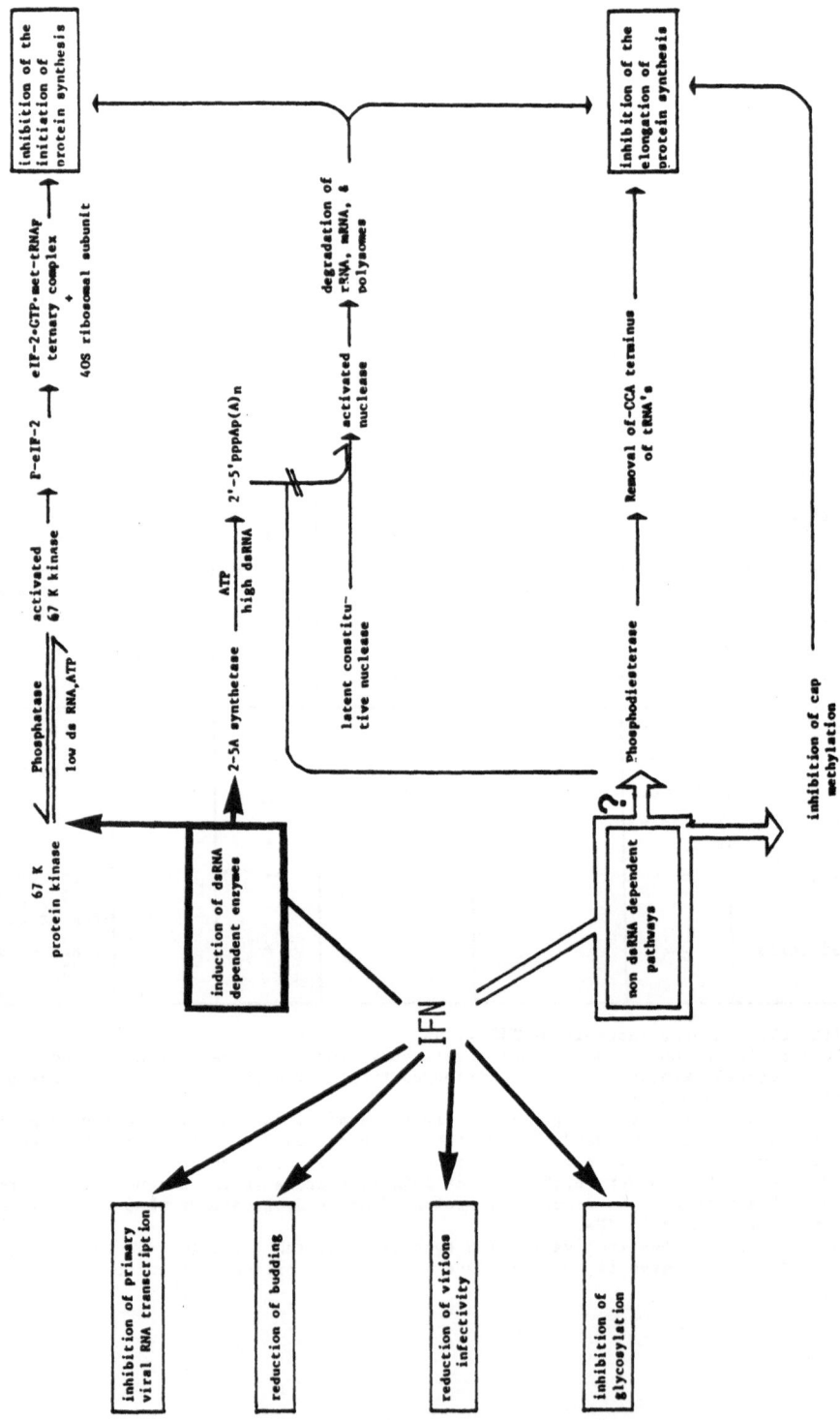

Fig. 1. Mechanism of the antiviral action of interferon.

mitogen-stimulated mouse splenocytes (Kimchi et al., 1979a). In human
lymphoblastoid cells there is a strong inverse parallelism between the
level of DNA synthesis and the rate of 2-5A synthesis induced by corti-
costeroid treatment (Krishnan and Baglioni, 1980b). In other cell lines
(Namalwa, or 3T3) there is often an inverse correlation between the
balance or ratio 2-5A synthetase/2',5' phosphodiesterase and the rate
of DNA synthesis (Krishnan and Baglioni; Kimchi and Revel, personal
communications).

      Nobody has to our knowledge directly measured intracellular 2-5A
concentration in different physiological situations and in different
phases of the cell cycle in syncronized cells. Another approach to the
possible role of 2-5A synthetase has been our work with IF^r cells.
These mutant cells derived from RSa (wild type) cells are exclusively
resistant to the anticellular action of IFN. However, they are normally inducible by IFN for both dsRNA activated enzymes (kinase and
2-5A synthetase). Their content in nuclease is perfectly normal and
this nuclease has a normal sensitivity to subnomolar concentrations of
2-5A when compared with the same enzyme from wild type RSa cells. Consequently either 2-5A is not responsible or not sufficient for the expression of an anticellular effect of IFN in these cells or another
target different from the nuclease is missing in this case as a consequence of the mutation responsible for IFN resistance. (Vandenbussche et al., 1981).

8.   OTHER BIOCHEMICAL ACTIONS OF INTERFERON

tRNA Effects

      Besides the two dsRNA-dependent enzymatic pathways involved in
the inhibition of mRNA translation, there is, in extracts of IFN-
treated cells, a third pathway of translation inhibition. This latter
which does not require dsRNA, affects polypeptide chain elongation,
and is reversed by the addition of tRNA (Content et al., 1975; Sen et
al., 1976; Zilberstein et al., 1976a). The mechanism that brings about
this tRNA deficiency in IFN-treated cell extracts is far from being
completely elucidated and does not seem to occur in vivo (Colby et al.,
1976; Sen et al., 1976). According to the recent observation of Revel
and his coworkers (Schmidt et al., 1979) the agent responsible for inactivating certain tRNA species in IFN-treated cell extracts could be
identical to the 2',5'-phosphodiesterase, if IFN-inducible, since this
enzyme also degrades the CCA terminus of tRNA's. Although 2',5'-
phosphodiesterase is not selective for a specific tRNA, minor species
present in low amounts in the cell would be more affected by a general-
ized decrease in tRNA activity.

Impairment of mRNA Methylation

      Most eukaryotic cellular and viral mRNA's have "cap" structures
at their 5'-termini. The methylation of the cap on a mRNA increases

its stability and the efficiency of its translation. Impairment of
reovirus mRNA cap methylation has been observed as well in extracts of
IFN-treated Ehrlich ascites (Sen et al., 1975) or Hela cells (Shaila
et al., 1977) as in intact cells (Desrosier and Longyel, 1979). In
these cases, there is an inhibition of the methylation of the mRNA
5'-terminal guanosine. Decreased methylation of the cap has also been
observed for both vaccinia mRNA and host cell mRNA's in IFN-treated
chick fibroblasts (Kroath et al., 1979). On the other hand, there is
little or no impairment of the cap methylation, and even an overmethy-
lation, both internally and in the caps, in SV40 mRNA and cellular
mRNA's isolated from IFN-treated monkey cells (Revel, 1979). How to
conciliate the results from the reovirus and vaccinia virus systems
on the one hand and those concerning the SV40 system on the other
hand remains to be clarified as well as the mechanism by which the
changes in mRNA methylation are induced.

Induction of New Proteins

     Another approach for the mechanism of action of interferon is the
search for new induced proteins synthesized upon IFN-treatment.
Recently, several groups have investigated by different methods IFN-
treated mouse, human or chicken cells.

     In the murine system, a number of proteins are selectively induced
by homologous IFN (Farrell et al., 1979; de Ley et al., 1979; Hovanessian et al., 1980).

     In several strains of human fibroblasts, treatment with either
human leukocyte or fibroblast IFN's induces the synthesis of at least
five proteins with M_r 120,000, 88,000 (both retained on dsRNA columns),
80,000, 67,000 and 56,000. The induction of these proteins occurs
only with human IFN's and is blocked by actinomycin D and cycloheximide,
indicating the requirement for de novo RNA and protein synthesis (Gupta
et al., 1979). In human FS4 fibroblasts, four new proteins with molecular weights ranging from 44,000 to more than 68,000 are synthesized
upon IFN-treatment (Knight and Korant, 1979).

     The appearance of newly synthesized proteins in IFN-treated cells
seems to parallel the appearance of the antiviral state as far as the
kinetics, the susceptibility to actinomycin D, the species specificity
for interferon and the dependence on interferon concentration are concerned. Moreover, it is possible that some of these IFN-induced proteins are related to the enzyme activities that are induced by IFN
treatment i.e. the protein kinase and the 2-5A synthetase. In fact,
Ball (1979), treating chick embryo fibroblasts by interferon, has
reported the induction of a protein of M_r 56,000 which co-eluted with
the 2-5A synthetase activity from a gel filtration column and also
bound to poly(I).poly(C) agarose, indicating that the protein and the
enzyme might be identical. However, it should be stressed that the
mentioned molecular weight is at variance with that found for 2-5A
synthetase from other cells. (see section 5).

Inhibition of Glycoprotein and Membrane Protein Biosynthesis

Interferon may also, in a non-dsRNA dependent pathway, inhibit the biosynthesis of glycoproteins and this mechanism might be implicated in the inhibition of proliferation of viruses that bud from the cell membrane (i.e. retroviruses). The production of infectious vesicular stomatitis virus (VSV) by IFN-treated L cells is reduced 30-200 fold whereas the production of virus particles is reduced 6 fold. (Maheshwari and Friedman, 1979). Biochemical and morphological studies (Maheshwari et al., 1980) have shown that the low infectivity of the VSV particles produced upon IFN-treatment might be related to the reduced amount of glycoprotein (G) and membrane protein (M) incorporated in such particles. Under these conditions the effects of IFN are mimicked by the antibiotic tunicamycin which is known to inhibit glycosylation of proteins.

9. IS THE 2-5A METABOLIC PATHWAY IMPLICATED IN OTHER FUNCTIONS THAN THE EXPRESSION OF ANTIVIRAL ACTIVITY?

As indicated in the introduction, IFN exerts multiple important non antiviral actions on the animal cells and in the entire organism. None of these, except the anticellular (section 7), have been explored in the context of enzyme induction. Superinduction and shut-off of IFN synthesis in human fibroblasts does not seem to correlate with 2-5A mediated phenomenon (Sehgal and Gupta, 1980). Conversely our recent study of IFN priming in NDV-induced mouse L 929 cells indicate that priming is accompanied by the induction of 2-5A synthetase activity in these cells and that the 2-5A system may play a role in the altered distribution of IFN-mRNA observed in IFN-primed cells (Content et al., 1980). Krishnan and Baglioni, (1980a) similarly suggested that treatment with interferon, inducing 2-5A synthetase, might increase interferon production after virus infection. They further suggested that 2',5'A synthetase might be part of the recognition mechanism for dsRNA of viral origin since structural requirement of double-stranded RNA for synthetase activation and IFN induction are similar (Minks et al., 1979b; Content et al., 1978). Among other normal non-infected, non-IFN-treated tissues containing appreciable level of 2-5A synthetase, one should cite mouse and human lymphocytes, the rabbit reticulocyte, the Krebs II ascites tumor cells, mouse erythroleukemia Friend cells, chick oviduct after removal of estrogen stimulation, lactating guinea pig mammary gland (Review in Torrence, 1981), dog liver and dog striated muscle tissue (Etienne et al., 1981). The significance of these diverse observations is not known. Either IFN or 2-5A synthetase (if induced through a non-IFN pathway, section 5 g) or both, could play a regulatory role at different stages of the differentiation of specialized animal tissues (Keay and Grossberg, 1980), but this is certainly not proved.

10. COMPARISON OF THE MECHANISMS OF ACTION AND EFFICACIES OF THE
DIFFERENT INTERFERONS

     All interferons confer resistance against viral infections to
animal sensitive cells, but different interferons are distinguished
on the basis of their types and subtypes (α_1 .... α_8,β,γ) as well as
of the cells which produce them (animal or bacterial) (section 1).

     The mechanisms of action of HuIFN's α and γ have been compared by
studying the kinetics of induction of 2-5A synthetase upon treatment
of Hela cells with different interferons in combination with inhibitors
of RNA and protein synthesis (Baglioni and Maroney, 1980). The kinetics of induction is faster for HuIFN than for HuIFN-γ . Moreover,
in the presence of cycloheximide, HuIFN-α seems to be able to activate
transcription of 2-5A synthetase mRNA which can be subsequently trans-
lated upon removal of the inhibitor. In the same situation, HuIFN-γ
is unable to induce 2-5A synthetase. These results suggest that
HuIFN-α might be a "direct" inducer of 2-5A synthetase whereas HuIFN-γ
would be an "indirect" inducer (Baglioni and Maroney 1980).

     Also in order to determine whether the actions of HuIFN-α and γ
have some common mechanism of action, Rubin and Gupta (1980) have
compared the differential efficacies of these interferons as inducers
of new proteins (see section 8) and as antiviral and antiproliferative
agents. HuIFN-γ induces four out of five proteins which are induced
by the HuIFN-α and is more effective for two of them. As antiproliferative agent, HuIFN-γ appears to be more efficient in inhibiting cell
growth of FS-4 (fibroblasts), Hela and osteogenic sarcoma cells. Moreover, HuIFN-γ seems to have a reversible cytostatic effect on normal
human fibroblast, whereas it provokes extensive death of the transformed cells tested.

     It is worthwhile to note that partially purified HuIFN-γ prep-
arations may be contaminated with some α- or β-like IFN (de Ley et al.,
1980) and that minimal amounts of such IFN have been shown to considerably potentiate the action of IFN-γ (Fleischmann et al., 1979).

     The various α and β HuIFN's have been cloned and expressed in
bacteria (see section 1). Bacterial α-interferons share with eukaryotic cell interferon a number of biological properties, such as antiviral and anticellular activities, antimigration activity, ability to
induce 2-5A synthetase and to stimulate natural killer cells. Cloning
and high performance liquid chromatography have allowed the identification of the HuIFN-α subtypes. When these are expressed in bacterial
cells, they exhibit different spectra of antiviral activities on
various animal species. This can also be reproduced <u>in vivo</u> (Stebbing,
personal communication).

     Moreover genetic engineering allows to construct and express
chimeric hybrids of bacterial interferons, whose various biological

activities may be radically different from those expected on the basis of the activities of the "parental" molecules. Those observations could present a real interest for clinical use of IFN's.

11. CLINICAL APPLICATIONS

As suggested a few years ago, (Content, 1978) it is possible to take advantage of these IFN-induced dsRNA dependent enzymes as biochemical markers of interferon action. In mouse infected with EMC virus or injected with IFN or with poly(I).poly(C), 2-5A synthetase appears in the serum within 1-3 days (Krishnan et Baglioni, 1980a).

In herpes virus infected mouse, 2-5A synthesis also appears in the trigeminal ganglion in parallel with the progression of the HSV viral infection (Sokawa et al., 1980).

Normal mouse lymphocytes as well as normal human lymphocytes also contain detectable level of 2-5A synthetase (Shimizu and Sokawa, 1979; Kimchi et al., 1979a; Revel et al., 1980; Johnston et al., 1980).

More pragmatically, Revel suggested that in acute viral infections the contents of lymphocyte 2-5A synthetase could reflect the degree of interferon impregnation of these cells (Personal communication, 1980). This could serve either as a mean to monitor IFN treatment (instead of measuring IFN level in the serum) or as a presumptive diagnosis of acute viral infection since this increased level of 2-5A synthetase in lymphocytes was found in patients with acute viral disease but not in those with acute bacterial disease.

ACKNOWLEDGMENTS

We wish to acknowledge Mrs. J. Herinckx for invaluable help in the preparation of this manuscript and Dr. L. Thiry, Mr. R. Olislager and Mr. P. Vandenbussche for critical reading of this paper.

This work was supported by grant 2.9006-79 from the Fonds de la Recherche Fondamentale Collective (Belgium).

REFERENCES

Aguet, M., 1980, Nature(London), 284 : 459-461.
Baglioni C., 1979, Cell, 17 : 255;264.
Baglioni, C. and Maroney, P.A., 1980, J. Biol.Chem., 255 : 8390-8393.
Baglioni, C., Minks, M.A. and Maroney, P.A., 1978, Nature(London), 273 : 684-687.
Ball, L.A., 1979, Virology, 94 : 282-296.
Ball, L.A. and White, C.N., 1978, Proc. Natl. Acad. Sci. U.S.A, 75 : 1167-1171.

Ball, L.A. and White, C.N., 1979, in :"Regulation of macromolecular synthesis by low molecular weight mediators", eds. Koch, H. and Ritters, D., Academic Press, New York : 303-317.

Blalock, J.E. and Baron, S., 1977, Nature (London), 269 : 422-425.

Blalock, J.E. and Harp, C., 1981, Archives of Virology, 67 : 45-49.

Blalock, J.E. and Smith, E.M., 1980, Proc. Natl. Acad. Sci. U.S.A, 77 : 5972-5974.

Blalock, J.E. and Stanton, J.D., 1980, Nature(London), 283 : 406-408.

Cantell, K., 1979, Interferon, 1 : 1-28.

Clemens, M.J. and Williams, B.R.G., 1978, Cell, 13 : 565-572.

Colby, C., Penhoet, E.E. and Samuel, C.E., 1976, Virology, 74 : 262-264.

Content, J., 1978, Ann. Méd. Vét., 122 : 243 - 255.

Content, J., Johnston, M.I., De Wit, L., De Maeyer-Guignard, J. and De Clercq, E., 1980, Biochem. Biophys. Res. Comm., 96 : 415-424.

Content, J., Lebleu, B. and De Clercq, E., 1978, Biochemistry, 17 : 88 - 94.

Content, J. Lebleu, B., Nudel, U., Zilberstein, A., Berissi, H. and Revel, M., 1975, Eur. J. Biochem., 54 : 1-10.

De Clercq, E., 1980, in : "Virus infections : Modern concepts and status", ed. Olson, L.C., Marcel Dekker, Inc., New York, N.Y. (in press).

de Ley, M., van Damme, J., Claeys, H., Weening, H., Heine, J.W., Billiau, A., Vermylen, C. and de Somer, P., 1980, Eur. J. Immunol., 10 : 877-883.

Derynck, R., 1981, Ph. D.Thesis,Ghent University, Ghent.

Desrosier, R.C. and Lengyel, P., 1979, Fed. Proc., 36 : 812.

Dougherty, J.P., Samanta, H., Farrell, P.J. and Lengyel, P., 1980, J. Biol. Chem., 255 : 3813 - 3816.

Epstein, D.A., Torrence, P.F., and Friedman, R.M., 1980, Proc. Natl. Acad. Sci. U.S.A., 77 : 107-111.

Etienne-Smekens, M., Vassart, G., Content, J., and Dumont, J.E., 1981, FEBS Lett. (in press)

Farrell, P.J., Balkow, K., Hunt, T., Jackson, R.J., and Trachsel, H., 1977, Cell, 11 : 187-200

Farrell, P.J., Broeze, R., and Lengyel, P., 1979, Nature (London), 279 : 523-525.

Farrell, P.J., Sen, G.C., Dubois, M.F., Ratner, L., Slattery, E., and Lengyel, P., 1978, Proc. Natl. Acad. Sci. U.S.A, 75 : 5893-5897.

Fleischmann, W.R. Jr., Georgiades, J.A., Osborne, L.C., and Johnston, H.M., 1979, Infect. Immun., 26 : 248-253.

Friedman, R.M., 1977, Bacteriol. Rev., 41 : 543-567.

Gresser, I., De Maeyer-Guignard, J., Tovey, M.G., and De Maeyer, E., 1979, Proc. Natl. Acad. Sci. U.S.A., 76 : 5308-5312.

Gupta, S.L., 1979, J. Virol., 94 : 282-296.

Gupta, S.L., Rubin, B.Y., and Holmes, S.L., 1979, Proc. Natl. Acad. Sci. U.S.A., 76 : 4817-4821.

Hovanessian, A.G., 1979, Differentiation, 15 : 139-151.

Hovanessian, A.G., and Kerr, I.M., 1979, Eur. J. Biochem., 93 : 515-526.

Hovanessian, A.G., Meurs, E., Aujean, O., Vaquero, C., Stefanos, S., and Falcoff, E., 1980, Virology, 104, 195-204.

Hovanessian, A.G., Meurs, E., and Montagnier, L., 1981, J. Interferon Res., (in press).

Hunt., T., 1980, in : "Recently discovered systems of enzyme regulation by reversible phosphorylation", ed. Cohen, Elsevier/North Holland Biomedical Press, Amsterdam : 175-202.

Isaacs, A., and Lindenmann, J., 1957, Proc. Roy. Soc., B147 : 258-267.

Johnston, M.I., Zoon, K.C., Friedman, R.M., De Clercq, E., and Torrence, P.F., 1980, Biochem. Biophys. Res. Commun., 97 : 375-383.

Justesen, J., Ferbus, D., and Thang, M.N., 1980, Proc. Natl. Acad. Sci. U.S.A., 77 : 4618-4622.

Keay, S., and Grossberg, S.E., 1980, Proc. Natl. Acad. Sci. U.S.A., 77, 4099-4103.

Kerr, I.M., and Brown, R.E., 1978, Proc. Natl. Acad. Sci. U.S.A., 75, 256-260.

Kerr, I.M., Brown, R.E., and Ball, L.A., 1974, Nature (London), 250 : 57-59.

Kimchi, A., Shure, H., and Revel, M., 1979a, Nature (London), 282 : 849 - 851.

Kimchi, A., Zilberstein, A., Schmidt, A., Shulman , L., and Revel, M., 1979b, J. Biol. Chem., 254 : 9846-9853.

Knight, M., Cayley, P.J., Silverman, R.H., Wreschner, D.H., Gilbert, C.S., Brown, R.E., and Kerr, I.M., 1980, Nature (London), 288 : 189-192.

Knight, E. Jr., and Korant, B., 1979, Proc. Natl. Acad. Sci. U.S.A., 76 : 1824-1827.

Krishnan, I., and Baglioni, C., 1980a, Nature (London), 285 : 485-488.

Krishnan, I., and Baglioni, C., 1980b, Proc. Natl. Acad. Sci. U.S.A., 77 : 6506-6510.

Kroath, H., Janda, H.G., Hiller, G., Kuhn, E., Jungwirth, C., Gross, H.J., and Bodo, G., 1979, Virology, 92 : 572-577.

Lebleu, B., Sen, G.C., Shaila, S., Cabrer, B., and Lengyel, P., 1976, Proc. Natl. Acad. Sci. U.S.A., 73 : 3107-3111.

Maheshwari, R.K., Demsey, A.E., Mohanty, S.B., and Friedman, R.M., 1980, Proc. Natl. Acad. Sci. U.S.A., 77 : 2284-2287.

Maheshwari, R.K., and Friedman, R.M., 1979, J. Gen. Virol., 44 : 261-264.

Marcus, P.I., and Salb, J.M., 1966, Virology, 30 : 502-516.

Marti, J., Vandenbussche,P., Silhol, M., Milhaud, P., Verhaegen, M., Content, J., and Lebleu, B., 1981, J.Interferon Res. (in press).

Minks, M.A., Benvin, S., and Baglioni, C., 1980, J. Biol. Chem., 255 : 5031-5035.

Minks, M.A., Benvin, S., Maroney, P.A., and Baglioni, C., 1979a, Nucleic Acids Res., 6 : 767-780.

Minks, M.A., West, D.K., Benvin, S., and Baglioni, C. 1979b , J. Biol. Chem., 254 : 10180-10183.

Miyamoto, N.G., and Samuel, C.E., 1980, Virology, 107 : 461-475.

Nilsen, T.W., and Baglioni, C., 1979, Proc. Natl. Acad. Sci. U.S.A.,
    76, 2600-2604.

Nilsen, T.W., Weissman, S.G., and Baglioni, C., 1980a, Biochemistry,
    19, 5574-5579.

Nilsen, T.W., Wood, D.L., and Baglioni, C., 1980b, Nature (London),
    286 : 178-180.

Revel, M., 1979, Interferon, 1 : 101-163.

Revel, M., Wallach, D., Merlin, G., Schattner, A., Schmidt, A.,
    Wolf, D., Shulman, L., and Kimchi, A., 1980, in :"Methods in
    Enzymology", Academic Press, New York (in press)

Roberts, W.K., Hovanessian, A., Brown, R.E., Clemens, M.J., and Kerr,
    I.M., 1976, Nature (London), 264 : 477-480.

Rossi, G.B., Dolei, A., Cioe, L., Benedetto, A., Matarese, G.P., and
    Belardelli, F., 1977, Proc. Natl. Acad. Sci. U.S.A., 74 : 2036-
    2040.

Rubin, B.Y., and Gupta, S.L., 1980, Proc. Natl. Acad. Sci. U.S.A.,
    77 : 5928-5932.

Samanta, H., Dougherty, J.P., and Lengyel, P., 1980, J. Biol. Chem.,
    255 : 9807-9813.

Schmidt, A., Chernajovsky, Y., Shulman, L., Federman, P., Berissi,
    H., and Revel, M., 1979, Proc. Natl. Acad. Sci. U.S.A, 76 : 4788-
    4792.

Scott, G.M., and Tyrrell, D.A., 1980, Br. Med. J., 280, 1558-1562.

Sehgal, P.B., and Gupta, S.L., 1980, Proc. Natl. Acad. Sci. U.S.A.,
    77 : 3489-3493.

Sen, G.C., 1981, in :"Progress in Nucleic Acid Research" ed. Davidson,
    J.N., and Cohn, W.E., Academic Press, New York, N.Y. (in press).

Sen, G.C., Gupta, S.L., Brown, G.E., Lebleu, B., Rebello, M.A., and
    Lengyel, P., 1976, J. Virol., 17 : 191-203.

Sen, G.C., Lebleu, B., Brown, G.E., Rebello, M.A., Furuichi, Y.,
    Morgan, M., Shatkin, A.J., and Lengyel, P., 1975, Biochem. Bio-
    phys. Res. Commun., 65 : 427-434.

Sen, G.C., Taira, H., and Lengyel, P., 1978, J. Biol. Chem., 253 :
    5915-5921.

Shaila, S., Lebleu, B., Brown, G.E., Sen, G.C., and Lengyel, P.,
    1977, J. Gen. Virol. : 37, 535-546.

Shimizu, N., and Sokawa, Y., 1979, J. Biol. Chem., 254 : 12034-12037.

Shulman, L., and Revel, M., 1980, Nature (London), 288 : 98-100.

Sikora, K., 1980, Br. Med. J. : 281, 855-858.

Sokawa, Y., Ando, T., and Ishihara, Y., 1980, Infect. Immun., 28 :
    719-723.

Stark, G.R., Dower, W.J., Schimke , R.T., Brown, R.E., and Kerr, I.M.,
    1979, Nature (London), 278 : 471-473.

Stewart II, W.E., 1979a, "The Interferon System", Springer Verlag,
    New York, N.Y.

Stewart II, W.E., 1979b, Interferon, 1 : 29-51.

Taylor, J., 1964, Biochem. Biophys. Res. Commun., 14 : 447-541.

Torrence, P.F., 1981, in : "Molecular Aspects of Medicine", (in press)

Torrence, P.F., and De Clercq, E., 1977, Pharmac. Ther., A2 : 1-88.
Tovey, M.G., Rochette-Egly, C., and Castagna, M., 1979, Proc. Natl. Acad. Sci. U.S.A., 76 : 3890-3893.
Vandenbussche, P., Content, J., Lebleu, B., and Wérenne, J., 1978, J. Gen. Virol., 41, 161-166.
Vandenbussche, P., Divizia, M., Verhaegen-Lewalle, M., Fuse, A., Kuwata, T., De Clercq, E., and Content, J., 1981, Virology, (in press).
Verhaegen, M., Divizia, M., Vandenbussche, P., Kuwata, T., and Content, J., 1980, Proc. Natl. Acad. Sci. U.S.A., 77 : 4479-4483.
Wallach, D., and Revel, M., 1980, Nature (London), 287 : 68-70.
West, D.K., and Baglioni, C., 1979, Eur. J. Biochem., 101 : 461-468.
Williams, B.R.G., Gilbert, C.S., and Kerr, I.M., 1979, Nucl. Acids Res., 6 : 1335-1350.
Williams, B.R.G., and Kerr, I.M., 1978, Nature (London) 276 : 88-90.
Williams, B.R.G., and Kerr, I.M., 1980, Trends in Biochem. Sci., 5 : 138-140.
Wood, J.N., and Hovanessian, A.G., 1979, Nature (London), 282 : 74-76.
Wreschner, D.H., Mc. Cauley, J.W., Skehel, J.J., and Kerr, I.M., 1981, Nature (London), 289 : 414-417.
Zilberstein, A., Dudock, B., Berissi, H., and Revel, M., 1976a, J. Mol. Biol., 108 : 43-54.
Zilberstein, A., Federmann, P., Shulman, L., and Revel, M., 1976b, FEBS Lett., 68 : 119-124.
Zilberstein, A., Kimchi, A., Schmidt, A., and Revel, M., 1978, Proc. Natl. Acad. Sci. U.S.A., 75 : 4734-4738.

CANCER AND TRANSFORMATION BY RETROVIRUSES

J. Deschamps[*], A. Burny[*o] and D. Portetelle[*o]

*Department of Molecular Biology, University of Brussels
1640 Rhode-St-Genese, Belgium, and
oFaculty of Agronomy, 5800 Gembloux, Belgium

INTRODUCTION

Development of an individual is a very complex phenomenon under continuous genetic control from the one cell stage to adulthood, senescence and death. Each important stage along the differentiation pathway is characterized by expression of a given array of genes, each one of them at a given and well controlled rate. Control mechanisms of gene expression are mediated by specific molecules, products of certain genes. These molecules act as inducers, repressors, modulators, effectors, catalysts .... of specific metabolic reactions. Any event whose consequence would be an alteration, a block of synthesis, a dramatic decrease or increase in concentration of one of the crucial specific molecules, might, in principle, break cell harmony and lead to cancer. It is evident, within the frame of such a concept, that many physical, chemical and virological agents are candidates as disruptors of differentiation programmes. It is equally evident that cellular parameters will be of paramount importance to allow or prevent cell disturbance by outside factors.

To paraphrase G. Klein's presentation in the introductory chapter to the book "Viral Oncology" [1] we can conveniently divide the family of retroviruses into:

1. The directly transforming RNA tumor viruses.
2. The non-transforming but nevertheless oncogenic leukemia and mammary tumor viruses.

Proviral DNA from group no 1 viruses carries a piece of cellular genome responsible for cell transformation[1-4].

The way by which viruses from group n° 2 interfere with normal cell program is not clear yet [1-4]. Obviously they lack the extra sequences required for a direct transforming effect. They might act through expression of a protein product of the viral information[6-7], through their site of integration, through mechanisms dependent on their own multiplication[9] or through modifications of regulatory sequences (the "hit and run" hypothesis)[5,8].

GENERAL ARCHITECTURE OF RETROVIRUSES

The general scheme of assembly of RNA and protein components of retroviruses has been reviewed[10,11]. We take here a particle of Rous Sarcoma Virus (RSV) as a prototype.

Each particle has a polyploid genome made of 2 identical molecules of RNA with a total number of ca 9500 nucleotides. These 2 RNA molecules are associated with several thousand molecules of a 12000 MW protein (p12) and several molecules of enzyme reverse transcriptase, to form the ribonucleoprotein complex. In turn, the complex is surrounded by several thousand molecules of p27 and p15 to achieve an icosahedric structure called the virus core. Core particles may be produced from entire virus particles by treatment with appropriate concentrations of non-ionic detergent. Core structures are packaged inside a first layer made of p19, a phosphoprotein, itself surrounded by an outer· membrane made of a lipid bilayer entirely covered by approximately one thousand protruding spikes made of two glycoproteins gp85 and gp35, held together by disulphide bonds.

Based upon morphological features, biological activities, structural antigens .... retroviruses can be divided into A, B, C or D-type particles. Their overall diameter approximates 100 nm, their buyant density being around 1.16 g/ml in sucrose and 1.12 in metrizamide solutions.

Remarks

1. Specific names of retroviruses always qualify the main neoplasm induced by that virus. Most viruses if not all, however, can induce a spectrum of neoplastic growths, affecting cells of different embryonic origins.

2. Viruses that are replication-defective need a helper that provides structural proteins and/or reverse transcriptase activity.

LIFE CYCLE OF RETROVIRUSES[11-13]

The replication cycle of retroviruses comprises the following steps:

1°. Adsorption and penetration of the virus in a permissive cell.

2°. Reverse transcription of the viral RNA genome into linear double stranded DNA. This occurs in the cytoplasm.

3°. Penetration of cell nucleus, circularization of the proviral DNA and its integration into the host DNA.

4°. Eventual transcription and translation of the proviral information by the normal cell machinery.

Different types of experiment and especially cloning and sequencing of unintegrated - and integrated - retrovirus proviruses have revealed striking similarities between these structures and moveable genetic elements of bacteria, yeast and drosophila[14].

The genome on non-defective retroviruses contains, from 5' to 3', the following genes:

1°. GAG - for group specific antigens. It codes for internal structural proteins of the virion.

2°. POL - for polymerase. It contains the information for reverse transcriptase.

3°. ENV - for envelope. It codes for the protein backbone of virion envelope glycoproteins.

In RSV, the most complete retrovirus, one finds a fourth gene, the "src" gene, responsible for cell transformation and a "c" region, made of a nucleotide sequence that is highly conserved among avian tumor viruses.

The functional viral RNA molecule harbors a CAP structure at the 5' end and a poly A stretch at the 3' terminus.

Reverse transcription of this RNA structure leads to appearance of a direct repeat of ca 600 deoxyribonucleotides at both ends of the unintegrated provirus. Integration itself induces appearance of 2 short (a few bases) inverse repeats at the host DNA site accomodating the provirus (fig. 1).

Defective viruses have altered, non functional (or deleted) GAG and/or POL and/or ENV genes.

TRANSMISSION OF RETROVIRUSES

Retroviruses are conveniently divided into exogenous and endogenous entities. Viruses are said endogenous to a given animal species

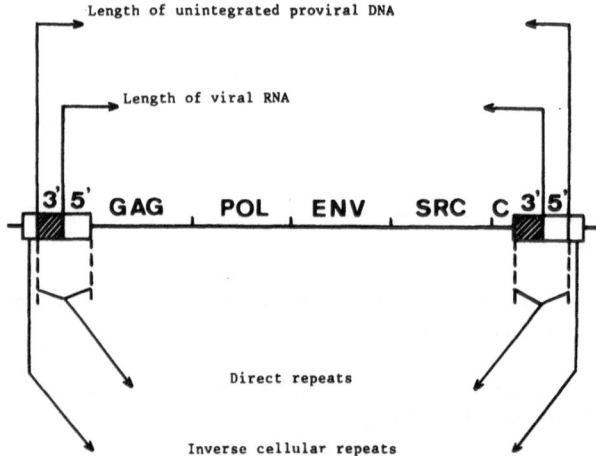

Fig. 1.

when their entire proviral information is present in the genome of
normal cells of that species. By contrast, exogenous viruses have no
sequence in common with their animal host. Their proviral sequences
are present only in target cells of affected organs. It follows that
endogenous viruses are vertically transmitted whilst horizontal transmission is the rule for exogenous viruses. No known physiological or
pathological function has been recognized so far for endogenous
viruses. They are always present as multiple copies in the genome and
seem to evolve more rapidly than the unique sequence DNA[15]. Exogenous
viruses have been identified as the cause of leukemias, lymphomas and
sarcomas.

VARIOUS HOST-VIRUS SYSTEMS

Table 1. Avian retroviruses

I.   Leukemia viruses.

  - They are competent for replication.
  - They do not transform cells in culture.
  - They require a long latent period before inducing
    neoplasms in vivo.

II.   Defective acute leukemia viruses.

  - They are defective for replication.
  - They transform their target cells and fibroblasts in
    culture.
  - They induce tumors, in vivo, with short periods of
    latency.

III.   Sarcoma viruses.

  - They are capable of replication and transformation
    of fibroblasts in culture.
  - They induce tumors in vivo.

Avian Leukemia Viruses

They cause predominantly lymphatic leukemia after a latency of months to years, following the inoculation in susceptible strains of chickens. They have not been shown thus far to contain any "onc" or transforming gene and the molecular basis of their oncogenic potential is therefore still obscure.

Recently transfection experiments have been reported, which consisted in introducing high molecular weight DNA extracted from ALV-induced neoplasms in NIH 3T3 mouse cells : they have revealed a transforming potential of a yet unidentified part of the infected cell genome towards those 3T3 cells. Moreover, it turns out that the viral genome itself is not responsible for the transforming effect; it was shown to be no longer present in some of the transformed mouse cells[8].

Aside from this, it has been shown that DNA fragments of chemically transformed and normal avian and murine cells, induce transformation of NIH 3T3 mouse cells with low efficiency, but that high molecular weight DNA of the resulting transformed cells induce transformation with high efficiency in secondary tranfection assays.

It thus seems that normal cells contain genes that are capable of inducing transformation once they have been activated.

Avian defective leukemia viruses

They are classified in 3 groups according to both their target cell specificity and the homologies of their specific genomic content: indeed, an internal portion of their genome originates from the host cellular DNA[16]. They are highly oncogenic and, _in vivo_, transform

- either erythroblasts (AEV)
- or macrophages (MC 29, CM II, MH 2, OK 10)
- or myeloblasts (AMV, E 26).

The hematopoietic disease they induce is correspondingly erythroblastosis, myelocytomatosis and myeloblastosis, respectively.

All those defective leukemia viruses seem to have been generated by recombination between an ALV helper and a cellular nucleotide sequence. Through the extensive study of some of those viruses, the following data could be collected: they code for a high molecular weight polyprotein (90.000 for CM II, 110.000 for MC 29, 100.000 for MH 2) presumed to be the fusion product of the recombinant viral-cellular genetic informations. The polyprotein bears determinants of the ALV 5' gag-region. Therefore those acute leukemia viruses are devoid of polymerase activity and lack env genes. They owe their "onc" potential presumably to the non structural part of their hybrid gene product. The non structural nucleotide sequences indeed are common to various strains with similar transforming spectra (for example, MC 29, MH 2, CM II) and absent from others whose hematopoietic target cells are different (such as AEV and AMV)[17]. For members of a same pathogenic family (MC 29, MH 2, CM II), only minor differences in the 3' part of the hybrid viral-cell sequence are detected, leading to a difference of MW for the gene product of this "onc" information[18]. A demonstration of the direct involvement of the gag gene-related polyprotein in the oncogenicity of those viruses has been provided for AEV[19] and MC 29. In the former case viral mutants have been described[20], the gag-related polyprotein of which happens to be smaller than that of the original transforming parent: those mutants show an altered ability to transform their bone marrow target cells.

The cellular homology of the onc gene of MCV (called "proto onc$_{MCV}$" gene) has recently been partially characterized[21]. It turns out to be a single locus in the chicken genome and consequently to be unrelated to the endogenous retroviral genes, the number and location of which being highly variable among the chicken species[22]. That "proto onc$_{MCV}$" is unlinked to the sarc information (the cellular homology of the ASV oncogene, cf hereunder). It is transcribed as a

2.8 Kb polyadenylated RNA also found in normal avian cells, suggesting that this proto onc$_{MCV}$ is a normal genetic locus in the vertebrate genome, the product of which could be involved in normal cellular growth and/or development.

An hypothesis concerning the mode of leukemogenesis of those viruses arose from the observation of simultaneity between the onset of the transformed phenotype and the block of differentiation of the infected cell (an AEV ts mutant was shown to confer a ts character to the transformed phenotype of the cells it infected, together with an induction of differentiation - i.e. haemoglobin synthesis - at the non-permissive temperature). The onc gene product could conceivably be thought to operate by direct competition with a physiological regulator of differentiation. This hypothesis also accounted for the specificity of the acute leukemia virus towards its target cell: only those possessing the corresponding non structural information would be able to compete with the regulatory element involved in the differentiation of a cell lineage[23].

A mutual exclusion between cell proliferation (mitogenesis) and cell differentiation might also explain the facts observed. Once a cell has sensed a mitogenic signal (the virus), it, no longer can be committed to differentiate. It has, indeed been reported that many of the chemical inducers of differentiation also inhibit lymphocyte mitogenesis in the mouse Friend leukemia system[24].

At any rate, what remains to be elucidated is the relationship between the sensitivity of a hematopoietic cell to a particular virus and the "onc" non structural information carried by that virus.

The avian sarcoma virus (or Rous Sarcoma Virus)

This non defective sarcoma inducing virus transforms fibroblasts both in vivo and in vitro. The viral genetic information proven to be responsible for oncogenicity is a 3' end located "src" gene, which has been both genetically and biochemically characterized in wild type and mutant ASV[25,26,27].

The src gene product is a 60.000 MW phosphoprotein (p60^{src}) which has been shown to possess protein kinase activity. p60^{src} contains two major sites of phosphorylation, a serin residue located in the amino-terminal portion of the polypeptide and a tyrosine residue is cAMP independent whilst phosphorylation of the serine residue is a cAMP-stimulated process. The src gene product being able to carry two phosphorous atoms is designated pp60^{src}.

The intracellular viral src gene product is found associated with the cell plasma membrane with a preference for areas of cell - cell and cell - substrate contact (gap junction)[28]. This localization

together with the fact that a ts pp60^{src}can reversibly provide the
cells with a transformed phenotype even after enucleation is strongly
suggestive of a direct cytoplasmic action of this protein kinase.
Interestingly, the pp60^{src} seems to be preferentially located at the
precise points where both the attachment sites of actin microfilament
bundles and the start of the extracellular fibronectin fibrils are
known to be located: it is therefore attractive to consider that
phosphorylation by pp60^{src} might directly influence both cell morphology and adhesion, and so confer the transformed phenotype to the
infected cell.

The viral src gene product has a cellular homolog expressed in
normal cells, product of a cellular sarc gene, situated at a single
constant locus in the chicken genome.

Interestingly enough, starting from the deleted transformation-
defective (td) ASV mutants, Hanafusa et al.,[29] have recovered transformation competent viral strains from chicken tumors produced after
long latency periods: those "recovered ASV$_s$" are presumed to result
from recombination of the td virus with cellular sequences.

In relation with this possible capture of a "src-like" cellular
information from the host genome, a molecular mechanism has recently
been suggested which is based upon the nucleotide sequence peculiar-
ities of the ASV genome[30]. An homologous region is reiterated on both
ends of the src gene and expected to allow site specific deletion or
integration of this src region[31]. This peculiarity would also account
for the frequent deletion of the src information, occurring after successive passages of the virus in cell cultures. The original generation of the "transforming virus genome might have followed this same
recombination scheme".

The src information which is unrelated to any other "onc" information known thus far appears to be required and sufficient for the
transformation capacity of an ASV strain: a cloned src genetic fragment, reintroduced into fibroblasts results in their acquiring the
transformed phenotype[32].

MAMMALIAN RETROVIRUSES

The Murine System

Murine leukemia viruses are classified according to their host
range. Depending on their ability to replicate on mouse cells and/or
on non-mouse cells, they are called:

    - ecotropic viruses (whose targets are mouse cells)

   - xenotropic viruses (whose targets are non-mouse cells, for
     example, mink cells)

   - dual tropic viruses (they can replicate on both type of cells).

     Among the ecotropic viruses are the N tropic and the B tropic
strains, the susceptibility of mice to one and/or the other of those
strains depending upon both the genetic identity of the host (mainly
through a genetic locus named $F_v - 1$) and viral gag gene determinates.
It has been shown that the block preventing a N ecotropic virus of
replicating in a F_v^{BB} (homozygous) mouse strain is located inside the
cell, at the level of DNA metabolism. The difference in host specificity between ecotropic and xenotropic viruses is related to adsorption of virus onto the cell and is located within the viral env gene.

A. Exogenous ecotropic murine leukemia viruses

     Three replication proficient, non transforming MuLV strains have
been discovered a long time ago, differing from each other by their
host spectrum:

   - the Mo.MuLV (Moloney Murine leukemia virus) is responsible for
     B type lymphoid leukemia

   - the F.MuLV (Friend) is the causative agent of erythroleukemia

   - the R.MuLV (Rauscher).

Their mode of action is still obscure.

B. Endogenous murine leukemia viruses

     The Gross virus, the first murine leukemia virus, was isolated
from the AKR mouse, a strain with high incidence of spontaneous
leukemia. It is endogenous to the mouse. Infection of young AKR mice
with Gross virus reduces the length of the latent period required for
overt leukemia development. It is believed that recombination events
between the Gross virus and a xenotropic virus, also endogenous to the
same mouse, is the cause of overt disease (see here under MCF and SFFV).

C. Transforming MuLV derived viruses

     Mo.MuLV and F.MuLV have been the native helper viruses for
replication-defective, highly oncogenic viruses.

   1°. Ab.MuLV (Abelson murine leukemia virus) was isolated sub-
        sequently to passage of Mo.MuLV in a BALB/C mouse.

2⁰. Mo.MuSV (Moloney murine sarcoma virus), which is sarcoma-inducing, was also generated after successive passages of Mo.MuLV in mice.

3⁰. Ha.MuSV (Harvey murine sarcoma virus) and Ki.MuSV (Kirsten murine sarcoma virus) were derived from a MuLV passed, into rats: they were shown to contain MuLV sequences together with rat cellular genetic information.

4⁰. In vivo recombination between F.MuLV and an endogenous xenotropic virus mostly in the env region is presumed to have given rise to the MCF transforming viruses (MCF = mink cell focus; these viruses exert cytopathic effects in mink cell cultures).

5⁰. A subsequent deletion within this recombinant MCF non-defective genome is thought to have generated the defective highly oncogenic SFFV (Spleen Focus Forming Virus) which is part of the so-called "Friend Virus Complex".

1⁰. - Ab.MuLV. This Mo.MuLV-derived virus lacks the required functions for replication as an infectious particle. It transforms lymphoid cells and fibroblasts in culture and induces rapid B cell lymphoma in vivo. It is presumed to result from a recombination between Mo.MuLV and normal cellular genes: there is 25% homology between both genomes. The Mo.MuLV related region is flanking an insert of 3.6 Kb. Ab.MuLV specific, which has an homologous counterpart in the normal mouse cell genome[33].

The only gene product of Ab.MuLV appears to be a p120 polyprotein possessing MuLV 5' gag determinants (p15, p12 and a part of p30) and a non structural part (MW = 90.000) which is presumably encoded by the newly acquired cellular sequences[34]. This polyprotein has to occupy a transmembrane position, because it can be detected outside the infected cells by a Ab.MuLV- specific anti-serum. It does not seem to be glycosylated.

p120 seems to be highly phosphorylated (in vitro, it is phosphorylated on a tyrosine residue) and to possess a protein kinase activity[35,36]. Several arguments strongly support the possibility that p120 possesses the transforming potential.

- p120 is the only known Ab.MuLV translational product.

- a number of morphological revertants of Ab.MuLV transformed mink cells have been isolated and found to lack detectable p120 expression.

- transformation defective Ab.MuLV mutants have recently been characterized that were found to be non-tumorigenic in vivo. Their p120 was shown not to be phosphorylated in vitro[37].

2°. - Mo.MuSV. Reported studies are consistent with the interpretation that a 85.000 M.W. polyprotein is involved in transformation. Transformation ts mutants are impaired in the synthesis of this p85 at the non permissive temperature. They also are consistent with the idea that the non structural sequences in p185 are derived from the acquired sequence of Mo.MuSV relatively to Mo.MuLV.

A transfection assay using in vitro synthesized Mo.MuSV DNA has shown that the Mo.MuSV DNA fragment containing the unique Mo.MuSV sequence is capable of inducing foci in mouse fibroblast cultures[38]. More recently, it was demonstrated[39] that an endogenously reverse transcribed Mo.MuSV is capable of transforming NIH - 3T3 fibroblasts in culture and that the resulting MSV DNA-transformed non producer cells are highly tumorigenic in newborn mice. An antiserum raised against helper Mo.MuLV proteins does not precipitate any gene product in these cells, suggesting that helper virus information is not involved in the transformation of NIH 3T3 cells by Mo.MuSV.

3°. - Ha.MuSV and Ki.MuSV. The Harvey and Kirsten MuSV, which are presumed to result from recombination between MuLV helper and rat cellular sequences have recently been shown to code for a phosphorylated (21,000 MW) gene product which could be the transforming protein of those viruses[40]. This p21 polypeptide indeed appears to be thermolabile in a ts transforming Ki.MuSV mutant. Moreover, NIH 3T3 cells transformed by moleculary cloned subgenomic DNA fragments of Ha.MuSV possess the p21 protein[41].

The p21 protein, which has a GDP and GTP binding activity[40] has been convincingly shown to be concentrated at the inner surface of the plasma membrane[41], as is the case for pp60^{src}; unlike pp60^{src} however, it does not accumulate at the gap junctions (junctions connecting adjacent cells). Knowing that adenylate cyclase, which is a plasma membrane component, is altered in transformation[42] and is regulated by a GTP binding protein, it seems an attractive hypothesis to speculate that the mechanism of Ha and Ki.MuSV-induced cell transformation would involve processes associated with guanine nucleotide binding proteins in the plasma membrane. In possible correlation with this, it is presumed that the effect of growth promoting hormones is mediated through a transmembrane signal which affects proteins on the inner surface of the plasma membrane: normal endogenous amounts of p21 (and pp60^{src}) would be normal components of a membrane growth promoting system that are magnified in transformation.

4°. - MCF and SFFV. They are both helper MuLV derived. Their env gene products have been shown to share antigenic determinants with their MuLV helper, but also to possess dissimilarities between each other and towards their helper parent. Those dissimilarities most probably account for their target cell specificity (lymphoid system for MCF, erythroid system for SFFV).

A MCF isolate has allowed the identification of a precursor polyprotein containing gag gene - coded proteins p15 and p12, and a non structural component for which no function has yet been assigned[43].

From heteroduplex analyses, it appears that the F-SFFV genome has incurred deletions in the pol and env gene regions. It has furthermore undergone sequence substitutions within the gag and env genes. It must thus code for altered and perhaps defective gag, pol and env gene products. Such modifications are perhaps related to the leukemogenic properties of the virus[44].

D.   The Radiation Leukemia Virus Isolates

They are unrelated to any known ecotropic or xenotropic retroviruses and are presumed to result from a radiation-provoked induction of the expression of latent oncogenic sequences capable of initiating a neoplastic transformation in susceptible lymphoid cells.

No satisfactory explanation for the mode of action of these agents is available at present.

E.   The Mouse Mammary Tumor Virus (MMTV)

Mu.MTV viruses form a closely related group of retroviruses which can be transmitted by two different modes[1,45].

- All strains of Mus musculus contain 6 to 10 MTV proviral DNA copies endogenous MTV information[45]. These copies are thus present in all cells of the mouse and seem to play no role in early mammary tumor occurrence. They are presumed to have arisen from an early infection of the germ line[46].

- In certain inbred strains of mice, the MMTV virus is exogenously transmitted to the offspring via the milk and appears to be responsible for a high incidence of mammary tumor occurence. The tumor cell genome has been shown to contain in this latter case, additional integrated MTV copies.

In some cases, spontaneous mammary tumors occur in low tumor incidence BALB/c mice rather late in the life of the animals. No enhanced amounts of integrated MMTV information is to be observed in this case. This could suggest that the endogenous copies of MMTV are in a repressed state in the normal cellular genome.

Hormones such as glucocorticoids, estrogen, prolactin and insulin, which are involved in the differentiation of the normal mammary gland, also seem to play a role in the regulation of viral expression and to contribute to the induction of mammary tumors in the mouse. No new viral DNA is integrated in this case, but the hormone treatment results in a transitory increased expression of MTV RNA in the lactating gland.

The molecular basis of the difference between the incidence of endogenous and exogenous copies of MMTV on mammary tumor induction is not understood nowadays.

The Feline System

Feline Leukemia Virus (FeLV) is a replication competent leukemia virus which induces leukemia at low frequency in cats[1]. Three subgroups of FeLV have been characterized so far according to interference properties[47]. Additional FeLV subgroups might exist as endogenous viruses in cat cells.

FeLV infected cells express Feline Oncornavirus Membrane Associated Antigen (FOCMA) at their surface. Antibodies to FOCMA confer immunity to the infected animal towards FeLV of FeSV tumor induction[48]. FeSV is a sarcoma inducing, FeLV-related type C retrovirus: it is presumed to result from a recombination event between FeLV gag sequences and a cellular genetic region. (At least two of the three FeLV subgroups were present in the original isolates of the known FeSV strains)[49].

FeSV synthesizes a hybrid FeLV gag-X protein, which is endowed with kinase activity[50]. A comparison between the tryptic peptides of the gag-X, p115 polyprotein synthesized in independently originated Gardner and Snyder-Theilen FeSV and their FeLV helper, lead to the conclusion that the potentially transforming sequences are mutually related. SM-FeSV, on the contrary, codes for a gag-X protein, the X part of which is unrelated to that coded for by ST or GA.FeSV (M. Essex, personal comm.). Of obvious, is the suspected relatedness or even identity between X and FOCMA. From recent data, it comes that "At least some cats make antibodies to the X portion of "gag-X" and such antisera show a typical FOCMA reaction by immunofluorescence when using either cat lymphoma cells or FeSV-transformed fibroblasts" (M. Essex, person. comm.).

Interesting to mention is the fact that infected cells from some tumor-bearing animals, carry this gag-X polyprotein without containing any other structural FeSV protein, nor showing any helper dependent FeSV production. The possibility that the transforming potential conferred by FeSV could be stably transferred without requiring the viral genome any longer, is as intriguing as the transforming capacity of lymphoid tumor DNA fragments (with no viral sequence) in the case of avian leukosis viruses.

The Bovine System

Enzootic bovine leukemia, a lymphoproliferative disease of cattle, is induced by Bovine Leukemia virus (BLV), a retrovirus exogenous to the bovine genus[51]. The infection spreads from herd to herd mostly through commercial exchanges of virus carriers. The origin of the virus as well as its mode of action are still unknown.

Goats, rabbits, sheep, pigs, chimpanzees ....can be infected. Sheep are so sensitive to infection (and tumor induction) that they are used in bioassays aiming at detection of the infectious agent.

The Primate System

Type C and Type D retroviruses, exo or endogenous to the host species have been described in a number of primates from the old and the new world[1,15].

A. Exogenous Viruses

The virus complex SSV-1/SSAV-1 (Simian Sarcoma Virus 1 / Simian Sarcoma associated Virus 1) consists of a defective sarcoma virus and its helper. The sarcoma component transforms fibroblasts in vitro and can induce tumors after inoculation to marmosets. It has been recently shown by blotting hybridization, that a DNA sequence, SSV-1 specific, is present in human and chimpanzee DNA. Is this the genetic information for cell transformation? No transforming protein has been identified, so far.

Several viral isolates have been characterized from gibbon apes with hematopoietic neoplasias. They are termed "Gibbon Ape Leukemia Viruses" (GALV). These viruses are closely related to one another and to SSV/SSAV. They can stimulate growth of human lymphocytes and immortalize them. These immortalized cells are tumorigenous in nude mice[52]. The mode of action of GALV is understood.

Of general biological interest too is the observation that GALV are closely related antigenically to an endogenous virus of an Asian mouse, Mus caroli.

B. Endogenous Viruses

A number of them have been isolated. They exhibit properties analogous to those of their counterparts in lower mammals.

One of them, the baboon endogenous virus provides an example of interspecies transmission of viral genes. It is closely related to an endogenous virus of many cat species, the RD 114 virus. From a number of experiments, it could be concluded that the baboon virus had been acquired by the cats, but solely by those that were in contact with African primates.

Another member of this group, endogenous to Presbytis obscurus; gave rise to the Mason Pfizer Monkey Virus (MPMV) exogenous to- and horizontally transmitted in Rhesus Monkeys. In contrast to the general rule, MPMV is not oncogenic in vivo.

These two cases stress the fact that viral information, even perhaps when endogenous, can escape the host cell genome and behave as moveable genetic element. Moreover, it has become evident that virogenes, even when endogenous, evolve more rapidly than the entire unique-sequence DNA, in which they are embedded, thus allowing a fine degree of discrimination among closely related primate species. Detailed and extensive studies along this line led to unsuspected and exciting conclusions such as, that man's evolution must have occurred outside Africa.

V......and man?

No firm and undisputable evidence exists today to cite retroviruses as causative agents of human neoplasia. An array of observations has been reported by a number of groups, tending to demonstrate that retroviruses are indeed present in the human genome and can be expressed. A nice series of experiments, recently disclosed by Reitz et al.[53] is most suggestive in that respect.

Irreproducible and controversial results has been so far, most of the harvest of investigators devoted to "human retroviruses". We find it, however, highly improbable that man's genome would have kept away from the otherwise widespread moveable genetic elements. Now that animal genes, coding for transforming proteins are being thoroughly characterized, they constitute new and powerful probes to investigate enigmas of human cancer.

CONCLUSIONS

From all the studies aiming at understanding how retroviruses transform cells, it emerges that, most probably, the host virus interplay leads to overexpression of information coded for by one (or more) normal cellular gene(s). The gene product is a phosphoprotein.

It exhibits in all detailed systems a membrane or transmembrane subcellular location, thus stressing the point that the cell membrane may be a major site of action for transformation.

When a normal counterpart to the transforming protein has been identified, it is not sure yet whether both polypeptides are functionally identical.

ACKNOWLEDGEMENTS

The work performed in the authors' laboratory was helped financially by the "Fonds Cancérologique de la Caisse Générale d'Epargne et de Retraite" and the State Contract "Actions concertées".

REFERENCES

1.   G. Klein (ed.)(1980). Viral Oncology, 842 pp. Raven Press, New York.
2.   Viral Oncogenes (1980). Cold Spring Harbor Symposia on Quantitative Biology, Vol. XLIV, 1400 pp. Cold Spring Harbor Laboratory, Cold Spring Harbor, N. Y.
3.   M. Essex, G. Todaro and H. zur Hausen, (eds.)(1980). Viruses in naturally occuring Cancers, 1440 pp. In: "Cold Spring Harbor conferences on Cell Proliferation", vol. 7. Cold Spring Harbor Laboratory, Cold Spring Harbor, N. Y.
4.   J. Deschamps, A. Burny and J. M. Provost, (1980). Factors regulating cellular and viral genes expression in leukemogenesis. In: "Advances in comparative leukemia research 1979" (D. S. Yohn, B. Lapin and J. R. Blakeslee ed.) pp. 75-81. Elsevier, North-Holland, N. Y.
5.   W. D. Hardy, A. J. McClelland, E. E. Zulkerman, H. W. Snyder Jr., E. G. MacEwen, D. Francis and M. Essex, (1980). Development of virus non-producer lymphosarcomas in pet cats exposed to FeLV. Nature, 288:90.
6.   J. Ghysdael, R. Kettmann and A. Burny, (1979). Translation of Bovine Leukemia Virus virion RNA in heterologous protein-synthesizing systems. J. Virol. 29:1087.
7.   J. G. Sutcliffe, T. M. Shinnick, I. M. Verma and R. A. Lerner, (1980). Nucleotide sequence of Moloney Leukemia Virus: 3' end reveals details of replication, analogy to bacterial transposons and an unexpected gene. Proc. Natl. Acad. Sci. U.S.A. 77:3302.
8.   G. M. Cooper and P. E. Neiman, (1980). Transforming genes of neoplasms induced by avian lymphoid leukosis viruses. Nature 287:656.
9.   M. S. McGrath, E. Pillemer, D. A. Kooistra, S. Jacobs, L. Jerabek and I. L. Weissman, (1980). T-lymphoma retroviral receptors and control of T-lymphoma cell proliferation. Cold Spring Harbor Symp. Quant. Bio., 44:1297.
10. R. C. Montelaro and D. P. Bolognesi, (1978). Structure and morphogenesis of type-C retroviruses. Adv. Cancer Res., 28:63.

11. J. Ghysdael, R. Kettmann and A. Burny, (1980). M-lecular aspects of RNA tumor virus biology. Pharmac. Ther., 9: 147.

12. R. A. Weinberg, (1977). Structure of the intermediates leading to the integrated provirus. Biochim. Biophys. Acta, 473:39.

13. A. Panet, (1980). Replication of murine leukemia viruses. In: "Viral Oncology". (G. Klein, ed.), pp. 109-134. Raven Press, New York, N.Y.

14. H. M. Temin, (1980). Origin of retroviruses from cellular moveable genetic elements. Cell, 21:599.

15. G. Todaro, (1980). Interspecies transmission of mammalian retroviruses. In: "Viral Oncology". (G. Klein, ed.), pp. 291-310. Raven Press, New York, N.Y.

16. M. Roussel, S. Saule, C. Lagrou, C. Rommens, H. Beug, T. Graf and D. Stehelin, (1979). Three new types of viral oncogene of cellular origin specific for haematopoietic cell transformation. Nature, 281:452.

17. D. Sheiness, K. Bister, C. Moscovoci, L. Fanshier, T. Gonda, M. Bishop, (1980). Avian retroviruses that cause Carcinoma and Leukemia: identification of nucleotide sequences associated with pathogenicity. J. Virol., 33:962.

18. K. Bister, H. C. Loliger and P. H. Duesberg, (1979). Oligonucleotide map and protein of CM II: detection of conserved and non-conserved genetic elements in avian acute leukemia viruses CM II, MC 29 and MH 2. J. Virol., 32:208.

19. H. Beug, G. Kitchener, G. Doederlein, T. Graf and M. J. Hayman, (1980). Mutant of avian erythroblastosis virus defective for erythroblast transformation: deletion in the erb portion of p75 suggests function of the protein in leukemogenesis. Proc. Natl. Acad. Sci. U.S.A., 77:6683.

20. G. Ramsay, T. Graf, M. J. Hayman, (1980). Mutants of avian myelocytomatosis virus with smaller gag gene-related proteins have an altered transforming ability. Nature,288:170.

21. D. K. Sheiness, S. H. Hughes, H. E. Varmus, E. Stubblefield and J. M. Bishop, (1980). The vertebrate homolog of the putative transforming gene of Avian myelocytomatosis virus: characteristics of the DNA locus and its RNA transcript. Virology, 105:415.

22. S. H. Hugues, F. Payvar, D. Spector, R. T. Schimke, H. L. Robinson, G. S. Payne, J. M. Bishop and H. E. Varmus, (1979). Heterogeneity of genetic loci in chickens: Analysis of endogenous viral and non-viral genes by cleavage of DNA with restriction endonuclease. Cell, 18:347.

23. T. Graf, N. Ade and H. Beug, (1978). Temperature-sensitive mutant of avian erythroblastosis virus suggests a block of differentiation as mechanism of leukemogenesis. Nature, 275:496.

24. K. H. Stenzel, R. Schwartz, A. L. Rubin and A. Novogrdsky, (1980). Chemical inducers of diffe-entiation in Friend leukemia cells inhibit lymphocyte mitogenesis, Nature, 285:106.

25.   R. L. Erikson, (1980). Avian sarcoma viruses: molecular bi-
      ology. In: "Viral Oncology". (G. Klein, ed.), pp. 39-53.
      Raven Press, New York, N.Y.

26   B. M. Sefton, T. Hunter, K. Beemon, (1980). Temperature-
      sensitive transformation by Rous sarcoma virus and tempera-
      ture sensitiveprotein kinase activity. J. Virol., 33:220.

27.   B. M. Sefton, T. Hunter, K. Beemon and W. Eckart, (1980).
      Evidence that the phosphorylation of tyrosine is essential
      for cellular transformation by Rous sarcoma virus. Cell,
      20:807.

28.   R. O. Hynes, (1980). Cellular location of viral transforming
      proteins. Cell, 21:601.

29.   L. H. Wang, C. C. Halpern, M. Nadel and H. Hanafusa, (1978).
      Recombination between viral and cellular sequences generates
      transforming sarcoma virus. Proc. Natl. Acad. Sci. USA,
      75:5812.

30.   T. Yamamoto, J. Sivaswami Tyagi, J. B. Fagan, J. Gilbert,
      B. de Crombrugghe and I. Pastan, (1980). Molecular Mechanism
      for the capture and excision of the transforming Gene of
      Avian Sarcoma Virus, as suggested by Analysis of Recombinant
      Clones. J. Virol., 35:436.

31.   A. P. Czernilofsky, A. D. Levinson, H. F. Varmus, J. M. Bishop,
      E. Tischer and H. M. Goodman, (1980). Nucleotide sequence
      of an avian sarcoma virus oncogene (src) and proposed amino-
      acid sequence for gene product. Nature, 287:198.

32.   D. G. Blair, W. L. McClements, M. K. Oskarsson, P. J. Fishinger
      and G. F. Van de Woude, (1980). Biological activity of
      cloned Moloney sarcoma virus DNA: Terminally redundant
      sequences may enhance transformation efficiency. Proc. Natl.
      Acad. Sci. USA, 77:3504.

33.   A. Shields, S. Goff, M. Pasking, G. Otto, D. Baltimore, (1979).
      Structure of the Abelson Murine Leukemia Virus genome.
      Cell, 18:955.

34.   W. J. M. Van de Ven, F. H. Reynolds, Jr., P. Nalewaik and
      J. R. Stephenson, (1979). The nonstructural component of
      the Abelson Murine Leukemia Virus Polyprotein p120 is encoded
      by newly acquired genetic sequences. J. Virol., 32:1041.

35.   O. N. Witte, A. Dasgupta, D. Baltimore, (1980). Abelson Murine
      leukaemia virus protein is phosphorylated in vitro to form
      phosphotyrosine. Nature, 283:826.

36.   W. J. M. Van de Ven, F. H. Reynolds, Jr., and J. R. Stephenson,
      (1980). The nonstructural components of polyproteins encoded
      by replication-defective mammalian transforming retroviruses
      are phosphorylated and have associated protein kinase
      activity. Virology, 101:185.

37.   F. H. Reynolds, Jr., W. J. M. Van de Ven and J. R. Stephenson,
      (1980). Abelson murine leukemia virus transformation-
      defective mutants with impaired p120 - associated protein
      kinase activity. J. Virol., 36:374.

38. P. Andersson, M. P. Goldfarb and R. A. Weinberg, (1979). A defined subgenomic fragment of in vitro synthesized Moloney sarcoma virus DNA can induce cell transformation upon transfection. Cell, 16:63.

39. M. H. T. Lai and I. M. Verma, (1980). Genome organization of retroviruses: VIII. Nonproducer cell lines of mouse fibroblasts transformed by Moloney murine sarcoma virus DNA synthesized in vitro. Virology, 104:407.

40. T. Y. Shih, A. G. Papageorge, P. E. Stokes, M. O. Weeks and E. Scolnick, (1980). Guanine nucleotide binding and autophosphorylating activities associated with the p21src protein of Harvey murine sarcoma virus. Nature, 287:686.

41. M. C. Willingham, I. Pastan, T. Y. Shih and E. M. Scolnick, (1980). Localisation of the src gene product of the Harvey strain of MSV to plasma membrane of transformed cells by electron microscopic immunocytochemistry. Cell, 19:1005.

42. I. Pastan and M. Willingham, (1978). Cellular transformation and the "morphologic phenotype" of transformed cells. Nature, 274:645.

43. R. A. Bosselman and I. M. Verma, (1980). Genome organization of retroviruses: V. in vitro synthesized Moloney murine leukemia viral DNA has long terminal redundancy. J. Virol., 33:487.

44. R. A. Bosselman, L. J. L. D. Van Griensven, M. Vogt and I. M. Verma, (1980). Genome organization of retroviruses: IX. Analysis of the genome of Friend Spleen Focus-forming (F-SFFV) and helper murine leukemia viruses by heteroduplex-formation. Virology, 102:234.

45. D. H. Moore, C. A. Long, A. B. Vaidya, J. B. Sheffield, A. S. Dion and E. Y. Lasfarques, (1979). Mammary tumor viruses. In: "Advances in Cancer Research" (G. Klein and Weinhouse, eds.), vol. 29, pp. 347-418. Academic Press, New York.

46. B. Groner and N. Hynes, (1980). Number and location of mouse mammary tumor virus proviral DNA in mouse DNA or normal tissue and of mammary tumors. J. Virol., 33:1013.

47. P. S. Sarma and T. Log, (1973). Subgroup classification of feline leukemia and sarcoma viruses by viral interference and neutralization tests. Virology, 54:160.

48. C. J. Sherr, A. Sen, G. J. Todaro, A. Sliski and M. Essex, (1978). Pseudotypes of feline sarcoma virus contain an 85.000 dalton protein with feline oncornavirus associated cell membrane antigen (FOCMA) activity. Proc. Natl. Acad. Sci. USA. 75:1505.

49. P. S. Sarma, A. C. Sharar and S. McDonough, (1972). The SM strain of feline sarcoma virus biologic and antigenic characterization of virus. Proc. Soc. Exp. Biol. Med., 140:1365.

50. W. J. M. Van de Ven, F. H. Reynolds, Jr., and J. R. Stephenson,
     (1980). The nonstructural components of polyproteins
     encoded by replication defective mammalian transforming
     viruses are phosphorylated and have associated protein
     kinase activity. Virology, 101:185.

51. R. Kettmann, D. Portetelle, M. Mammerickx, Y. Cleuter,
     D. Dekegel, M. Galoux, J. Ghysdael, A. Burny and
     R. Chantrenne, (1976). Bovine leukemia virus: an exogenous
     RNA oncogenic Virus. Proc. Natl. Acad. Sci. USA, 73:1041.

52. P. D. Markham, F. Ruscetti, S. Z. Salahuddin, R. E. Gallagher
     and R. C. Gallo, (1979). Enhanced induction of growth of
     B lymphoblasts from fresh blood by primate type C retro-
     viruses (gibbon ape leukemia virus and simian sarcoma virus).
     Int. J. Cancer, 23:148.

53. M. S. Reitz, N. R. Miller, F. Wong-Staal, R. E. Gallagher,
     R. C. Gallo and D. H. Gillepsic, (1976). Primate type C
     virus nucleic acid sequences (woolly monkey and baboon types)
     in tissues from a patient with acute myelogenous leukemia
     and in viruses isolated from cultured cells of the same
     patient. Proc. Natl. Acad. Sci. USA., 73:2113.

INTRACELLULAR SIGNALS IN PITUITARY HORMONE SECRETION

Barry L. Brown and *Pauline R.M. Dobson

Dept. of Human Metabolism & Clinical Biochemistry,
University of Sheffield Medical School,
Beech Hill Road, Sheffield, S10 2RX, U.K.
*Dept. of Biochemistry, University of Nottingham
Medical School, Clifton Blvd., Nottingham NG7 2UH,UK

INTRODUCTION

The regulation of the secretion of hormones from the anterior pituitary gland depends, to a large degree, on the release of substances from the hypothalamus into the hypophysial portal vessels of the median eminence. These neural substances act in conjunction with other blood-borne substances from the periphery to regulate the function of the various anterior pituitary cells. Thus far, the only hypothalamic substances that have been unequivocally demonstrated are the peptides; thyrotropin-releasing hormone (TRH), luteinizing hormone-releasing hormone (LHRH) and somatostatin (or GHRIH). In addition, most evidence favours the supposition that dopamine is the major prolactin inhibitory factor (PIF), although there have been sporadic reports of a separate PIF (notably GABA). Dopamine acts directly on pituitary cells to inhibit prolactin secretion, it is found in the portal blood in concentrations sufficient to affect prolactin release, and its release from the hypothalamus appears to be influenced by effectors of prolactin secretion in vivo. There is also evidence for the existence of corticotropin-releasing factor (CRF), growth hormone releasing factor (GRF), and a prolactin releasing factor (PRF). This latter activity may be distinct from TRH which also stimulates prolactin secretion.

The actions of the three characterised peptides are not completely specific : TRH stimulates the release of prolactin as well as TSH; LHRH stimulates the release of LH and FSH, and somatostatin has inhibitory effects on TSH secretion, in addition to affecting GH secretion. Moreover, all three hypothalamic peptides, like many

other hormonal peptides, have been found in extra-hypothalamic
locations.   Somatostatin is also synthesised in the pancreas and has
profound effects on gastrointestinal function. In addition, LHRH has
been shown, recently, to have direct effects on the gonads.

Within the anterior pituitary gland there are, at least five
different cell types secreting prolactin (mammotroph), TSH
(thyrotroph), growth hormone (somatotroph), the gonadotropins - LH and
FSH (gonadotroph), ACTH and other peptides of the pro-opiocortin
family (corticotroph).   To some degree this cellular heterogeneity and
the relative non-specificity of control by hypothalamic hormones has
complicated the elucidation of the intracellular mechanisms involved
in the control of pituitary hormone secretion.   Various approaches
have been taken including the measurement of the secretion of a
specific pituitary hormone (both <u>in vivo</u> and <u>in vitro</u>) in response to
substances known or suspected to influence the regulation of putative
intracellular mediators, and the direct measurement of these putative
mediators in response to specific hormonal (or other) stimuli in
tissue slices, isolated cells, separated (ie partially purified) cell
types and in hormone secreting tumour cells.   Furthermore, numerous
attempts have been made to compare the effects of secretagogues and
inhibitors on the dynamics and dose response relationships of
secretion and of accumulation of cyclic nucleotides or of ion fluxes.

Our aim, in this short review, is to give an overall impression
of the state of the art with respect to the relative roles of the
various proposed intracellular mediators of hormone secretion.   We
will deal in turn with the evidence for and against a role for cyclic
AMP, cyclic GMP and Ca^{++} in anterior pituitary function.   It
should be borne in mind that the often contradictory results found in
this field may reflect in part the above-stated problems of
heterogeneity and specificity.   The bibliography has been selected to
reflect the controversies in this field.   In addition, papers have
been chosen which have extensive references.   The bulk of the
discussion in this review will centre on the control of the secretion
of TSH, prolactin, LH and FSH and, to a lesser extent, growth
hormone.

Receptors

The initial step in the mechanism of action of TRH, LHRH,
dopamine and somatostatin appears to be their interaction with
receptors on the plasma membrane of their respective target cells.
Many groups have shown selective and saturable binding of the hormones
to pituitary tissue, cells and membrane fractions.   In the case of
TRH, binding to GH_3 cells (a clonal strain of rat pituitary tumour
cells secreting prolactin and growth hormone) and to a mouse pituitary
TSH-secreting tumour (TtT) has also been observed.   For both TRH and
LHRH, the subcellular fractions possessing the receptors were co-
purified with adenylate cyclase activity and other plasma membrane

enzymes. Studies have suggested that only a fraction of the receptors need to be occupied to elicit a maximal secretory response.   The dopamine receptors in the anterior pituitary have been designated as being type D2 on the basis of the lack of dopamine-stimulated adenylate cyclase and on the pharmacological profile of dopamine agonists and antagonists. Recently it has been shown that the displacement of tritium labelled-spiroperidol from pituitary dopamine receptors by the benzamides (dopamine antagonists) was sodium dependent.   Typical neuroleptic activity was not influenced by sodium and the conclusion was reached that there is a subpopulation of dopamine receptors with which benzamides react.

The number of TRH receptors on both mammotrophic and thyrotrophic cells is not constant. Chronic exposure of GH_3 cells to TRH leads to a decrease in the number of TRH receptors. Thyroid hormones also decrease the number of receptors, but not the affinity for TRH, on these tumour cells, on the mouse pituitary tumour cell (TtT) and in membrane fractions of pituitary tissue from rats treated with thyroid hormones in vivo. Moreover, oestrogens, which increase the prolactin and TSH response to TRH, increase the number of TRH receptors but not their affinity, in homogenates of pituitaries from rats treated with oestradiol, and in mammotrophic tumour cells treated in culture for 48 hours with oestradiol. Oestrone was about 10% as potent, and tamoxifen, an antioestrogenic compound, inhibited the increase in receptors induced by oestradiol. This effect of physiological levels of oestradiol may be mediated through cytoplasmic oestrogen receptors. Glucocorticoids also increase the number of TRH receptors on the GH cells.   There is also evidence for internalisation of TRH into the GH cell. In one study it was reported that, after being taken up by the cells, TRH is slowly metabolised to its constituent amino acids. However, other workers reported that, after internalisation, the peptide was spontaneously unloaded without being degraded.

In the case of LHRH it has been demonstrated that the number of LHRH receptors increases in diestrous in advance of the proestrous rise in LHRH secretion.   There is also evidence that LHRH induces its own pituitary receptors and that the increase in receptor number after castration may reflect increased endogenous secretion of LHRH.

Thus, there are a number of mechanisms whereby both the hypothalamic peptides and peripheral hormones regulate receptor numbers. It may be that several regulating mechanisms operate at the receptor level to modulate the response to each hypothalamic hormone.

Adenylate cyclase

Although it has long been clear that normal pituitary cells and pituitary tumour cells possess the adenylate cyclase enzyme, there have been few reports of investigations of its response to hormones and other putative effectors. This may, in part, be a reflection of the cellular heterogeneity of the normal gland mentioned earlier.

Activation of adenylate cyclase in the rat pituitary by crude hypothalamic extract and LHRH have been reported. However, other workers have reported binding of LHRH to the plasma membrane without activation of adenylate cyclase. GH (both GH_1 and GH_3) cells also contain an adenylate cyclase which responds to effectors. The adenylate cyclase of a clone of GH_3 cells was stimulated by low concentrations of chlorpromazine, a dopamine antagonist. Homogenates of normal pituitary tissue showed no response to chloropromazine unless guanylimidodiphosphate (Gpp(NH)p) was present. Although in this particular study, no response to dopamine was observed, homogenates of human prolactin-secreting adenomas have been shown to possess an adenylate cyclase which was inhibited by dopamine. Similar results have been reported with normal rat pituitary adenylate cyclase, although some investigators have been unable to find such an effect. These results are interesting in the light of the wealth of evidence for dopamine-stimulated (and chlorpromazine-inhibited) adenylate cyclase activity in areas of the brain. It has been shown that TRH has no effect on the adenylate cyclase of TtT cells either in the presence or absence of Gpp(NH)p.

Somatostatin inhibited both basal and chlorpromazine-stimulated adenylate cyclase activity in GH_1 cells, but had no effect on fluoride or Gpp(NH)p-stimulated activity. Somatostatin reduced the Vmax of the reaction without altering the apparent Km for ATP or the requirement for Mg^{++}. Addition of EGTA or varying the concentration of Ca^{++} had no effect on somatostatin inhibition of adenylate cyclase. However, whether this inhibitory action of somatostatin on adenylate cyclase can account for its effect on hormone release is unclear, since others have shown that somatostatin can inhibit growth hormone release in response to exogenous derivatives of cyclic AMP.

Intracellular cyclic nucleotides

In the early 1970's, a great deal of evidence accrued which was interpreted as suggesting an obligatory role for cyclic AMP in the mediation of the control of pituitary hormone release. Cyclic AMP and its derivatives, and phosphodiesterase inhibitors were shown to increase the secretion and often to potentiate the effects of either hypothalamic extracts or synthetic releasing hormones. Furthermore, several investigators found that LHRH increased intracellular cyclic AMP in anterior pituitary tissue or cells, and that there was a close parallelism between stimulation of cyclic AMP accumulation and gonadotropin release with a wide range of LHRH analogues. Moreover, LHRH antagonists were reported to lead to parallel inhibition of LH and FSH release and cyclic AMP concentration. Additionally, TRH was shown to have a small effect in increasing cyclic AMP in normal pituitary tissue. It has also been reported that somatostatin lowered both basal and stimulated levels of cyclic AMP. With the advent of the various clonal cell lines of rat pituitary tumour cells, further studies indicated that TRH caused a dose-dependent increase in cyclic AMP in both the GH cells and the TtT cells.   The application of cell separation techniques to the   anterior pituitary has led to reports of a TRH-stimulated accumulation of cyclic AMP in both thyrotroph-rich and mammotroph-rich preparations.   Furthermore, it was observed that micromolar concentrations of dopamine prevented the rise in cyclic AMP induced by TRH in a mammotroph-rich but not in a thyrotroph-rich preparation.   It has also been shown that these concentrations of dopamine also inhibit basal and IBMX-stimulated cyclic AMP accumulation in pituitary glands.

However, the interpretation of the data outlined above as indicating an obligatory role for cyclic AMP has been questioned. Firstly, not all investigators have been able to demonstrate changes in pituitary cyclic AMP accumulation in response to LHRH or TRH, although this discrepancy may be a reflection of the cellular heterogeneity of the tissue.   Nevertheless, several investigators have reported increases in hormone secretion occurring before measurable increases in cyclic AMP were observed.   In cultured rat pituitary cells, LHRH induced a rapid release of LH without significant alterations in intracellular or extracellular cyclic AMP, or in occupancy by cyclic AMP of the regulatory subunit of cyclic AMP-dependent protein kinase.   A   number of studies have indicated that neither cholera toxin nor prostaglandin E_1 (both of which increase cyclic AMP accumulation) had any effect on the release of LH. However, in one report these agents did appear to increase LH release, but not as markedly as did LHRH.   In one study in which cyclic AMP production was observed to be stimulated by LHRH, it has been suggested that LHRH-stimulated prostaglandin biosynthesis and cyclic AMP accumulation may be separate (and not sequential) physiological events, since aspirin lowered prostaglandin synthesis in response to LHRH, but had no effect on cyclic AMP levels.   Others have also concluded that the ability of LHRH to raise cyclic AMP concentrations is independent of its ability to increase LH release.   Pituitary glands from adult female rats or from immature animals responded to LHRH with an increase in LH release, but no change in cyclic AMP concentration was observed. Pituitaries from immature males did produce changes in cyclic AMP, but only with high doses of LHRH and only after 90 minutes incubation.   Lower doses of LHRH only enhanced cyclic AMP accummulation in pituitaries from adult male rats or from castrate male rats with testosterone replacement.   These findings were taken to indicate that cyclic AMP cannot be regarded as the exclusive mediator of LHRH action.

There also appeared to be no correlation between TSH release and intracellular cyclic AMP levels in TtT cells, nor was there any effect of TRH on protein-bound cyclic AMP in these cells.   Moreover, in some of those studies reporting that TRH does cause an increase in cyclic AMP accumulation, there is an obvious discrepancy between the doses of the hormone causing cyclic AMP accumulation and the secretory response.   In addition, experiments with GH_3 cells indicated that low doses of TRH induced an increase in prolactin release without an accompanying change in cyclic AMP concentrations.   Studies on prolactin and TSH secretion in response to cholera toxin and prostaglandins have given conflicting results.   While both of these agents generally cause an increase in the release of TSH, cholera toxin has been reported to either stimulate or have no effect on prolactin secretion.   Furthermore, most studies indicate that the prostaglandins do not induce an increase in prolactin release.

A number of studies have indicated that both cholera toxin and prostaglandins are potent secretagogues for growth hormone in vitro and that this effect is accompanied by, and may be causally related to, an increase in cyclic AMP.   Despite the observations of inhibition of adenylate cyclase activity and cyclic AMP production by somatostatin, recent reports have indicated that this peptide does not modify intracellular cyclic AMP concentrations under conditions in which growth hormone release is inhibited.   Recent studies have shown that nanomolar concentrations of dopamine, which inhibit prolactin secretion, have no effect on the cyclic AMP content of pituitary glands in vitro.   Furthermore, the inhibition of prolactin secretion by dopamine was unaffected by cholera toxin and it was concluded that the inhibitory effects of dopamine were probably not mediated by lowering of cyclic AMP concentrations.

The presence of a cyclic AMP-dependent protein kinase has been demonstrated in the adenohypophysis, and phosphorylation of plasma membrane and secretory granule proteins has been reported. There is some evidence that LHRH can increase the incorporation of phosphate into protein regardless of the presence or absence of increased cyclic AMP concentrations.   There is evidence for an effect of TRH in activating protein kinase activity, but only at high concentrations. The evidence relating to whether somatostatin affects kinase activity is also contradictory with reports indicating either inhibition or a lack of effect.   Thus, there is insufficient data at the present time to make any definitive statements regarding the role of phosphorylation in the release process.

The possible role of cyclic GMP in the control of pituitary hormone secretion has also been investigated, but here again the results are somewhat contradictory.   While some workers have suggested that cyclic GMP may act as a mediator of LHRH action, others have concluded that this nucleotide does not appear to act as a mediator of LHRH-induced gonadotropin secretion.   It has been reported both that somatostatin causes an increase in cyclic GMP levels in whole pituitary glands, and that it has no effect in bovine pituitary cells.   In contrast there is a report indicating that cyclic GMP is a potent growth hormone secretagogue.   Thus, it would appear that at this time the contradictory data prevents an understanding of a possible role for cyclic GMP in this tissue.

The role of calcium ions

The secretion of anterior pituitary hormones is dependent on the presence of extracellular calcium. While stimulated hormone secretion does not occur in the absence of extracellular calcium, it has been suggested that this pool may not necessarily play a crucial role in the release process. Evidence has been presented that mobilisation of intracellular calcium may be a critical event in the secretory process. Measurements of calcium flux in response to stimuli of secretion have given equivocal results. Thus, while there are a number of reports showing that TRH and analogues of cyclic AMP have no effect on Ca^{2+} influx, others have reported increased Ca^{2+} uptake in response to theophylline, and to a purified GRF preparation. It is probable that these conflicting results are due, at least in part, to the cellular heterogeneity mentioned earlier, and that the positive result with the purified GRF reflects the larger number of somatotrophs in the gland. Measurements of calcium efflux have shown that both TRH and LHRH cause increased efflux from normal cells and that TRH stimulates calcium release from GH_3 cells. Moreover, parallel LHRH dose response relationships of Ca^{2+} efflux and LH release have been noted.

Studies on the electrical characteristics of pituitary cells also implicate calcium flux as a major factor in the control of hormone secretion. Spontaneous depolarising action potentials have been recorded in normal pituitary and in GH_3 cells. Moreover, TRH increases the frequency of these calcium-dependent action potentials. In one study, TRH elicited spiking in approximately 10% of the cells, which was taken as an indication that it was the thyrotroph and mammotroph cells that were responsive. There is also some evidence that dopamine exerts an inhibitory effect on the action potentials. A number of investigators have shown that both the action potential and secretion were blocked by inhibitors of calcium channels such as La^{3+}, Co^{2+} and methoxyverapamil, but not by tetrodotoxin or by the absence of sodium ions. The general conclusion has been that the action potentials involve calcium influx which participates in stimulus-secretion coupling in the pituitary cells. In GH_3 cells and some non-functioning human pituitary tumours, calcium-dependent action potentials were generated or accelerated by agents which caused hormone release, and slowed by those which decreased secretion. However, very few cells from human prolactinomas were excitable, suggesting that the control mechanisms in different cell types may not be the same.

Various other studies, using both calcium ionophores and calcium "antagonists", including methoxyverapamil, flunarizine, Mn^{2+}, Co^{2+} and La^{3+}, have also suggested that calcium ions are important in the mediation of hormone release. Pituitary hormone secretion stimulated

by agonists or by depolarisation is subject to inhibition by calcium
antagonists. However, it appears that flunarizine is less effective
in inhibiting K^+-induced secretion of prolactin than in inhibiting
TRH-induced secretion (unpublished observation). Interestingly,
methoxyverapamil inhibits basal prolactin secretion, but is without
effect on basal TSH secretion. This difference may be a reflection
of the high level of prolactin secretion _in vitro_ due to the release
from the tonic inhibitory control by PIF. In one study it was
reported that calcium ionophores, particularly A23187, did not affect
basal prolactin release but reversed the dopamine-mediated inhibition
of prolactin secretion. This was taken as evidence in favour of an
inhibitory effect of dopamine on calcium flux. However, other
studies have shown that A23187 does indeed cause prolactin release in
the absence of dopamine. Both A23187 and increased extracellular K^+
concentration caused an increase in growth hormone release which was
inhibited by somatostatin. It was also found that growth hormone
release (but not cyclic AMP accumulation) was inhibited by depletion
of tissue calcium. However, it appears that somatostatin does not
affect calcium efflux and it has been suggested that somatostatin acts
at a late stage in the secretory process. One group who had
previously concluded that neither prostaglandins nor cyclic
nucleotides were involved in the mechanism of action of LHRH, have
recently demonstrated that LHRH induced a calcium-dependent release of
arachidonic acid. Phospholipase inhibitors prevented the effect of
LHRH on arachidonate formation and LH release. Exogenous
arachidonate stimulated LH release, an effect that was not dependent
on calcium. So it is possible that phospholipid turnover is involved
in LHRH-induced secretion of LH, and that the calcium-dependent step
is at a post-receptor locus, probably at the activation of
phospholipase A2.

It is possible that Ca^{2+} and cyclic nucleotides act cooperatively in the control of hormone secretion from both normal and
tumour cells. One suggestion has been that calcium ions play a
pivotal role in mediating secretion and that cyclic AMP may modulate
the calcium fluxes. Another suggestion has been that cyclic AMP
causes translocation of Ca^{2+} from a bound pool to a free pool and
that secretogogues can bypass cyclic AMP by causing Ca^{2+} influx.
Thus, it seems that Ca^{2+} and cyclic nucleotides should not
necessarily be considered separately in relation to hormone secretion.
It is of interest to note in this regard that calcium ions with the
participation of calmodulin (the calcium-dependent regulator) can have
profound effects on cyclic nucleotide phosphodiesterase activity and
on adenylate cyclase activity in some tissues. Indeed, calmodulin-
activatable phosphodiesterases have been identified in pituitary
tissue. It has recently been demonstrated that, in the brain, raised
cyclic AMP levels can overcome the inhibitory effects of calcineurin
on calmodulin activated phosphodiesterase. The biphasic effect of
several dopamine antagonists on prolactin secretion provides some
evidence for a role for calmodulin in secretion. At low

concentrations these antagonists, including pimozide, metoclopramide and domperidone, overcome the inhibitory effects of dopamine, whereas at higher (μmolar) concentrations they inhibit prolactin secretion. It has been known for some time that pimozide is a potent inhibitor of calmodulin-activated phosphodiesterase activity. Furthermore, it has been shown recently that domperidone inhibits calmodulin-activated phosphodiesterase activity and also that the phenothiazines trifluoperazine and prochlorperazine inhibit prolactin secretion at μmolar concentrations (unpublished observations). These results suggest that calmodulin may play a key role in the mediation of the secretion of prolactin (and possibly of the other pituitary hormones). It is also tempting to suggest that some, at least, of the other drugs which cause a biphasic effect on secretion, or are only inhibitory, are also effective inhibitors of calmodulin.

SUMMARY

While a substantial amount of information exists on the various proposed intracellular mechanisms governing pituitary hormone secretion, considerable controversy still remains. It would appear that the consensus of opinion is currently swinging towards the conclusion that calcium is the major intracellular mediator of the action of LHRH on the secretion of the gonadotropins in particular, but also of TRH and dopamine action on TSH and prolactin secretion. It is clear that the cellular heterogeneity of the anterior pituitary gland has been a contributory factor in many of the contradictory results obtained and has, for the most part, precluded direct determination of the putative intracellular mediators. Although the use of tumour cell lines secreting just one or two hormones appears to have helped in clarifying these mechanisms, it must be remembered that these are not normal cells. As an example of the potential problems, it has recently been reported that GH_3 cells do not possess high affinity receptors for dopamine. The establishment of relatively simple techniques for the separation of the different cell types from the anterior pituitary will be of considerable value for the elucidation of the intracellular control mechanism.

Selected bibliography

Azhar, S., and Menon, K.M.J., 1977, Cyclic nucleotide phosphodiesterases from rat anterior pituitary: Characterisation of multiple forms and regulation by protein activator and Ca^{2+}, Eur. J. Biochem., 73:73-82.
Barnes, G.D., Brown, B.L., Gard, T.G., Atkinson, D., and Ekins, R.P., 1978, Effect of TRH and dopamine on cyclic AMP levels in enriched mammotroph and thyrotroph cells, Mol. Cell Endocr., 12:273-284.
Conn, P.M., Morrell, D.V., Dufau, M.L., and Catt, K.J., 1979, Gonadotropin-releasing hormone action in cultured pituicytes: Independence of luteinizing hormone release and adenosine 3',5'-monophosphate production, Endocrinology, 104:448-453.

DeCamilli, P., Macconi, D., and Spada, A., 1979, Dopamine inhibits adenylate cyclase in human prolactin secreting pituitary adenomas, Nature, 278:252-254.

Faure, N., Cronin, M.J., Martial, J.A., and Weiner, R.I., 1980, Decreased responsiveness of GH_3 cells to the dopamine inhibition of prolactin, Endocrinology, 107:1022-1026.

Gautvik, K.M., Iversen, J-G., and Sand, O., 1980, On the role of extracellular Ca^{2+} for prolactin release and adenosine 3',5'-monophosphate formation induced by thyroliberin in cultured rat pituitary cells, Life Sci., 26:995-1005.

Gershengorn, M.C., Marcus-Samuels. B.E., and Geras, E., 1979, Estrogens increase the number of thyrotropin-releasing hormone receptors on mammotropic cells in culture, Endocrinology, 105:171-176.

Gershengorn, M.C., Rebecchi, M.J., Geras, E., and Arevalo, C.O., 1980, Thyrotropin-releasing hormone (TRH) action in mouse thyrotropic tumour cells in culture: Evidence against a role for adenosine 3',5' monophosphate as a mediator of TRH-stimulated thyrotropin release, Endocrinology, 107:665-670.

Heindel, J.J., and Clement-Cormier, Y.C., 1981, Regulation of adenylate cyclase activity in GH_1 cells by chlorpromazine and a heat-stable factor, Endocrinology, 108:310-317.

Kraicer, J., and Spence, J.W., 1981, Release of growth hormone from purified somatotrophs: use of high K^+ and the ionophore A23187 to elucidate interrelations among Ca^{2+}, adenosine 3',5' monophosphate and somatostatin, Endocrinology, 108:651-657.

Labrie, F., Borgeat, P., Drouin, J., Beaulieu, M., Lagace, L., Ferland, L., and Raymond, V., 1979, Mechanism of action of hypothalamic hormones in the adenohypophysis, Ann. Rev. Physiol., 41:555-569.

Naor, Z., Zor, U., Meidan, R., and Koch, Y, 1978, Sex difference in pituitary cyclic AMP response to gonadotropin-releasing hormone, Am. J. Physiol., 235:E37-E41.

Naor, Z., Snyder, G., Fawcett, C.P., and McCann, S.M., 1980, Pituitary cyclic nucleotides and thyrotropin-releasing hormone action: The relationship of adenosine 3',5' monophosphate and guanosine 3',5' monophosphate to the release of thyrotropin and prolactin, Endocrinology, 106:1304-1310.

Ozawa, S., and Kimura, N., 1979, Membrane potential changes caused by thyrotropin-releasing hormone in the clonal GH_3 cell and their relationship to secretion of pituitary hormone, Proc. Natl. Acad. Sci, USA, 76:6017-6020.

Ray, K.P., and Wallis, M., 1981, Effects of dopamine on prolactin secretion and cyclic AMP accumulation in rat anterior pituitary gland, Biochem. J., 194:119-128.

Schrey, M.P., Brown, B.L., and Ekins, R.P., 1978, Studies on the role of calcium and cyclic nucleotides in the control of TSH secretion, Mol. Cell. Endocr., 11:249-264.

Stefanini, E., Clement-Cormier, Y.C., Vernaleone, F., Devoto, P., Marchisio, A.M., and Collu, R., 1981, Sodium-dependent interaction of benzamides with dopamine receptors in rat and dog anterior pituitary glands, Neuroendocrinology, 32:103-107.

Taraskevich, P.S., and Douglas, W.W., 1977, Action potentials occur in cells of the normal anterior pituitary gland and are stimulated by the hypophysiotropic peptide thyrotropin-releasing hormone, Proc. Natl. Acad. Sci. USA, 74:4064-4067.

Taraskevich, P.S., and Douglas, W.W., 1978, Catecholamines of supposed inhibitory hypophysiotropic function suppress action potentials in prolactin cells, Nature, 276:832-834.

Thorner, M.O., Hackett, J.T., Murad, F., and MacLeod, R.M., 1980, Calcium rather than cyclic AMP as the physiological intracellular regulator of prolactin release, Neuroendocrinology, 31:390-402.

West, B., and Dannies, P.S, 1979, Antipsychotic drugs inhibit prolactin release from rat anterior pituitary cells in culture by a mechanism not involving the dopamine receptor, Endocrinology, 104:877-880.

INDEX

Adenylate Cyclase, 3, 7–61, 87–94, 117–135
  and calcium in the adrenal gland, 254
  and calmodulin, 224–225
  and cholera toxin, 10, 13, 18
  and fluoride, 13, 23, 26, 34, 43
  inhibition of, 78–81
  and prostaglandins, 110–111
  in the pituitary, 321–322
  solubilization of, 31–45
Acetyl-CoA carboxylase
  regulation of, 181
ATP-citrate lyase
  regulation of, 181–182

Binding
  and guanine nucleotides, 76, 78
  of hormone, 3, 8–30, 50, 54, 56–59

Calcium, 4
  and ACTH action, 254
  and calmodulin, 219–231
  and cell growth, 237
  and guanylate cyclase, 66–68
  and phosphodiesterase, 102
  and secretion, 209–218, 325–327
  and smooth muscle contraction, 187–189
Calmodulin, 4
  and adenylate cyclase, 43
  and calcium dependent ATPases, 227–228
  and guanylate cyclase, 66
  and myosin light chain kinase, 189

  in pituitary secretion, 326–327
  processes regulated by, 222–228, 241
  properties of, 219–222
  structure-function relationship, 228–231
Cancer,
  and retroviruses, 299–314
Cell
  astrocytoma (132 INI), 19, 25
  erythrocyte, 11–14, 18, 23, 25, 90, 129
  fibroblast (WI 38), 23, 120–133, 235
  glioma (C6-2B), 16–18
  hepatoma (HC-1), 23
  Leydig, 261–262, 264
  lymphoma (S49), 11, 13, 19, 21–22, 93
  parietal mutants in the cyclic AMP system, 137–145 269–273
  pituitary
    GH(1,3), 321–322
    TtT, 321–323
Chromaffin Cell
  and secretion, 211, 212, 213
Cholesterol
  and testosterone synthesis, 260, 261
Contractile Proteins
  in leukocytes, 216
  properties and functions of, 244–245